Victoria Island
Banks Island
Cape
Bathurst
BEAUFORT SEA
Herschel
Island
PACIFIC OCEAN

DOROTHY ADDAMS BROWN

Whales, Ice, and Men

The History of Whaling in the Western Arctic

WHALES, ICE, and MEN

The History of Whaling in the Western Arctic

JOHN R. BOCKSTOCE

University of Washington Press *Seattle and London*

in association with the **New Bedford Whaling Museum** *Massachusetts*

First Edition, 1986

Composition by the University of Washington Department of Printing, Seattle
Printed in the United States of America

Library of Congress Cataloguing in Publication Data

Bockstoce, John R.
Whales, ice, and men.

Bibliography: p.
Includes index.
1. Whaling—Arctic Ocean—History. I. New Bedford
Whaling Museum. II. Title.
SH382.B63 1986 338.3′7295 85-91266
ISBN 0-295-96318-2

Designed by Audrey Meyer
Map illustrations by Robert D. Hutchins

JAMES BARTON,

SHIPSMITH,

AND

Manufacturer of Whale Craft

OF EVERY DESCRIPTION.

☞ Orders solicited, and all such executed promptly and satisfactorily. ☜

City Wharf, - - - - NEW BEDFORD.

for Romayne

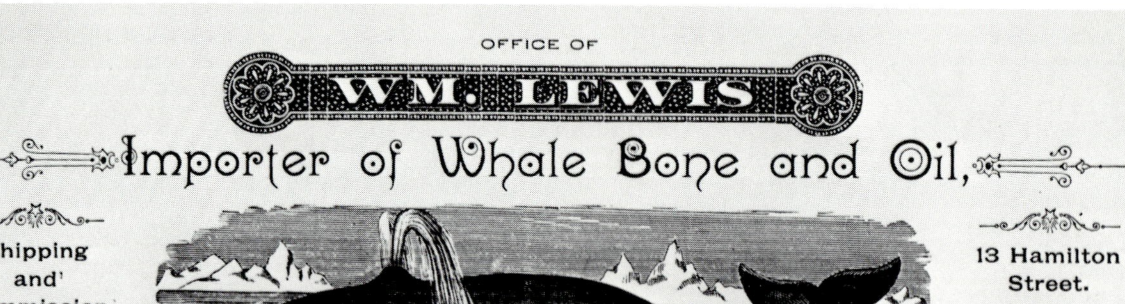

Shipping
and
Commission
Merchant.

13 Hamilton
Street.

P. O. Box 66.

Contents

MAPS

Foreword

Nowhere does the hunt for whales extend further back in time than in the western Arctic. Since at least 500 B.C., native peoples along the Siberian shore regularly pursued marine mammals, including the largest that came within their reach, the bowhead whale. By 1,000 A.D., the techniques and implements employed along the Asian coast had crossed the Bering Strait, enabling the Eskimos to occupy northern Alaska, then Arctic Canada and Greenland. The essential elements of the technology were simple: a boat to bring a hunter and his crew within striking distance, a toggle-head harpoon, and a sealskin float to hinder the whale's escape. Within the limits of the hunters' strength and skill, the gear was effective enough to take perhaps a hundred whales a year from a population thought to have numbered 30,000.

In 1848, an entirely different enterprise interrupted the older pattern of native peoples. In that year, an American whaleship, the bark *Superior,* sailed through Bering Strait and discovered the largely unexploited stock of bowheads. As with California gold—that other discovery of 1848—a "rush" ensued; within two years, two hundred whaleships were cruising in the western Arctic. During the summer of 1850 their crews took over 1,700 whales. By the end of the decade, the number taken had risen to 6,400, with another 1,200 mortally wounded but not recovered.

The pursuit continued for another fifty years, although with diminishing results as the stock of whales declined. By 1910, when commercial whaling in the Arctic had come to an end, the survival of the bowhead was in doubt.

Despite their vastly different impact on the stock of whales, American and Eskimo whaling shared elements of similarity. The prey was certainly the same: the large and unaggressive bowhead. The basic implements of capture were much alike, at least in function: toggling harpoons to fasten to the whales and floats to prevent them from escaping, whether inflated sealskins or Yankee whaleboats towed behind. The differences were those of motive and mobility. With the whaleship as a floating base, the Americans could chase the bowheads far from shore wherever the ice permitted. For the Eskimo, the walrus-skin umiak restricted the hunt to open leads between the offshore pack and land. Even more significant, however, was the differing purpose of the hunt. For the Eskimos, the bowhead meant survival in a harsh climate. For the Americans, the whale was sought for commercial gain, and only the calculations of profit and loss set limits to the hunt.

In the pages that follow, the story of commercial whaling in the western Arctic is told by a scholar intimately acquainted with the terrain—not only as it can be found in the historical records or at archaeological sites, but from long experience on the shores and waters where the great adventure was played out. His book is written with such mastery and vigor that we confidently greet it as the finest history yet written on any aspect of American whaling.

Richard C. Kugler
New Bedford Whaling Museum

Preface

This book began on the ice of the Chukchi Sea one night in May 1971. I was a green hand on an Eskimo whaling crew, learning to hunt one of the largest animals on earth, the bowhead whale. Despite the fact that I had already been on the ice for several weeks, each time I gazed around I was struck by the exotic scene before me. That night the midnight sun hung low on the northern horizon, washing the jumbled ice fields with a viscous red light. I shifted uneasily on the freight sled and tried to keep the cold wind from finding its way into my parka. Judging from their nervous movements, the six Eskimos seated beside me were equally unsuccessful in fighting the cold.

Our whaleboat, an eighteen-foot sealskin-covered umiak, sat ready on the ice, its harpoon projecting over the bow. The captain and the harpooner paced before us in a curious syncopated shuffle, tapping one foot against the other to force circulation back into their feet.

For the past five days we had had little sleep. Offshore winds had opened a wide ribbon of water between the shore-fast ice and the moving pack and in this water lead the whales had been migrating past us toward the Beaufort Sea. Fruitlessly we had been chasing them round the clock.

Despite the fatigue, hunger, and cold, several Eskimos, as always, were alertly scanning the inky lead—now 400 yards wide—for the telltale V-shaped spout of a bowhead. The harpooner, as he had many times that day, checked to see that the safe pins were in the darting guns and the bomb-lance shoulder gun, a twenty-seven-pound brass smooth-bore.

Both weapons were developed in the nineteenth century by Yankee whalers to reduce the loss of wounded bowheads that escaped into the pack ice. The darting gun, mounted on the harpoon shaft beside the toggle iron, fires a small bomb into the whale the instant the iron strikes; the bomb explodes a few seconds later. The shoulder gun can throw a similar bomb accurately as far as ten yards or more. Because they were far more effective than hand lances, these weapons were quickly adopted by Eskimo whalers.

Also in the boat was a large inflated sealskin float with 180 feet of three-quarter-inch manilla rope coiled around it. The harpoon float has been a part of Eskimo whaling gear for more than a thousand years, an invention of marvelous simplicity that slows the whale's flight, tires the animal by the drag, and, when the whale is submerged, signals its location to the hunters above.

Suddenly the captain stopped pacing; his eyes were riveted on a spot far down the lead. The crew snapped alert. Hanging in the shimmering Arctic mirage was a ball of vapor. Seconds later the sound carried to us: PAHHHHH, a great rush of air exhausting from the whale's huge lungs.

We forgot how cold and tired we were. We scrambled into the umiak and set off, driving the boat quickly and quietly with short, powerful strokes, the captain pointing the bow 400 yards ahead of where the whale had surfaced. Four other boats also launched, each aiming for a spot where the whale might rise next.

The whale surfaced in ten minutes very near one of the boats. I saw a crew straining to overtake it. The harpooner stood abruptly, raised his weapon, and as the umiak grazed the whale's back, he thrust it deep between the animal's shoulder blades. The darting gun went off with a roar, throwing the shaft high in the air.

Instantly the whale dove, and one of the Eskimos threw the float overboard. At first it spun crazily as the carefully coiled line was drawn after the sounding whale. The thirty-fathom rope quickly ran out, however, and the float was pulled under. A moment later we felt the dull WHUMP of the exploding bomb. Soon the float popped up nearby.

By now eight boats were converging on the scene. As the dying whale erupted, its flukes thrashing, another harpooner fired his shoulder gun, putting a bomb at the base of its skull, killing it instantly.

My mind was febrile but I realized I was witnessing two forgotten eras at once. The hunt straddled two cultures and two ages. The basic elements of the hunt had developed a thousand years ago; yet some of the tools and procedures had come from New England only a little more than a century ago.

Why had this hybrid developed?

My search for an answer to that question would consume more than a decade.

I had been sent to Point Hope by my friend and teacher, Professor Froelich Rainey. Rainey, an archaeologist, had lived there in the early 1940s while he and his Danish colleague, Dr. Helge Larsen, were excavating along the shore, probing for the origins of Eskimo culture. Whaling was considered to be one of the fundamental parts of the culture, and Rainey believed that a study of the contemporary whaling society could shed light on his and Larsen's archaeological discoveries. In addition to their several summers of digging, he spent most of one year at Point Hope with his family. His whaling experiences and his research into the community contributed to the success of the archaeological work and resulted in a significant study in its own right.

Thirty years later I was working for Rainey near Bering Strait, excavating the remains of an ancient Eskimo whaling village. Rainey advised me that the experience of whale hunting would give me a wider perspective on my own work. Each spring for more than a decade I went to Point Hope to hunt bowheads.

From the first, the origins of modern shore whaling fascinated me, but although I visited many libraries and archives searching for information, I found only a cold trail. I resolved then to write a comprehensive history of ancient and modern shore whaling in the western Arctic.

I soon discovered, however, that the history of modern shore whaling is inextricably linked to the history of pelagic whaling in the Arctic; that the whaling industry, both pelagic and shore-based, was the region's most powerful agent of change. Even today many of the area's biological, economic, and social conditions bear the indelible tincture of the industry. The whaling industry introduced the outside world to the natives and, by nearly eliminating the whales and walruses, it changed the natives' world as well. Paradoxically, these two hunting peoples for the most part co-existed peacefully and—in the perception of each—with mutual benefit.

When I began my research into this essentially unstudied topic, each question seemed only to lead to another. It soon became clear that before I could attempt to write a thorough history of shore whaling, it was necessary first to carry out a comprehensive investigation of its origins: the pelagic whaling industry in the western Arctic.

While my archival research was underway I continued my fieldwork. In 1972 I began a series of surveys, searching along the continental coast of western arctic America for the remains of the whaling industry. I had been on many walrus hunting expeditions with the Eskimos of Bering Strait and I had come to admire their big, walrus-hide umiaks. They were tough, fast, flexible, and extremely seaworthy even when loaded with a ton or more of supplies. I used a thirty-two-foot umiak to navigate the triple impediments of sinuous coasts, ice, and shallows, and in all I traveled back and forth about six thousand miles. The work consumed nearly ten summers. Having drawn on information in logbooks and other manuscripts, I was able to locate a number of wrecks, graves, houses, and other structures from the whaling era.

One unforeseen benefit of those surveys was that I learned about the physical conditions in which the whalemen operated. The ice, the gales, and the shoals were constant challenges. At most places in the western Arctic I have been rained or snowed on, pummelled by the wind, deceived by the fog, frustrated by the ice, or grounded on the shoals. This was an easy—if not entirely pleasant—way of getting to know the area and the problems that faced the whalemen.

Most important, however, is that I was indeed fortunate to be hired by the New Bedford Whaling Museum—for in 1971 I could not have realized that this book could be written nowhere else. The Whaling Museum contains the world's largest collections of logbooks, printed books, charts, manuscripts, documents, whalecraft, and fine arts on the history of the American whaling industry. In this environment I was equally fortunate to receive the generous and rigorous tutelage of Richard C. Kugler, Director, and Philip F. Purrington, Senior Curator. In entirely complementary ways they broadened and refined my understanding of the American whaling industry.

There have been many periods of American history that have been romanticized. To this day most writers of the whaling industry have glorified or damned it beyond reality: jingoism, xenophobia, missionary zeal, political and social perceptions, environmental concern, and many other points of view have colored their interpretations.

At the same time most writers have ignored the industry's essential element. It is entirely unsurprising that almost everyone who was involved with the whaling industry—owners, captains, officers, sailors, Yankees, Eskimos, Portuguese, Hawaiians, and many others—entered it solely to make a profit. Whalemen did not set out on four-year voyages to the world's most dangerous and remote waters for the joys of national expansion, geographical discovery, religious proselytization, debauchery, scientific inquiry, or a number of other motives which modern writers have attributed to them. Nor were whalemen as a whole especially brave, craven, curious, dull, careless, careful, restless, lazy, callous, greedy, noble, stupid, intelligent, or cruel. The whaling industry simply represented the industrial revolution's expansion to the most remote waters of the world. This expansion occurred at a time when the acquisition of natural resources was considered to be a God-given right. Whaling was a rational economic endeavor for most of the people involved in the industry.

Modern writers have tended to portray both the industry and its men in crude polarities. Were the whalemen bold, world-wandering mariners who, seeking to support their families, sailed in dangerous and uncharted seas while they did battle with the largest animals on earth; or were they lupine exploiters who raped the world's whale stocks and debauched the natives in a greedy search for profits? Were they locked in an archaic, dying industry while doing their cooperative best to bring home a catch for the whole crew; or did the owners and captains mercilessly exploit the working men who were unable to find better employment? Were the whalemen acquisitive expansionists who cared nothing for international boundaries; or were they merely hunting whales, in the manner of the nineteenth century, where they found them? Did they harm the arctic natives; did they help the arctic natives? These polarities are simplistic. I hope this book will help the reader to understand the complexity of these issues.

The history of this book is also linked to the history of the western Arctic bowhead whale population. In 1977 the issue of the bowheads' survival reached critical proportions. Conservationists claimed that at the current level of the Eskimos' annual kill the population would soon be extinct and, therefore, that the Eskimos' hunt must be stopped. The Eskimos retorted that such action would cause irreparable harm to the social fabric of their communities. The U.S. government was in a quandary, drawn between its stands in favor of both conservation and native rights.

At the core of this controversy lay two questions: how many bowheads are alive today and how many were alive before the whaling fleet began to reduce the population? With these answers it would be possible to calculate the degree to which the population had been reduced today; hence the degree of the modern population's fragility. Having established this, it would be possible to determine the number of whales that could safely be taken by the Eskimos.

By using aircraft and shore-based spotters the government was eventually able to determine quite accurately the size of today's population—somewhere around 4,500 whales. The question remained, however, of the size of the original population. A number of quick and sloppy estimates had been made, but nothing comprehensive or rigorous had been attempted. My colleague, Daniel Botkin, professor of environmental studies at the University of California, Santa Barbara, and I devised a method of using the daily information in whaleships' logbooks to calculate the annual reduction of the population and to estimate the size of the population before the beginning of the commercial hunt.

This required that first I identify all whaleships that cruised in the western Arctic year by year. At the Whaling Museum my assistants and I consulted historical shipping newspapers from around the world, printed documents and books, and insurance records. We compiled a list of the 2,700 annual cruises in the western Arctic from 1848 to 1914.

With that completed I convened a group of researchers at the Museum and moved on to our primary data, the logbooks. We surveyed manuscript repositories throughout the world and found logbooks covering one-third of all the cruises to the western Arctic, a surprisingly large percentage. Most fortunately, 20 percent of the cruises were covered by documents which were complete, legible, and sufficiently detailed for our purposes.

We then began the work of extracting the data from these logbooks. We worked primarily from microfilm and recorded the data on computer forms. For each day, for each vessel, we recorded the following information: the ship's position, the visibility and weather conditions, the amount of ice cover, the type and number of animals seen. We also recorded whether the whales had been seen only, chased and escaped, wounded and escaped likely to live, wounded and escaped likely to die, found dead, or captured and processed. For those whales that were processed we recorded their size as the whalemen expressed it: in barrels of oil and pounds of baleen.

In all we extracted data covering 66,000 days of observations, equal to 19 percent of all the cruises made to the western Arctic—one of the longest continuous bodies of historical data compiled on any mammal species.

Aided by a computer, Daniel Botkin analyzed our data and found that whaleships had killed 20,000 bowheads. He estimated that about 30,000 were alive before the fishery began in 1848. The difference between the two figures is accounted for by the shore-based kill, by a suppression in the birth rate, and by the population that remained at the end of the commercial fishery.

This project allowed me to survey comprehensively the holdings of those manuscript collections throughout the world that related to the western Arctic fishery. In all I probably took notes on a thousand logbooks as well as many other holographic documents. This work gave me an understanding of the day-to-day events of the fishery. During my research for this book many colleagues from several disciplines have generously given me their assistance and suggestions. While each may have hoped for a different emphasis, I must point out that this is not an anthropological study of the changes that the whaling industry vested upon the native peoples, nor a social history of the men who sailed aboard the ships, nor a biological history of the faunal populations that the industry exploited. I intend to return to these topics in the future, but this book is simply a history of the whaling industry in the western Arctic—the topic that must be treated first.

I must set forth a few more definitions. The western Arctic fishery was a unified search for the bowhead. Few other whales were taken. For the whalemen this fishery included most of the waters north of the Aleutian and Commander islands. In the area known as Bristol Bay—east of the Pribilof Islands and south of Saint Matthew's Island—right whales were hunted, but bowheads rarely found. Whalemen viewed the Bristol Bay fishery as an extension of the right whale fishery of the Gulf of Alaska. Similarly, I have not discussed the shore-based whaling operations of the twentieth century in the Aleutian Islands and Gulf of Alaska. No bowheads were taken in this fishery. Consequently I have used the phrase "western Arctic" to include all of the waters of the Bering, Chuckchi, and Beaufort seas, excluding Bristol Bay.

For units of measurement I have used degrees Fahrenheit and statute, not nautical, miles. Times I have expressed in civil time, converting the information from those logbooks that were kept in nautical time (in which the new day was recorded as beginning at noon). Some vessels frequently crossed the 180th meridian and recorded the changes in date properly. In the interest of clarity I have, for the most part, altered their dates to being *west* of Greenwich.

In using quotations from manuscripts I have usually added punctuation and standardized spelling. Specialized whaling terminology is defined in the glossary. For place names I have retained the whalemen's own terminology because it is far more evocative and descriptive than modern equivalents, especially for geographical features on the Asian coast: for instance, False East Cape is used in preference to *Mys Intsova*, and Indian Point for *Mys Chaplina*. All obsolete and obscure place names are also defined in the gazetteer (Appendix 4). Definitions of other nautical terms and geographic names can be found in standard marine dictionaries and gazetteers. Modern dollar equivalents are based on *The Historical Statistics of the United States* (1975), volume 1, pages 182-83. A chronological summary of important events in the western Arctic whale fishery will be found in Appendix 1.

The book contains an enormous number of references. Had the citations been incorporated in the body, they would have overwhelmed the text with thousands of superscript numbers. I have therefore grouped the references by topic near the end of the book. Citations for manuscript repositories are given according to Library of Congress symbols. A key for these symbols is at the beginning of the notes section.

The illustrations that help bring the story of Arctic whaling to life, unless otherwise noted, come from the rich historical photography collection of the New Bedford Whaling Museum.

John R. Bockstoce
New Bedford, Massachusetts

May 1985

Acknowledgments

I am deeply indebted to Richard C. Kugler, Director of the New Bedford Whaling Museum, and to Philip F. Purrington, Senior Curator. My colleagues at the Museum have been no less generous with their time, assistance, and considered advice; each has contributed substantially to this book. I am very grateful to Elton Hall, former Curator of Collections, Virginia M. Adams, Librarian, Nicholas S. Whitman, Curator of Photography, and David A. Henderson, Visiting Scholar.

This book could not have been written without the aid of my assistants, Elizabeth Rex and Gail Scott Sleeman. Their consistent and conscientious support in all phases of the research and writing have been of inestimable value. I am also very grateful to Ernest S. Burch, Jr., and to Stephen Braund for their advice and assistance in many areas and to Elizabeth Kugler for her fine index.

Over the past decade many other people have generously helped with the research: Patricia Albright, Fred Alt, Douglas D. Anderson, Willits Ansel, Herbert Appossingok, Terence Armstrong, Serghei Aroutiounov, Jacqueline Astor, Michael Astor, Walt Audi, Barbara Austin, Lucy Avakanna, Mrs. John Backland, Gregory Baker, Rosalie Baker, the late William A. Baker, William S. Baker, Albert M. Barnes, Bruce Barnes, Don Barnett, William Barr, Don Berlo, Jane Bernert, the late Mrs. John Bertonccini, Nigel Blackwell, Rene Blahuta, Gale Blosser, James Bodena, George Bodfish, Waldo Bodfish, Serghei Bogojavlensky, Ed Booth, Daniel Botkin, Howard Braham, Willard Brooks, Noel Broom, Thomas Brower, Sr., Nicholas Bruen, William and Linda Bucklin, the late Col. Laurence Bunker, Franklin Burch, John Burns, Ed Chapman, William Cockney, Jr., William R. Collins, Agnes Conrad, Alan Cooke, Nan Cooper, Paul Fenimore Cooper, Jr., Bruce Courson, Adam Weir Craig, Alan R. Crane, Thomas B. Crowley, Paul Cyr, Dr. C. A. Dana III, Andrew David, Phyliss DeMuth, Wayne Donaldson, Judith Downey, Hugh Downs, Lynne Dunne, Frank and Ursula Ellanna, Linda Ellanna, Francis Fay, Richard S. Finnie, Raymond Fisher, Douglass C. Fonda, Jr., Mark Fraker, Stuart Frank, David Frankson, Sr., Milton Freeman, Kathy Frost, Benjamin Fuller, Stephen Gamsby, Natalie Garfield, Karen Garnett, James H. McM. Gibson, the late Mrs. Flavel Gifford, William Gilkerson, Robert Gill, Carl Grauvogel, William Graves, David Gray, James Griffith, Richard Griffith, Iona Grimston, Bonnie Hahn, Kenneth Hahn, Patrick Hahn, Edwin S. Hall, Jr., William Hanable, Archibald Hanna, Gordon Harrison, Don Harvie, Robert Hauser, George Hobson, Wybrand Hoek, Clive Holland, the Reverend Mark Hollingsworth, the late Warren Howell, David Hull, Frank Hunt, Gale Huntington, Harold Huycke, Rob Ingram, Melvin H. Jackson, Todd Jackson, Whitney Jackson, Jonathan Jensen, Barry Wayne Jesse, Carl Jewell, Sven Johansson, Barbara Johnson,

Rene Jussaud, Youssuf and Estrellita Karsh, John Kendall, Brina Kessel, Aziz Kheraj, Bernhard Kilian, the late Jimmie Killigivuk, Harry King, Laurie Kingik, the late Reverend Herbert Kinneeveauk, Luke Koonook, Sr., Karl Kortum, Michael Krauss, Sheppard Krech III, Tony and Jeannette Krier, Igor Krupnik, Amos Lane, Sr., Craig Fisher, Robin Lane Fox, Homer Langlois, Dinah Larsen, Willy Lasarich, William Laughlin, Molly Lee, Robert Leet, Edward Lefkowicz, Hamish Leslie Melville, Father Robert LeMeur, Jean-François and Maryke Le Mouel, John Lennie, Margaret Lennie, Francis B. Lothrop, Tom Lowenstein, Lloyd Lowry, Paul McCarthy, Priscilla McCarthy, John McDonough, the descendants of E. A. McIlhenny, Bob Mackenzie, Hoss MacKenzie, John Maggs, Willman Marquette, Kenneth R. Martin, Dwight Milligrock, Lionel Montpetit, Pat Murray, Evelyn Nef, Virginia Newhouse, Felix Nuvuyayoak, Kenneth Okolski, the late Reverend Donald Oktollik, John Oktollik, Mrs. Harold Osborn, Harold Gray Osborn, William O'Shea, Erica Parmi, Thierry du Pasquier, Laddy Pathy, Mrs. Jefferson Patterson, Harvey Pearce, Charles Pedersen, Ted Pedersen, Robert Pegau, Dale Perty, Victoria Pleydell-Bouverie, John and Lucy Poling, Donald and the late Dorothy Poole, George W. Porter, Sr., George W. Porter, Jr., Froelich Rainey, Elmer Rasmusson, Sr., Dorothy Jean Ray, Tad Reynales, Marian Ridge, Pat Ridzon, Tom Robinson, Derek Roe, W. Gillies Ross, Jean-Loup Rousselot, Graham and Diana Rowley, Diana Russell, Frederick Schmitt, Blayney Scott, Peter and Alma Semotiuk, Chester Sevek, G. Terry Sharrer, Leon Shelabarger, Stuart Sherman, Anne Shinkwin, Ann Shirley, Edward M. Simmons, Philip C. F. Smith, Shirlee Ann Smith, Benoit Souyri, Robert Spearing, William Starbuck, Douglas Stein, Gary Stein, Richard Stern, Sam Stoker, M. K. Swingle, William E. Taylor, Jr., Douglas Thomas, Cynthia Timberlake, Patrick Toomey, James VanStone, the Earl of Verulam, Roy Vincent, Sr., Robert C. Vose III, Shelly Wanger, Lincoln and Tahoe Washburn, George Wentzell, Alison Wilson, Frank Winter, Paul Wish, Helena Wright, Alexey Yablokov, David Zimmerly. To these and to many others I owe a sincere debt of thanks.

Whales, Ice, and Men

The History of Whaling in the Western Arctic

1 / *The First Whaleship to Bering Strait*

On July 23, 1848, Captain Thomas Roys stood on the deck of the whaling bark *Superior* with a revolver in his hand. His small ship was becalmed in Bering Strait, the shallow fifty-mile-wide gut that separates the Pacific's waters from the Arctic Ocean. The strong current was carrying his ship north at three or four knots. Ahead loomed menacingly the Diomede Islands, two black, flat-topped pillars of rock that spring vertically from the sea and rise more than a thousand feet. Twenty-five miles to either side, the massive headlands of East Cape, Siberia, and Cape Prince of Wales, Alaska, framed his view of the Arctic Ocean. He was headed there on a bold voyage of discovery, searching for whales in waters where only a handful of ships had ever been.

But Roys's gaze was directed much closer to his vessel. Seven umiaks, walrus-skin canoes, carrying more than two hundred and fifty Eskimos were heading toward the *Superior* from the American shore. Poised as he was at the threshold of the unknown, at the gates of the Arctic, a thousand miles beyond the nearest whaleship, Roys was naturally anxious about the natives who, for all he knew, might be hostile and who certainly outnumbered his crew by at least eight to one. To make matters worse, the *Superior* was entirely devoid of arms, excepting the "one Blunt & Sims revolver that would not go unless you threw it."

Fortunately a southeasterly breeze sprang up, allowing Roys to push the bark beyond their reach, but this was by no means the end of his troubles. His crew, from whom he had concealed his course, only three days before had discovered their location and were consumed by a fright that drove his first mate to tears. Near mutiny at the prospect of probing further into distant and hostile seas, they obeyed his orders only reluctantly.

They were soon enveloped by one of the chilling and depressing fogs that haunt Bering Strait in the summer. When the fog lifted a day later, Roys knew that he had made the greatest whaling discovery of the century. Whales appeared on all sides of the ship in vast numbers. His officers pronounced them to be humpbacks—fast swimmers that were both hard to catch and known to sink once killed.

Roys knew better. Three years earlier while he commanded the bark *Josephine* of Sag Harbor, Long Island, a fighting right whale had smashed him with its flukes and broken two of his ribs. Roys had been forced to take the *Josephine* to the town of Petropavlovsk on the Kamchatka Peninsula so that he could recuperate ashore while the ship continued its cruise under the first mate. During his convalescence he had met a Russian naval officer who had been north of Bering Strait and who mentioned the herds of strange whales he had seen in those waters. Roys began to suspect that these whales were the same as the "Greenland" whales that Europeans had been capturing for more than two centuries in the high latitudes beyond the North Atlantic.

Believing that there might be money to be made beyond Bering Strait, he ventured one hundred dollars to purchase the Russian charts of those waters. Shortly after Roys returned to the *Josephine,* his conviction was reinforced. During a gam with the Danish whaleship *Neptun* as they cruised "on Kamchatka"—the right whale grounds in the North Pacific off the east coast of the Kamchatka Peninsula—Captain Thomas Sodring told Roys of three strange whales he had recently taken near Petropavlovsk. At first Sodring thought the whales were the right whales they had been taking, but on closer inspection he noted some distinctive features, one of which was that its nib—the tip of its upper jaw—lacked the large barnacle-like callosity found on right whales. But most important, this whale was rolling in fat and it had an unusually large amount of whalebone (baleen) in its mouth. Whalebone—flexible and strong—was used for corset stays and a hundred other commodities that today are served by plastics.

As Roys sailed the *Josephine* toward her home port of Sag Harbor he had the opportunity to reflect upon what he had learned, and after his arrival home in May 1847, he began to consult the published works of the explorers who had sailed north of Bering Strait. Roys found reports of whales in the narratives of the British explorers who had pushed north of Bering Strait, Royal Navy Captains James Cook and Frederick William Beechey.

Captain Roys had grown up in the whaling industry and did not deceive himself with the idea that a ship's owner would allow him to sail to Bering Strait to test his hypothesis. Although phenomenally lucrative whale stocks might lie beyond the known whaling grounds, it was extremely unlikely that the owners would risk their capital on a radical departure into dangerous seas while there were plenty of whales to be taken on Kamchatka and in the Gulf of Alaska.

On Roys's next voyage the Grinnell Minturn Company assigned him the bark *Superior.* At 275 tons she was small for her rig, and the company outfitted her for a short and inexpensive voyage, planned to last less than a year, to Desolation (Kerguelen) Island in the Southern Ocean, southeast of the Cape of Good Hope.

He put to sea in July 1847 as if on a normal whaling voyage during which he would follow the strategy outlined to him by the owners. Once on the whaling grounds they had designated, it would be up to him to devise the best tactics for taking whales. He was to follow orders—apart from the usual discretion allowed to masters of ships when out of touch with their owners.

Roys headed directly to the Crozet and Desolation islands at the edge of the Antarctic but was unlucky there in the rugged weather of the Roaring Forties, taking only three whales. He continued east, reaching Hobart, Tasmania, with a cargo of only 123 barrels of oil. Roys then set out on a short, equally unsuccessful cruise to the South Pacific. He returned to Hobart to refit for the next leg of his voyage and, with his men already grumbling, he decided to play his wild card. He wrote to the

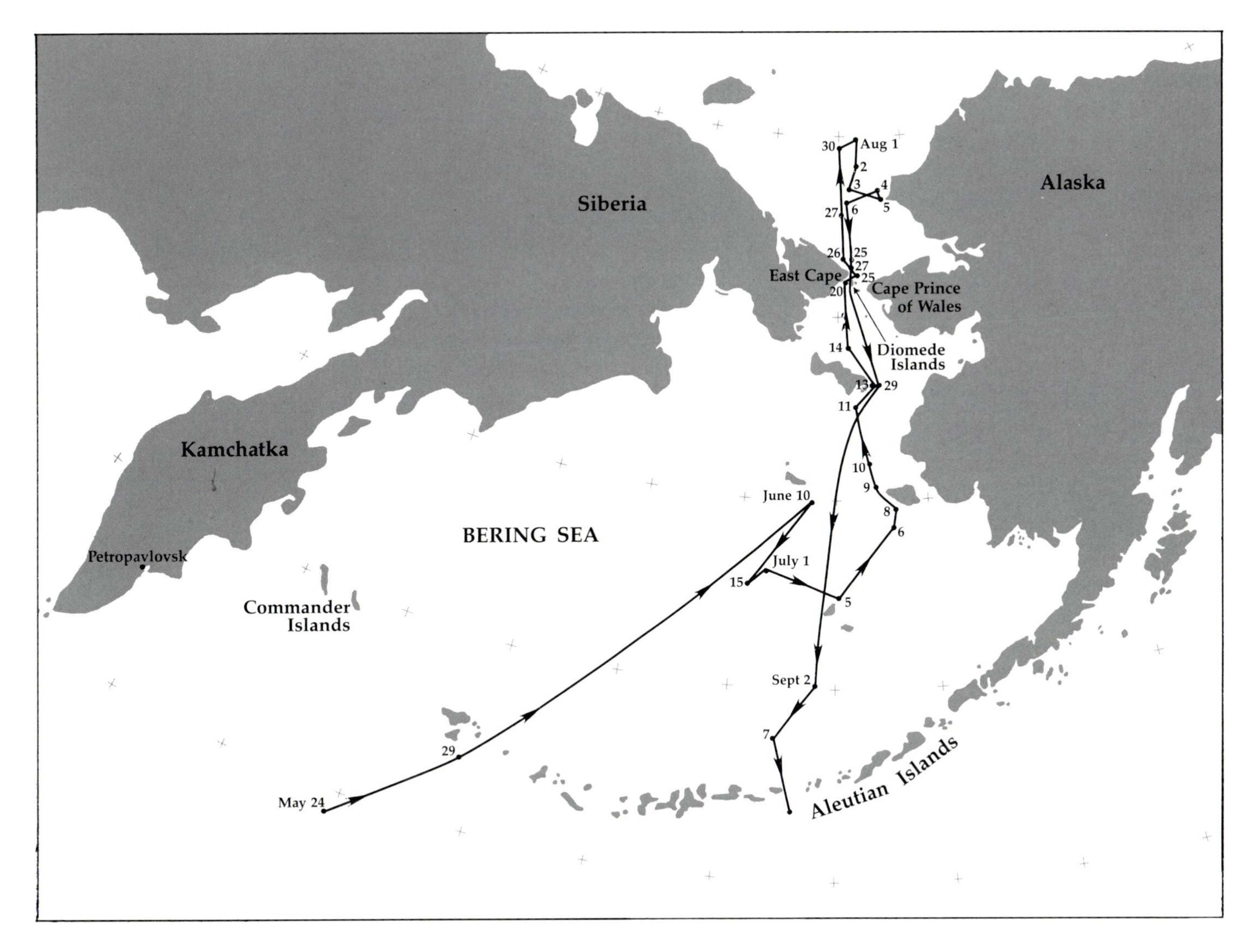

Route of Thomas Roys's cruise in the Superior, *1848*

owners from Hobart, informing them of his intention to go to Bering Strait, knowing that by the time they received his letter he would be in the Arctic Ocean.

Some have suggested that the news of the death of his young wife in August 1847, shortly after the birth of their only son, may have lowered his threshold of caution. But his decision was not made lightly. Several years later he wrote: "There is heavy responsibility resting upon the master who shall dare cruise different from the known grounds, as it will not only be his death stroke if he does not succeed, but the whole of his officers and crew will unite to put him down."

On July 25, when Roys ordered his boats lowered in Bering Strait, his crew, "who never expected to see home again," manned their boats and set off after the whale that some probably now realized was not a humpback, but a species they had never seen before. Roys reported that his men "were not inclined to meddle with the 'new fangled monster' as they called him." They must have proceeded as tentatively as possible toward the huge black shiny mass. Believing that this was the Greenland whale (hence relatively docile and easily killed), Roys drove his boat toward the creature through the twilight of midnight and his boatsteerer, the harpooner, fastened the whaling iron without difficulty. "The whale went down," Roys wrote,

> and ran along the bottom for a full 50 minutes and I began to think that I was fast to something that breathed water instead of air and might remain down a week if he liked. He then came up and was immediately killed. My officers now declared it was a humpback sure. We took the whale alongside, the weather thick and calm and only 25 fathoms of water. I told my mate to let go and anchor to keep from being drifted about in the current and not know[ing] where I was he looked at me with astonishment that I would think of anchoring in the open. . . .

Once the anchor bit, the bark swung around and surged on its cable in the three-knot current. As the fog lifted, they found themselves only three miles beyond Big Diomede Island in the middle of the strait. They had become the first commercial hunters to take a whale north of the Aleutians.

When they began to cut in their whale they found to their surprise that its baleen (the long keratinous plates that hang from the whale's upper jaw and allow it to filter its food, tiny shrimp and other plankton, from the water) was twelve feet long, far longer than most right whales' baleen. To their equal astonishment when they tried out the blubber, rendering it into oil, they found it yielded 120 barrels, a vast amount compared to most other whales.

Despite the continuing apprehension of his officers and men, Roys drove the ship another 250 miles north before turning back. He took whales all the way. His tryworks were never cool, so fast did he catch whales and melt oil from their blubber. Nevertheless, his "officers and crew were living in hourly expectation of some unforseen calamity and almost beside themselves with fear." "I actually believe," he added, "if they had any hope that open mutiny would have succeeded they would have tried it to get away from this sea."

North of Bering Strait he caught eleven whales that yielded 1,600 barrels of oil, thus filling his ship so that he could take no more. He passed south through the strait on August 27, and, no doubt to the relief of his men, he headed straight for Hawaii, arriving October 3 at Honolulu.

Roys immediately announced the news of his success. The *Superior*'s voyage set off a flurry of excitement. What news not spread by word of mouth was quickly broadcast by the Reverend Samuel Damon in his Honolulu missionary newspaper, *The Friend*, a mixture of temperance advocacy, proselytization, and marine intelligence, that was carried round the world by mariners. By early 1849 most of the marine journals of the world had picked up the story.

Roys's cruise was not only the most important whaling discovery of the nineteenth century, it was also one of the most important events in the history of the Pacific. Over the seven decades following his discovery, the richness of the western Arctic

The discoverer of the Bering Strait whaling grounds: Captain Thomas Roys in the mid-1860s. He lost his left hand in 1858 in an explosion while experimenting with a rocket harpoon Smithsonian Institution

Sketch of a bowhead whale by Albert Gaballi, a whaleman aboard the steam bark Bowhead *in the Arctic, 1903-6*

whale stocks lured the whaling vessels of the United States, France, Germany, Hawaii, and Australia through the shallow and treacherous waters of Bering Strait. More than 2,700 whaling voyages were made into Arctic waters at the cost of more than 150 whaleships lost and the near extinction of the bowhead whale, as Roys's whales came to be called.

The vast American presence in the western Arctic added impetus to the sale and purchase of Alaska. The support facilities generated by the whaling industry in Hawaii contributed to the American commercial presence in those islands that led, ultimately, to their annexation. Natives on both sides of Bering Strait, as well as in northern Alaska and western Arctic Canada, were supplied with western manufactured goods and their lifeways changed—for better and worse. As World War I approached, the Arctic whaling industry, in its death throes, spawned the maritime fur trade in the western Arctic, an activity that employed the Eskimos and a few remaining whalemen until the beginning of the Second World War.

Lastly, this first search for oil in the western Arctic touched off the second, for it was a whaleman who discovered the oil seeps that led to the Prudhoe Bay oil strike.

2 / *From New England to Bering Strait, 1750–1848*

Strangely enough it was the sperm whales of the Atlantic that drew the whaling fleets to the bowhead whales of Bering Strait.

In the first half of the eighteenth century the demand for lubricating and illuminating oils on both sides of the Atlantic was met for the most part by the whaling industry in the waters east and west of Greenland and off the coast of Labrador where right whales and bowheads were hunted. Increasingly during this period American whalemen became a force in the fishery. They sold their whale products both in the colonies and to England, and by the outbreak of the American War of Independence, the sale of oil and baleen had become one of the colonies' most important sources of credit in England.

Then, in the middle of that century, the American whaling industry began a period of dramatic growth—spurred by two innovations. The first was the development of tryworks aboard ship. Shipboard tryworks—large iron pots set in a brick oven built on deck—enabled the whalemen to render oil from blubber at sea. Before 1750 shipboard tryworks had not been systematically employed on American ships. As long as whale stocks were available near shore, the carcasses could be more simply towed to a shore base for flensing and rendering. In fisheries farther north the blubber could be packed into casks as the British did and kept without turning rancid until the ship reached a shore base or home port.

By 1750, however, nearby whale stocks were depleted. Whalemen were forced to range farther north and south on longer voyages. It seems likely that because of the expansion into the warmer waters of the south where spoilage was a problem, shipboard tryworks became a necessity. The change freed American whalemen from their fetters to shore and allowed them to become truly pelagic hunters.

The second innovation also propelled American whalemen into distant waters. A method was developed for making a superior grade of candle from spermaceti—the whalers' term for a waxy substance found in the sperm whale's head. The process was introduced possibly by Jacob Rodriguez Rivera, a Sephardic Jew who arrived in Newport, Rhode Island, in 1748. The candle manufacturers' added demand for spermaceti increased the value of sperm whale products, bringing the focus of the American industry onto the sperm whale. With their shipboard tryworks, the Americans quickly spread south in search of the species—to the Azores, the Brazil Banks, and, by 1775, to the Falkland Islands. By that year, with a fleet of three hundred ships, the Americans had a virtual monopoly in the sperm whale fishery.

Their achievement was short-lived. The outbreak of the War of Independence in 1775 marked the beginning of four decades of misfortune for the American whaling industry. The hostilities hamstrung the industry by cutting the Americans off from their European markets and putting the Royal Navy and British privateers on the lookout for American vessels. In the years following 1776, the British industry flourished, filling the vacuum left by the Americans.

In 1783, with the end of hostilities, as the Americans struggled to regain their former markets, Britain effectively embargoed American oil by placing a duty on the importation of whale oil roughly equal to its market value. The Americans were equally unable to compete in European markets. "The English," wrote Thomas Jefferson, "had now begun to deluge the markets of France with their whale oils and they were enabled by the great premiums given by their government, to undersell the French fisherman, aided by feebler premiums, and the American, aided by his poverty alone."

Furthermore, the Spaniards, who bitterly resented the arrival of British ships off their shores in South America, made the American vessels similarly unwelcome. During the "Quasi-War" with France in the last years of the eighteenth century, French privateers also took their toll of American ships, as did the British ships in the War of 1812.

Only after the end of the War of 1812 did the American whaling industry begin to expand rapidly once again. At the same time the British industry—America's only serious competitor—began to founder. Whale oil prices declined steeply in England as rapeseed oil replaced whale oil in textile manufacturing and the rising British coal oil and coal gas industry forced the price of whale oil below the cost of its acquisition.

The American industry, however, was riding the flood tide of a growing nation and a strong market for oil. Despite abundant supplies of cheap manpower and raw materials, industrialization in the United States lagged, retarding the effects of the coal and petroleum industries on the whale oil market. Vessels from New Bedford, New London, and many other ports in southeastern Massachusetts, Rhode Island, Connecticut, and Long Island spread throughout the whaling grounds of the North and South Atlantic, then into the Pacific, following a route the British whalers had sailed since the last years of the eighteenth century.

The American ships departed from the familiar "on shore" grounds near the coast of Chile, cruising near the Galapagos Islands, on the "off shore" ground south of the equator between 105 and 125 degrees west, and "on the line," the 5,000-mile corridor that spanned the Pacific at the equator. The first whaleships reached Hawaii in 1819 and soon after they were on the Japan grounds, the waters extending far to the east of that country.

During the 1820s and 1830s the whaling fleets cruised throughout the Pacific from 50 degrees south to 40 degrees north chasing the nomadic sperm whales. The length of these voyages gradually increased from about fifteen months to about thirty months as the numbers of sperm whales declined. By 1840 sperm whale stocks in the Pacific were severely diminished. On the assumption that they had already combed Pacific waters south of the fortieth parallel for their whale resources, the fleet moved north.

Another factor was also drawing the ships north: the price of whalebone had begun to rise. New women's fashions called for full skirts, buoyed out by whalebone hoops. The demand caused the price of whalebone to double between 1840 and 1844. At the same time substitute illuminants were causing a decline in the price of sperm oil. The total value of the sperm whale, which lacked baleen, was overtaken by the value of the right whale's combined yield of oil and whalebone.

From 1843 to 1845 a large number of whaleships from several nations were cruising on the rim of the North Pacific, hunting right whales along the Kurile Islands, on Kamchatka, or in the Gulf of Alaska ("on Kodiak" and the "Northwest" ground). We know, for instance, that Captain Mercator Cooper took the ship *Manhattan* briefly into the Okhotsk Sea in 1845 and that Roys's *Josephine* and Sodring's *Neptun* and eight other American and French whaleships touched at Petropavlovsk in the summer of 1845. By the season of 1844 the *Whalemen's Shipping List* was listing ships as "on Kamchatka," indicating that those waters were already being regularly cruised.

Although Roys's voyage in the *Superior* was an innovative departure, the belief that "polar whales" existed in the Bering Sea was shared among whaling masters and merchants alike. Polar explorations were a popular subject in the mid-nineteenth century and probably whaling merchants had made note of the reports of whales in the Bering Sea and Arctic Ocean.

As early as 1845 ships had taken bowheads in the waters between Petropavlovsk and the Commander Islands. In the three following years a large number of ships cruised on Kamchatka and probably also heard of or saw these whales. "Ships recently arrived at Lahaina, from Kamchatka," the *Friend* reported in October 1846, "give a most favorable account of the fishery on that ground. Our correspondent informs us that it is not impossible that there will be an average of fifteen hundred [barrels]." Most important, on June 10, 1848—a month before Roys had reached Bering Strait—Captain Cyrus Manter of the *Ocmulgee* cruising on Kamchatka, lowered his boats for a polar whale which he identified as such. In the first few years after the *Superior*'s voyage a number of ships encountered "polar whales" near the Commander Islands, confirming the presence of bowheads in waters near where the *Neptun* and *Ocmulgee* had traveled.

Thus it is likely that even had Roys not made his bold voyage in 1848, within the next few years the whaling industry, drawn to the Pacific a half century earlier in pursuit of sperm whales, would have pushed to Bering Strait and beyond in search of the bowhead.

3 / *An Arctic Whaling Voyage*

The preparations for a whaling voyage began as soon as a whaleship returned to its home port. Within hours of reaching the wharf the ship was empty of its officers and men, and the ship's agent immediately put the vessel in the care of a shipkeeper, who had complete authority over her until she went to sea again.

The first business at hand was unloading the cargo. The shipkeeper and a stevedore oversaw the longshoremen as they hoisted the heavy oil casks out of the hold and stored them on the wharf. The casks, with the ship's name marked on them, were arranged in close order and then covered with a thick mantle of seaweed to prevent the sun from shrinking the staves and thus causing leakage. The baleen was stored in a warehouse, as were the ship's sails, whalecraft (harpoons, lances, and boat gear), and other equipment.

As soon as the men had secured all the movable gear, attention was focused on any repairs the ship might need. If the bottom needed work, the ship was usually "hove down" to expose first one side and then the other. The men rove a strong chain around the mainmast head and winched it down as far as sixty degrees from vertical by using heavy blocks and tackle moved by a powerful windlass on the wharf. Ship carpenters then boarded their rafts and began stripping away the copper sheets and pine sheathing and then scraping the bottom smooth to expose the seams in her planking. The caulkers dug the old oakum out of the seams and hammered in new. Then the workmen covered the hull with hot tar and tarpaper or felt. This layer they covered with inch-thick pine planks to sheath the hull in such a way that each seam was straddled by a plank. The men then nailed overlapping "copper" sheets—often a cheaper metal alloy known as yellow metal—onto the sheathing everywhere below the ship's maximum waterline.

The caulking, tarring, sheathing, and coppering protected the hull from a number of problems, principally leaks and wear, but also from the teredo worm and marine growth. The teredo worm, a sort of sea termite that lives in temperate waters, invades hulls lacking sheathing and tarring; unchecked, the worms will eventually honeycomb wood planking. The electrochemical action of sea water on the metal plates retards marine growth by causing the exposed plate surface to slough away. Because of this sloughing, barnacles and weed could only get a poor hold on the hull. The ship could better maintain its speed without the friction against the water that these unwanted passengers caused. Although whaleships were known as slow sailers, particularly the bluff-bowed vessels built before 1840, the whalers made as much speed as possible when sailing to and from the whaling grounds.

Once the hull repairs were complete, the ship was returned to an even keel and

Whaleship Josephine *hove down in New Bedford*

work began on the topsides and rigging. A master rigger directed his men in replacing old masts and spars and worn standing rigging. The ship was also painted. The cooper measured the hold and began building variously shaped casks to fill the space as completely as possible; a "barrel of oil" (31 1/2 U.S. gallons) existed only as a unit of measurement.

The stevedore and his longshoremen then began loading the ship. First, the largest oil casks, holding fresh water for drinking and salt water for ballast, were stowed in the "ground tier," the lowest level. Some of the casks in the upper tier (called the "riding tier") also held fresh water, but others held provisions, knocked down oil casks, or equipment. When they began taking whales, the salt water would be pumped overboard and the casks filled with oil. As the food and water were consumed the empty casks would be used to hold whale oil as well.

Hazard's Wharf, New Bedford, 1868. Sails drying on the bark Massachusetts

At the same time the ship's sails, whalecraft, and other gear were being delivered to the wharf. Much of the whalecraft and lighter equipment, which would be needed shortly after putting to sea, was stowed on the middle deck in casks in the blubber room. This deck, often referred to as the "'tween decks," was usually divided into three general sections: the afterquarters and steerage where, respectively, the officers and boatsteerers berthed; the blubber room where the great strips of the whale's fat were stored until they were tryed out; and the forecastle, the dark triangular place in the bow where more than twenty men bunked "in an odor more or less mephitic," according to one observer. While the loading was under way,

Oil casks on the New Bedford waterfront, ca. 1870

a mason was at work building the tryworks, nesting two try pots, great iron cauldrons each with a capacity of 200 gallons or more, in a bed of bricks and mortar above a fire box.

As the date of sailing drew near, the ship's agent—the principal partner among her owners—notified the outfitters that he would need a crew. The outfitters, known universally as "landsharks" because of their methods, then notified agents in other towns. The agents, in turn, provided the landshark with men and received a share of the premium paid for procurement. In the early years of the nineteenth century it was not difficult to ship a crew from the nearby regions of New England and New York. As the world's whale stocks were depleted, however, voyages became longer and less profitable, and fewer men were willing to sail before the

Casks holding shooks that will be made into oil casks as needed on the voyage. Merrill's Wharf, New Bedford

mast on a whaleship for a voyage of three or four years. The agents had to search farther afield and the crews became more polyglot and generally of a lower calibre.

The officers provided the outfitters with a modest profit and little concern, for they stood to gain from a successful voyage. With the crew, the situation was reversed. The landshark had already paid for the man's transportation to the port, for his outfit of clothes for the voyage, for his straw mattress (his "donkey's breakfast"), and for his board and lodging in port. With the prospect of a long voyage and only a modest profit looming, the sailor's ardor could easily wane as sailing day approached. The landshark consequently kept a close eye on his charges, for he stood to gain not only his premium should the man ship out, but also from

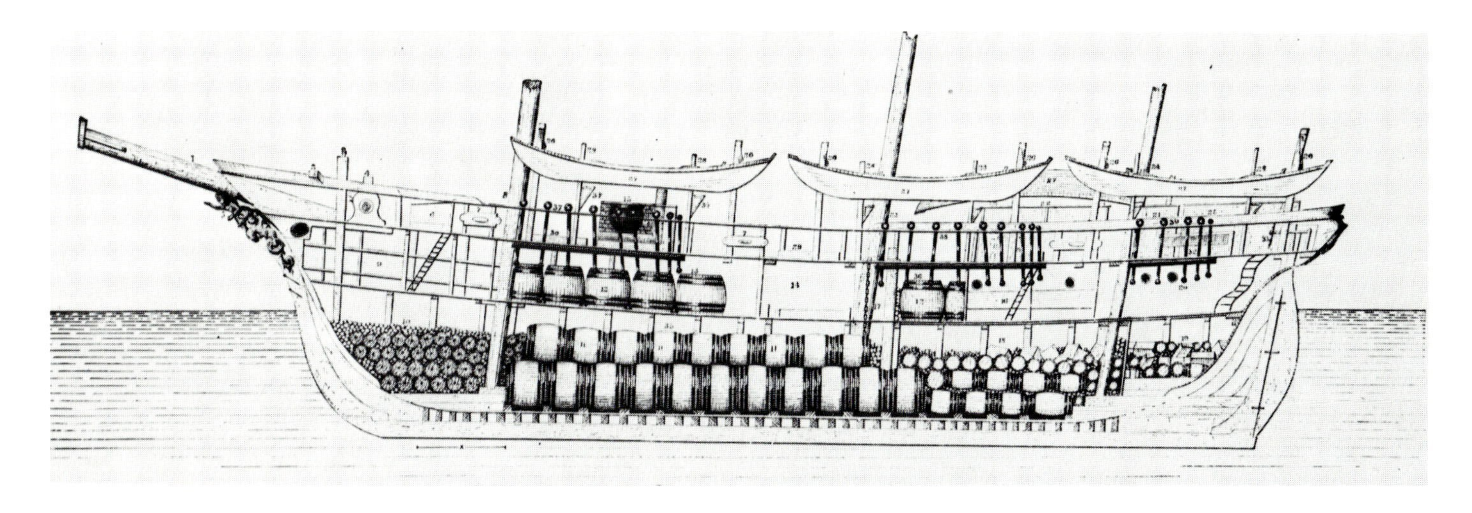

Sectional plan of the bark Alice Knowles. *From J. T. Brown 1887*

the exorbitant mark-up he had charged for the man's outfit. Walter Burns described his outfitting in San Francisco:

I was given a sailor's canvas bag, a mattress, a pair of blankets, woolen trousers, dungaree trousers, a coat, a pair of brogans, a pair of rubber sea boots, underwear, socks, two flannel shirts, a cap, a belt and a sheath knife, a suit of oilskins and sou'wester, a tin cup, tin pan, knife, fork and spoon. That was all. It struck me as a rather slender equipment for a year's voyage. A runner footed up the cost.

"Why," he said with an air of great surprise, "this foots up to $53 and your advance is only $50."

He added up the column of figures again. But he had made no mistake. He seemed perplexed.

"I don't see how it is possible to scratch off anything," he said. "You'll need every one of these articles."

He puckered his brow, bit the end of his pencil and studied the figures. It was evidently a puzzling problem.

"Well," he said at last, "I'll tell you what I'll do. Bring me down a few curios from the Arctic and I'll call it square."

I suppose my outfit was really worth about $6—not over $10.

These sums were paid to the landshark by the ship's agent within a few months of the ship's sailing; the agent, in turn, deducted them from the sailor's profits when the ship returned home.

The net profits from a whaling voyage were often dispersed in three roughly equal parts: one third for the owners; one third for maintenance of the ship; and one third for the officers and men. The whaling voyage was arranged as a cooperative venture with the officers and men receiving shares, or "lays," of the net profits. The captain, mates, boatsteerers, and cooper received lays from 1/8 of the net to 1/100, depending primarily on their rank but to some extent on the desire of the agent to secure their services. The able and ordinary seamen, steward, cook, and blacksmith took lays of

Building the tryworks on deck

1/100 to 1/160. The "green hands" (inexperienced seamen) and cabin boy received "long lays" from about 1/160 to about 1/200.

When the ship returned to port, the profits were often computed by determining the value of the ship's oil and baleen at the current market prices (although the agent could choose to sell them at any time in the future) and deducting from it the costs of outfitting, piloting, port charges, tow boats, wharfage, commissions on the sales, watchmen's fees, and many other items. In 1851 when the *Benjamin Tucker* returned to New Bedford from the Arctic on her fourth voyage with a cargo of 73,707 gallons of whale oil, 5,348 gallons of sperm oil, and 30,012 pounds of whalebone, at current prices the gross value of her catch was $47,682, the net, $45,320 ($525,712 in 1982 dollars), and an ordinary seaman on a 1/160 lay earned $283.25 ($3,285.70 in 1982 dollars) for two years' work.

But the seaman's actual profits were far less than that because his debts to the landsharks were deducted, as well as any charges he made during the voyage on the ship's "slop chest" (store) for clothing, knives, or other requisites. Often these prices were inflated. Many seamen ended their voyages in debt to the ship owners and were paid off with one dollar. And at times desertion and death during the voyage meant that only half of the original crew returned to the home port to claim their share of the profits.

Hohman has calculated that the average earning of an ordinary seaman on a whaling voyage in the period from 1840 to 1860 amounted to less than twenty cents per day, equal to one-half to one-third of the pay for a comparable job ashore. The owners usually did better—but not always. Quite apart from the cost of the vessel, the outfitting costs could amount to $25,000 to $35,000, depending on the size of the ship and length of the voyage. In 1837, during a profitable phase of the industry, 25 percent of the voyages were losing propositions; and by 1858, a time of slump for the industry, the number of losing voyages was greater than 64 percent.

One may ask why, with these odds, did owners risk their money in the whaling industry. In fact, there were wide fluctuations in the fortunes of whaling voyages, and some brought in big profits. As an example, the ship *Envoy* of New Bedford sailed in 1848 and with two seasons in the Arctic brought in a net of $138,000. To protect themselves against losing voyages most of the owners divided their capital among several ships and partially insured the ships and cargoes against loss or damage.

Despite these precautions there was no substitute for sound business judgment in the whaling industry. The agents paid close attention to the quality of their captains and officers and used a fine pencil to calculate the costs of outfitting their ships. They also took great pains to gather intelligence on the productivity of new whaling grounds to better direct their efforts. The New Bedford firm of Aiken and Swift, for instance, issued its captains a detailed instruction book on the potential of the various whaling grounds and on the best times to cruise, emphasizing that the information was to be kept in the strictest confidence. But beyond this, and beyond their ability to sell the cargo at the most favorable moment, lay only the abilities of the captain, officers, and crew. The captain had to be allowed great discretion, and without skillful officers, energetic, capable men, and a measure of luck, no voyage could be a success.

On sailing day a party of the owners and their friends went aboard the ship and the tug towed it out of the harbor while some of the seamen, who had been dumped aboard barking drunk by the landsharks, slept off their thick heads in the forecastle. When the ship was ready to sail, the well-wishers went aboard the tug, leaving the officers to cope with the men, many of whom were seasick, hungover, or totally inexperienced. No doubt it was then that the full reality of a four-year voyage began to dawn on the new hands.

The captain soon called the seamen aft and gave them their specific orders. On one voyage Captain Edward Davoll began:

Whaleship Niger *outward bound*

My men, you must be well aware that the object of this voyage is to get a cargo of oil. To do this, it becomes necessary that some system be established in order to effect our purpose, and it becomes my duty to lay it before you, which you will strictly observe and abide by through the voyage. I am here to conduct the voyage; the rest of you to aid and assist to the best of your abilities, according to my order, which will, after this, be made known by and through my officers.

After warning the men not to fight or be lazy and wasteful, he continued:

When you are at the masthead on the lookout for whales, sing out for everything that you see. If whitewater, sing out "There She White Waters!"; if a breach, sing out "There She Breaches!"; if a spout, sing out "There She Blows!"; if blackskin, sing out "There She Blackskins!"; if a sail, sing out "Sail ho!!" Always sing out at the top of your voices. There is music in it. It makes things lively and cheerful and we shall never run past any sperm whales then and call them something else. We have glasses on purpose to make out any and all things which may appear, and before passing anything of the whale species I want them brought into use, so as to make sure what it is we see. I shall have a bounty up for a man that raises the whales and none gets it that does not sing out. . . .

Don't let yourselves be heard to grumble in any way. I and the officers can do all that. Grumblers and growlers won't go unpunished. Your places when on deck is forward of the tryworks where you are to remain at all times unless you have orders to come aft. Remember what I have said to you from now to the end of the voyage. I now dismiss you. Go Forward.

With this preamble completed, the next business at hand was choosing the watches. On a bark with five whale boats—the most common size and rig in the mid-nineteenth century—there were about thirty-five men. Twenty seamen were berthed "before the mast" in the forecastle. The remainder included the captain, four or five mates, five boatsteerers (harpooners), a cooper, a carpenter, a steward, a cook, and a cabin boy. A number of captains brought their wives and children.

The first and second mates, the "watch headers," chose the men for their watches. One of the two watches was always on deck, the other off duty. There were seven watches in a twenty-four-hour period, five of four hours and two of two hours. The short watches, called the "dog watches," allowed the men to shift watches each day. All hands, however, were called to chase whales and to trim sails in heavy weather.

The captain and mates chose the crews for their whaleboats in order of rank. Each tried to get the best men from the pool and the fourth or fifth mate had to take the leftovers. In the first years of the Arctic fishery the captain, as well as the mates, commanded a whaleboat. When chasing whales, they left the ship under the guidance of "shipkeepers," the steward, cook, cabin boy, cooper, and carpenter. As the Arctic fishery wore on, however, the ships pushed farther north and the complexities of navigating amid the Arctic's fog and ice forced the captains to remain aboard to keep the ship free of danger and to direct the chase from the crow's nest while staying as close to the boats as possible. When this change occurred, the fifth boat went under the command of a fifth mate, or "boatheader," often a man of talent but limited experience who was likely to be "fleeted up" should a man of higher rank be disabled or derated.

At times the presence of whales spurred the captain to join the chase. In August 1893 Captain George Leavitt of the *Mary D. Hume* recorded:

(All boats already down.) At 3 A.M. I took the cook, cabin boy, carpenter, Smithers and lowered spare boat. At 5 A.M. I struck. . . . When I struck Smithers was thrown out of the boat and things were sort of mixed up for a minute or so. Got Smithers into the boat again. [The whale] took nearly all my line, as I could not hold on while Smithers was overboard. Boy and carpenter, scared, had to growl at them and finally got them at work. Whale was breaching when I struck and filled the boat half full of water. Fun there for a little while.

With the preliminaries over, while the ship headed south into the Atlantic, the officers began teaching the men the operations of the ship. Most important, of course, was the procedure for catching whales, and because whales might be raised any time after clearing port, the boats had to be rigged and the men had to be taught how to work them.

The American whaleboat stands as one of the supreme achievements of nineteenth-century watercraft design. Light, fast, seaworthy, inexpensive, and maneuverable, it contained nothing superfluous to its purpose. But had the green crewmen been aware of its noble history, it is unlikely they would have been greatly encouraged as they eyed the thirty-foot double-ender with its thin cedar planks and low freeboard, knowing they would often be in it far at sea, miles from the ship, amid ice, fog, murk, or fighting whales.

The whaleboats were carried in davits on the side of the ship. On an American bark there were generally three on the port side and two on the starboard side, with a gap between the starboard boats to allow space for processing the whale. The boats were carried in either of two positions (on the upper or lower cranes) depending on the state of the sea and the readiness of the ship for whaling. Arctic boats were usually thirty feet long and were often built slightly stronger than those used in temperate seas.

To the uninitiated the American whaleboat carried a bewildering amount of gear. It carried a sail, mast, and boom of any rig that appealed to the boat's officer. Sailing was the quietest way to approach a whale. Although the oarlocks were muffled, rowing was noisy; consequently, the oars were used principally to pull upwind of a whale or for towing. The paddles were quieter and the men used them to approach an unsuspecting whale or to gain extra speed under sail. The mate steered the boat under sail with a rudder, which he could quickly unship and draw up. When at close quarters with a whale he replaced the rudder with a twenty-foot steering oar that allowed him to pivot the stern to stay clear of the flukes.

In the early years of the Arctic fishery, before the advent of explosive weapons, whaleboats carried two harpoons in the bow. The harpoon was a composite implement consisting of a hardwood shaft five or six feet long, the "iron" (a razor-sharp barb on a two-and-a-half foot soft metal shaft) and a rope "iron strap" that lashed both pieces together. The term "harpoon" seldom entered whalemen's parlance; "iron" was used to indicate the entire instrument. Of the two irons carried in the bow, the "first iron" was fastened directly to the whale line; the "second iron" had a short warp that was looped around the whaleline. If the boatsteerer (who was the harpooner) did not have time to strike the whale with both the first and the second irons, he immediately threw the second iron overboard to prevent it from thrashing around wildly in the boat as the whaleline raced through its loop.

The whale line was made of strong, loose-laid manila rope, two inches in circumference with a tensile strength of 6,000 pounds. Two separate lengths of line were coiled in two tubs placed between the aft thwarts. The first tub held 225 fathoms (1,350 feet). The last few feet of the line in the first tub was led out over the edge so that if the whale took the length, the end could be quickly tied on to the line in the second tub, which held 75 fathoms (450 feet). If the whale dived deep or ran into the ice, the end of the line in the second tub was rigged the same way, ready to be bent onto lines from other boats. Because bowhead whales frequently ran for the safety of the pack ice, Arctic whaleboats often carried larger tubs, holding altogether more than 400 fathoms of line. Captain Hartson Bodfish once watched a running bowhead take more than two miles of line from five boats.

In both tubs the line was coiled carefully in tight spirals so that it could run out at blistering speed after a diving whale without kinking. A kink jamming in the bow chock could pull a boat under in a flash.

From the first tub the line was led aft to a post at the stern called a loggerhead. Two turns of line were taken around the loggerhead to allow the men to brake the speed of a running line. The line was led from the loggerhead forward through a rope bridle (the "kicking strap") fastened to the thick bow thwart, then through a chock at the bow, back to a small open compartment in the bow called "the box," and finally was attached to the two irons. Ten to twenty fathoms of line (the "box warp") were loosely coiled in the box to provide slack so that when an iron was thrown or darted, tension on the line would not deflect it from its course. The kicking strap prevented the line from towing the boat sideways or stern first, should it jump out of the bow chock when fast to a whale.

The whaleboat also carried three lances for killing the whale. The lance had a razor-sharp blade at the end of an iron shaft five or six feet long, and, like the harpoon, this piece was mounted on a hardwood pole. In the boat were also two or three extra harpoons; a toggle ready to be thrust through a dead whale's lip in order to attach a towline; a "boat spade" (a spatulate blade on a short shaft for cutting the hole in the whale's lip for the towing toggle); a bucket for wetting the whale line to keep it from catching fire as it rasped around the loggerhead behind a running whale; a piggin for bailing; a keg of drinking water; a drogue to attach to the whale line; several "waifs," flags for marking the whale's carcass; two knives; a grapnel; a hatchet to cut the whale line in case it became fouled on the boat and threatened to take the boat under; a fog horn, a boat hook; a lantern keg holding matches, candles, a lantern, some food, tobacco, and emergency equipment; as well as a few other items, depending on the tastes of the mate and boatsteerer.

After the advent of explosive weapons in the 1850s, the Arctic boats also carried two darting guns for wounding the whale while simultaneously fastening to it, a heavy, smooth-bore shoulder gun for firing a "bomb lance" into the whale, two darting bombs, and six shoulder bombs.

The crewmen's duties were well established. The mate, or boatheader, was the officer in command. He steered the boat to the whale and made all decisions until the moment of striking. Without a skillful boatheader who knew the behavior of the various species of whales, it was difficult to get close enough to strike. The boatsteerer was equally important. He was the harpooner, a sobriquet rarely used by American whalemen. The boatsteerer's skill in fastening the iron was essential. If he missed a fair chance to strike more than once he would face an irate captain and perhaps be sent forward to join the seamen while another man would be assigned his place.

Once the boat was fast to a whale, the boatsteerer and boatheader invariably changed places—hence the nomenclature—and it became the mate's duty to kill the whale. In the first years of the Arctic fishery (before explosive weapons came into general use), this was done by the crew pulling hand-over-hand up the whale line to the tiring whale. The mate would then pump the lance in and out of the whale ("churning it") until he had struck the whale's "life" (the heart or lungs) and it began to spout thick blood.

Behind the boatsteerer sat the bowman, who pulled the second oar. He was the most experienced crewman in the boat and took charge of raising and lowering the mast, which was stepped in a hinged wooden collar. He lowered the mast the instant they struck the whale. Behind him, in order, were the midship oarsman and tub oarsman. The latter took charge of pouring water on the whaleline at the loggerhead to prevent it from burning when a whale was running or sounding rapidly. The after oarsman was also the principal bailer and he assisted in lowering the mast.

These duties, as well as ship handling and many other skills, were drummed into the men as they headed across the Atlantic toward the Azores. By the 1850s many whaleships left port short-handed. The captains counted on completing the crew in the Azores or Cape Verde islands where labor was cheaper than in the United States. For their part, many of the islanders welcomed a chance to escape conscription and poverty by shipping aboard a whaleship. Many chose never to return to their homelands, taking up residence in the port cities of southeastern New England, where they formed the nucleus of the region's Portugese and Cape Verdean communities.

A few of the ships bound to the Arctic entered the Pacific via the Indian Ocean, but most chose their sailing dates so that they could take the shorter route, entering the Pacific via Cape Horn during the southern summer. They planned their routes to take them across the whaling grounds of the south Atlantic and the west coast of South America before they swung westward to run along the equator, then north to the Japan grounds, then southeast to refit in Hawaii in March.

Because they were situated centrally in the Pacific, the ports of the Hawaiian Islands were the most often visited by Pacific fleet whaleships for refitting, but the ships also touched at Sydney, Hobart, Hong Kong, Hakodate (after 1853), Ponape, Guam, San Francisco, Talcahuano, and Tumbes, among others, to take advantage of cheap provisions and labor. Many of the whaleships accumulated substantial cargoes of oil and whalebone during their winter cruises and used the ports to transship their products by freighter or returning whaleship. In these ports they also took on wood, water, and fresh food (vegetables, fruit, fowl, pigs, goats, sheep, and cattle) and new crew to make up for men lost through death, desertion, or disability. Many of the recruits—Hawaiians ("Kanakas"), Maoris, "Guamies," European beachcombers, and members of many other groups as well—shipped out for the remainder of a voyage, but others, particularly the Hawaiians, went as "seasoners" only for the northern cruise.

After a week or two in port, the ships put out to sea, bound to the Arctic. The livestock was penned on deck, to be slaughtered as needed over the next two or three months. When the animals were gone, the fare returned to a routine of salt beef ("salt horse"), salt pork ("salt junk"), beans, potatoes, and the like, with "duff" (a mixture of flour, lard, raisins, and yeast) to relieve the monotony. But even the duff could run out. "We get no duff now at all," wrote Orson Shattuck in Bering Strait, August 1851. "Nothing for dinner but hard bread & salt junk, with the exception of rice and bugs twice a week and some beans that were raised by old father Noah before the flood."

The voyage to the pack ice of the Bering Sea usually took four to six weeks. The first duties at sea were sawing and storing firewood. Then, with the temperature falling as the ship neared the Aleutians, the captain broke out the slop chest to allow the men to buy the warm clothing they would need for the high latitudes. A stove was set up in the forecastle. Without the warmth of the stove, hoarfrost formed on the forecastle walls from the moisture of the men's breath and, with the first warm weather, the thawing hoarfrost made the forecastle even damper and nastier.

The whaleboats had been unrigged before starting north and had been either moved inboard and stored upside down on the boat racks or put in the upper cranes to avoid damaging or losing them in the mountainous seas of the North Pacific. Ships from Hawaii usually entered the Bering Sea via the most direct route, through "Seventy-two Pass" (Amukta Pass). The men began to rig the boats once again, putting the upturned boats back on the cranes, checking the condition of the boat falls in the davits, and breaking out the whalecraft from the blubber room.

The men scrubbed and painted the boats, fitted them with their spars, checked the whalecraft carefully for any signs of weakness or wear, and sharpened the irons and lances. Any new whaleline they uncoiled and towed behind the ship for a day or so to work out any kinks that might cause it to foul as it ran from the line tub. They then coiled it loosely on deck and allowed it to dry before placing it in tight flemish coils in the tubs. The men also dismantled and cleaned the windlass, checking its pawls for fitness to haul up the massive pieces of blubber that could weigh many tons. They scrubbed out the try pots so that the oil would not be discolored by dirt or rust.

Next the cutting tackle had to be prepared. The massive blocks allowed the men to haul the blubber up from the whale's carcass. The axles of the sheaves, the block straps, and the line (the "cutting falls") were carefully inspected to make certain they could withstand the strain of heaving up the blanket pieces of blubber. The tackle was then sent aloft to the mainmast head, where it was chained or secured with very heavy cordage.

The men then put together the cutting stage, an arrangement of three strong eighteen-inch-wide planks and a hand rail, which would be rigged outboard on the starboard side, ready to lower over a whale. Once lowered in position amidships, the officers stood on it and from that vantage point, helped strip the blubber from the whale's carcass with their long-handled blubber spades.

Once the whaling gear was ready, the men turned to preparing the ship for the ice and seas they were approaching. Because they would not need speed while cruising on the whaling grounds, they often unrigged a number of the uppermost sails, spars, and masts, and sent them down. At the foretopmast crosstrees they set up a "crow's nest," a box-like framework of wood covered with a dodger of canvas. From the crow's nest the officers and boatsteerers kept watch for ice and whales, protected up to the shoulders from the piercing wind.

While making a passage across the deep waters from Hawaii to the Arctic, anchors were both unnecessary and an encumbrance on deck. The whaleships consequently carried both main anchors securely lashed to the inboard rail with their ninety fathom chain cables unbent and stowed below decks. The hawse holes were plugged against the sea. Once past the Aleutians, the anchors were rigged again.

Arctic ships carried heavier anchors and chain than other whalers. The waters they traversed north of the sixtieth parallel were uniformly shallow, less than thirty fathoms. The ships could anchor anywhere and often needed to because a lee shore was never far away. The main anchors weighed a ton or more. Often, however, instead of carrying two heavy and cumbersome anchors, they replaced one with a five- to seven-hundred-pound kedge anchor. The lighter ground tackle was easier to use during the frequent stops and starts that were necessary amid the drifting ice.

The whalemen knew that they would have to go into the pack ice of the Bering Sea if they were to stay near the whales as they migrated toward the Arctic Ocean. Although the rotting pack ice in the Bering Sea in the spring is flat and rarely thicker than six feet, the winds and currents can quickly close the leads of open water that marble the ice, completely surrounding ships in vast tightly packed fields of broken ice floes.

Unlike the conditions in the Chukchi and Beaufort seas, the ice did not often crush ships in the Bering Sea. The whalemen, however, feared the action of the ice floes on the rudder. Even a small drifting floe coming in contact with the rudder could rip the retaining chains away and splinter the rudder off its pintles. To make matters worse, if a storm were blowing in the southern Bering Sea, the swells it sent out could be felt far to the north, deep in the pack ice, where the violent heaving of the floes could be equally damaging to the steering gear.

To give themselves some measure of protection, the whalemen attached tackles to each side of the rudder, bringing the lines in over the ship's quarter to insure extra purchase against the force of the ice. Occasionally, however, despite these precautions, the force of the ice would be so great as to straighten out the hooks on the tackles and snap the pintles right off.

Once these preparations were completed the men confronted some of the world's most dangerous waters. Not only was ice a constant threat, but the area was the home of unpredictable currents, violent sudden gales, and shoals that were poorly charted at best. From 1848 to 1914 the gales, the ice, the shoals took their toll of more than 150 whaleships north of the Aleutians—and countless more were damaged. It is a testimony to the skillful seamanship of the whalemen that these losses amount to only about 5 percent of the total number of more than 2,700 whaling cruises in those waters.

The Bering and Chukchi seas together resemble a crude hourglass with the waist being the narrow fifty-mile-wide gut of Bering Strait. These names, however, did not come into standard usage until the 1950s, nor did our knowledge of their physical geography expand greatly until that time. The whalemen developed their own terminology for the North Pacific whaling grounds. They divided the Bering Sea into four major sections. The southeastern corner, east of the Pribilof Islands, they called "Bristol Bay." No bowhead whales were found in the region but right whales could be taken in July, August, and September. The "Kamchatka Sea"—the waters lying west of the Pribilofs and south of Cape Navarin—made up the vast majority of the present Bering Sea. The "Anadyr Sea" included Anadyr Gulf and the waters southeast. "Bering Straits" comprised the waters from Saint Lawrence Island to just north of Bering Strait; that is, all of the constricted passage between the two seas. Lastly, all waters north of "Bering Straits" were the "Arctic Ocean," or simply "the Arctic."

These distinctions in fact accord well with our knowledge of the region. The Bering Sea is basically divided into waters of two radically different depths. The entire area southwest of a line between Cape Navarin, Siberia, and the Alaska Peninsula is frequently as deep as 2,000 fathoms (12,000 feet), whereas the rest of the sea and all of the Chukchi Sea are part of the continental shelf and are generally less than thirty fathoms (180 feet). The shallow part comprises the submerged Bering land bridge, or Beringia, which during the Pleistocene was successively dried and flooded by the sea, the result of the expansions and contractions of the great continental glaciers.

Because of the last climatological warming and the consequent release of vast amounts of water from the glaciers to the oceans, the land bridge was submerged about ten thousand years ago. The rivers and waves and ocean currents then began their work, re-shaping the bottoms and coasts to their present form. Today the principal movement of the water in these seas is the great current that surges north out of the Pacific, entering the Bering Sea between the Commander Islands and the westernmost Aleutian Island, Attu. The current's main mass then runs northeast toward Bering Strait, while a small branch doubles back at Cape Navarin and travels southwest along the Siberian and Kamchatkan coasts and re-enters the Pacific. The main body, flowing at a rate of about 800,000 cubic meters of water per second, forks around Saint Lawrence Island, through Bering Strait, and is deflected by the Cape Thompson-Cape Lisburne peninsula. It flows over Herald Shoal and east of Herald Island before leaving the Chukchi Sea. A branch splits from the current at Point Hope and runs northeast along the Alaska coast to Point Barrow, where it forks again. This main branch continues north, but a weaker branch turns east along the shore into the Beaufort Sea. Another current, unrelated to the others and weaker, enters the Chukchi Sea from the east Siberian Sea via Long Strait and flows southeast along the north coast of Siberia to Bering Strait.

For the whalemen the great current from the Pacific was both their lifeblood and their peril. Transporting sediment, it built the low and dangerous sandspits and shoals at Indian Point, Cape Prince of Wales, Point Hope, Icy Cape, Point Franklin, and Point Barrow. Far more important, the entire chain of life in those waters is linked to that current.

The Bering Sea is the world's third largest, being smaller only than the Mediterranean and South China seas, but it supports one of the world's richest marine ecosystems in which flourish 170 plant species and 300 faunal species. The animal species, including the bowhead whales, are supported by the phytoplankton (microscopic algae), which in turn is nourished by this great ocean current that sweeps north out of the Pacific at abyssal depths and then rises over the continental shelf carrying a wealth of nitrates, phosphates, silicates, and trace elements. These mix with the oxygen-rich upper waters and produce a fertile broth that has been calculated to produce about 274,000,000 metric tons of phytoplankton per year. The mass of phytoplankton allows a concentration of zooplankton (tiny marine animals that feed on phytoplankton) of as many as 10,000 individuals per cubic meter of water. This phenomenal plankton bloom is seasonal and occurs for the most part at the edge of the melting pack ice. It is triggered by the action of the sunlight as the ice recedes in the lengthening Arctic days. As the ice melted, the whalemen hunted the bowheads grazing the lush marine garden at the edge of the pack ice.

The whalemen learned that where they saw "feed slicks"—oily slicks on the surface of the water indicating dense concentrations of plankton—bowheads were likely to be nearby. The whalemen learned to locate the slicks by watching for "bowhead birds," phalaropes that feed on the plankton.

The whales graze through these slicks. They also feed on blooms under water and on the bottom. On the surface they swim with their mouths wide open so that the water passes through the three hundred or more baleen plates that hang, lattice-like, from each side of the upper jaw. After a few minutes, the whale slowly closes its cavernous mouth, expelling the water and swallowing the mass of plankton ensnared by the hairy fringe on the inside edge of the baleen plates. The whale then often blows from six to nine times and dives. If undisturbed, it generally stays down for ten to twenty minutes before rising to feed again. Thus do the tiny organisms sustain bowheads that can grow to a length of seventy feet, weigh sixty tons, and carry a coat of blubber as much as a foot and a half thick.

These great, gentle animals, so perfectly adapted to their environment, move slowly in the Arctic, swimming about two miles per hour while feeding, following the plankton blooms and cutting great visible swaths through the surface slicks as they go. Their only natural enemy is the killer whale and their only impediment to movement is the ice, which at the same time provided protection from their unnatural enemy, whalemen.

The pack ice in the Bering Sea reaches it southernmost expansion in late March.

Its actual boundary may vary as much as three hundred miles depending on the winter's weather, but in most years the maximum margin is never far from the edge of the continental shelf. By late April, with the rising temperatures of spring, the pack ice begins to loosen as leads (lanes of open water through the pack ice) and polynyas (spaces of clear water) increase. In the spring the bowheads move north through these open spaces. The whalemen quickly learned that most whales disappeared into the pack ice in April, and if the ships were to make a respectable spring catch, they must follow the whales through the ice to intercept them in the polynyas at Bering Strait before they vanished once again into the temporarily impregnable fastnesses of the Chukchi Sea.

By mid to late May the ice in the northern Bering Sea is in an advanced state of disintegration, leaving only the largest and heaviest floes to persist until the end of June or early July. In the Chukchi Sea, however, the disintegration of the ice takes place later. A long lead develops along the Alaskan coast in March and April and the first bowheads use this narrow waterway on their migration to the Beaufort Sea. By late May or early June the warmer water entering from the Bering Sea creates substantial polynyas in the southern Chukchi Sea. Although abnormal periods of prolonged northerly winds can completely block Bering Strait, keeping the whales from moving north, by late May most of the whales have usually entered the Chukchi Sea or are already in the Beaufort Sea.

The Chukchi pack ice continues to disintegrate throughout June, July, and August. Its margin retreats progressively farther north under the influence of wind and current and generally reaches its northernmost retreat in early September. The limits of its retreat are determined by the weather conditions in any year. Logically enough, south winds create the greatest recession both by their own warmth and pressure and by raising the speed of the current, and north winds produce the opposite effect.

In the autumn the pack ice of the Chukchi Sea for the most part does not advance southward; rather, the deepening cold of September and October creates the growth of ice along shore and in the center of the sea. In addition, some young ice is carried down into the southern Chukchi Sea by the current that runs out of Long Strait (between Wrangel Island and Siberia), and with the strong northerly winds that prevail in the autumn some chunks or fields of old ice (ice that formed more than a year earlier and is consequently denser and heavier) are carried down as far as Bering Strait or occasionally to Saint Lawrence Island.

Most of the whales leave the Beaufort Sea in late August and September and move to feeding grounds near Herald Island and Herald Shoal. A few go farther and enter the East Siberian Sea. During September and October many are seen on the north coast of the Chukchi Peninsula. In late October and early November they are near Bering Strait and in the bays on the south coast of the Chukchi Peninsula. Some whales remain all winter on the south coast of the Chukchi Peninsula, but others move farther south and remain from December to March near the ice edge in the northwestern Bering Sea.

In the Bering Sea ice begins to form in October and November and advances until its maximum coverage is reached in early March on a line from Bristol Bay to near Cape Navarin and thence southward along the Siberian coast to Karaginski Island. The growth of the ice edge at this time occurs more from the action of the north wind pushing it south than because of low temperatures forming it outward.

It was amid this changing seascape that the whalemen sought their quarry. Arctic whaling cruises presented the whalemen with two different sets of problems: in April, May, and June the men were forced to work through the ice to intercept the bowheads in the northern Bering Sea; in July, August, and September they stayed near the pack ice (but not in it), avoiding the shoals and gales while following the whales in the northern Chukchi Sea.

In the Bering Sea the whaleships usually reached the edge of the pack ice in mid-April. They found that although the center of the pack seemed more dense, frequently there was more open water on one side or the other, depending on the prevailing winds. For the most part, however, the whalemen used the western route to the polynyas of the Straits. Their route usually took them toward Cape Navarin, a high black rocky mass that is visible for fifty miles at sea. When the ships came within sight of it, they often simultaneously found the ice edge, and when they found a promising lead, they "entered the ice" on their six-week race toward Bering Strait.

The ships were frequently becalmed on their passage through the ice, and just as frequently the ice would mysteriously close around them, imprisoning them in a vast field of broken floes, putting them at the mercy of the wind and current. The current rarely posed any danger to the ships, but if they were unlucky enough to get into the branch that runs southwest along the Siberian shore, they could be carried helplessly away from their destination while the rest of the fleet forged ahead after the whales.

The whalemen also found that they could make the current work for them. If they could find a big enough floe, one that was very thick, hence deep in the water, they could hitch themselves to it while it was driven along by the northerly current, moving through thinner ice on the way. In 1886 the New Bedford bark *Sea Breeze* furled all sail in a northeast gale, tied up to a "cruiser," and was carried forty-five miles to windward in three days.

But "it was oak against ice," as one Captain put it, describing the push toward Anadyr Gulf when some open water appeared before the ship. "Everything seemed very strange," wrote Herbert Aldrich aboard the *Young Phoenix* in 1887:

> On every side was ice; at the northeast was the pack; at the west was the shore—high and snow-clad—just south of the bight under Cape Navarin. The thermometer recorded only twenty degrees above zero. . . . Two men were aloft in the crow's-nest to pick out a path for the ship, and watch for whales; a third was on the bowsprit, or try-works, to steer clear, as much as possible, of large cakes of ice, and the watch on deck was ready, at any instant, to wear or tack ship. As we wormed our way along there was a constant flow of commands: "starboard!" "steady!" "port!" "steady!" "let her luff a little!" "steady!" as we passed in and out among the cakes of ice.

The Hunter *approaching Cape Navarin, Siberia*

Collisions with the ice floes were unavoidable, expecially when it was dark or foggy. Usually the ship's copper sheathing and reinforced bows were strong enough to withstand the impact, but a number of ships—especially in the early years of the fishery—were badly stove, or lost, when they rammed headlong into ice floes in the Bering Sea. Occasionally, too, a ship would be stove without the crew sensing it. When the *America II* hit a small cake of ice in 1851, the captain was not concerned until a few hours later, when the steward reported that water was coming over the combings of the after-hatch. The men hardly had time to grab a few provisions and lower the boats before the ship rolled over on her beam ends. The crew made their way to nearby vessels, and the captain sold what remained of his ship for fifty dollars.

The John P. West *becalmed in Arctic waters* Denver Museum of Natural History

If the ship was stove and the men discovered it in time, then the only recourse was to cover the weak point with a sail to slow the flow of water, to rig a protective mat over that and to keep the ship afloat with her pumps while making for a good harbor—generally Rudder Bay, Plover Bay, or Saint Lawrence Bay on the Chukchi Peninsula, or Port Clarence or Chamisso Island in Alaska. Once there they had to shift the vessel's cargo to expose the damage, or, if the hole was well below the waterline, they had to unload the ship, dismantle the rigging, and set up jury tackle to heave her down.

Captain George Bauldry stove the *Arnolda* near Point Barrow in 1873. His efforts to save his ship were not uncommon:

I stove in one bow, breaking four planks and four timbers. To press it back into place, it was necessary to cut off three more timbers. I was advised to abandon the ship, but declined to leave her until I was obliged to. By the time I got the last timber cut off the hold was full and the water had reached the between-decks. We had the blubber of three whales on board, so that the main hatches could not be got at to bail out and aid the pumps: but by working in the water, I got a tarred blanket over the break, then put boards over, and, with spars, wedged the bow back into place. Had she not been a live-oak ship, the pressure on the other bow would have broken that out. To add to the confusion, the boat-steerers got at a keg of rum, and all were dead drunk when I got on deck.

In this crippled condition I went down to Plover Bay [more than 700 miles from Point Barrow], and there, by the aid of Capt. W. H. Kelley, and others, got the bow out of water, filled up the cracks with sawdust, put on tarred canvas, then planked it all over. Thus a check was put on the leak. We got out of Behring Sea all right, but just below the Aleutian Islands encountered a gale which threatened to swamp us. It would have done so had it lasted long, for the leak gained on us, though the pumps were going all the time. When we arrived at Honolulu we were leaking 20,000 strokes in twenty-four hours.

If the men found themselves being trapped in a closing pack, they redoubled their efforts to get clear, knowing that the pack could act like a vise on their craft. "Our ship drifted into the ice," wrote the mate of the *Cossack*. "It made her groan." The only way out was by force, and many ships rigged rope mats over their bows, using old rigging, whale line, or whatever was available to soften the blows as they tried to ram their way out.

Other ships escaped by using their boats to tow them to safety, or, if the men had no room to maneuver, by "box-hauling," backing all their yards and sailing the ship stern-first. If the ice nearby was heavy enough, the men could attach their cutting falls to it and use the windlass to winch their way out. Other ships sent their boats ahead to drop their anchors and then used the windlass to kedge themselves out. Some captains as a last resort filled bottles with gun powder and shoved them under the ice on long poles to blast their way clear.

If the pack closed around them and there was no way out, the best they could do was to seek the protection of a bay in a particularly large piece of ice, hoping that it would save them from the pressure of the pack. As the floes closed about them, some captains rigged their anchors from the ends of their yards and used them like pile drivers, trying to break off the menacing points of ice before they could crush the hull.

If the ships were caught without the protection of a large floe, they were equally afraid of being pounded to pieces by the heaving and grinding floes. It is eerie to be surrounded by the pack ice with a strong wind blowing. Within the pack the ice deflects the wind from the water, leaving it often like a mirror. Sooner or later a light swell arrives from outside the pack ice. It moves the ice floes almost imperceptibly. The swell soon grows, driven in by a storm blowing far away. The noise on the hull rises to a crunching hiss, then a crashing and grinding shriek. Today, sitting out one of these storms in a steel ship, it is easy to understand how a wooden ship could be stripped of her copper and sheathing and have her rudder sprung or ripped right off as the ship heaved and rolled in the ice. You can feel the terror of the mate on the *Sea Breeze* who wrote, "a tremendous heavy sea. Ship pounding fearfully on thick ice." The next day he added, "We have found cutwater, copper & sheathing gone & rudder broke & ship badly damaged."

Blasting through the ice. The man at the right is lighting the fuse; the other two will push the charge under the ice on the pole Private collection

After the month-long passage to Anadyr Gulf, the ships usually sailed into loosely scattered ice near Cape Bering. For the next month and a half they would enjoy the triple luxuries of the best weather of the Arctic summer, twenty-four-hour daylight, and increasingly open seas. These pleasant conditions were, of course, a secondary matter in their search for whales, but, fortunately for the whalemen, they also enjoyed reasonable success in the waters near Bering Strait during the first twenty years of the fishery.

Especially at this time of year, with warm winds passing over cold seas, their main problem was the fog. Bering Sea fogs are famous for their dank density. Time and again the men cursed the fog as they tried to avoid the ice and the land. The mate of the *Minerva* wrote on Independence Day, 1865, "Celebrated the day by cutting the fog with a knife." Or the mate of the *Lydia,* "Thick as burgoo. You cannot see the jib boom end half the time." or on the *Condor,*"Thick fog. Tacking & re tacking and re re tacking to keep clear of the ice."

It was often easier to keep clear of the land than the ice. The rocky headlands of Bering Strait support colonies of millions of seabirds and whalemen learned that when they heard the incessant cries of the seabirds, the ships were usually approaching a cliff. They located headlands by firing a cannon into the fog and listening for an echo. Most important was the use of the lead line for soundings. The bottom is regular north of sixty-two degrees north and any abrupt rise signals approach to land. They also kept track of their position by the color of the water. The water on the eastern side of the Bering Strait region is often greenish and full of silt from the Yukon River.

The fog not only hid the land; it concealed nearby ships as well. When the men thought they were near other ships they constantly sounded their fog horns or fired guns. In one instance near the turn of the century the fog horn was surprisingly effective:

> Midway of our run down the Behring Sea a thick fog closed about us and we kept our fog horn booming. Soon, off our bows, we heard another fog horn. It seemed to be coming closer. Our cooper, an old navy bugler, became suspicious. He got out his old bugle and sounded "assembly" sharply. As the first note struck into the mist, the other fog horn ceased its blowing. We did not hear it again. When the mist lifted, no vessel was in sight, but the situation was clear. We had chanced upon a poaching sealer and when she heard our cooper's bugle, she concluded we were a revenue cutter and took to her heels.

In the southern Chukchi Sea the warm northerly currents usually forced the ice north and the ships were able to reach Cape Lisburne—225 miles north of Bering Strait—before the middle of July. A week or so later they could often make Icy Cape although much of the northwestern Chukchi Sea remained ice-covered. This part of the arctic cruise was relatively predictable, but beyond Icy Cape the whalemen could not count on the pack ice to continue its contraction: north of that point the ice was increasingly governed by the strength and direction of the wind. The whalemen could usually expect to get as far as Point Barrow in early August, but above Icy Cape they ran between the land and a large triangular tongue of ice that hangs in the center of the Chukchi Sea. The tongue, formed by the warm currents passing to the east and west of it in the Chukchi Sea, wags back and forth according to the wind direction. Unlike the ice in the Bering Sea, the ice in the northern Chukchi Sea does not disintegrate; a whaleship entering the pack ice north of Bering Strait could be surrounded and then swept north into the Arctic Ocean, far beyond the hope of rescue.

Wind is the dominant factor in the Chukchi Sea in the autumn and, to the whalemen, autumn began about the first of August when the gales begin to increase in frequency and violence. The entire Bering Sea has only a dozen good harbors but there are none at all in the Chukchi Sea for vessels drawing more than about six feet, as virtually all whaleships did. The whaleships, therefore, had to be ready to anchor on exposed and often lee shores. If the pack ice was nearby, it could rapidly come down on them. Although the men tried to work their way to Point Barrow as quickly as possible to be able to intercept the bowheads coming out of the Beaufort Sea, most sailing vessels left Point Barrow at the end of August, when the northerly winds began to make them increasingly vulnerable.

They usually ran south along the shore as far as Point Franklin, then followed the ice edge west until they reached "Post Office Point," the southern tip of the tongue of ice extending south in the center of the Chukchi Sea where ships often met and exchanged mail. From Post Office Point the ships changed course and went northwest to cruise on Herald Shoal, or farther, to "the Hole," the deep cleft in the pack ice carved by the warm current running out of Bering Strait. The sailing ships usually left these grounds by the beginning of October. In the ugly autumn weather and deepening cold the "eastern ice" (the tongue) and the "western ice" (the pack ice in the western Chukchi Sea) were usually growing and there was a real danger that the two ice masses might meet before the ships could escape to Bering Strait.

To avoid the western ice the ships steered directly for Cape Lisburne before turning south for Bering Strait. This passage was the most dangerous of the whole season. In the spring they had crept north through the Strait with good weather and the current behind them. In the fall they had to career to the south in growing wind and shrinking daylight, approaching what was, in effect, a vast lee shore to shoot through its sole, narrow gap. Once in the Strait, they could find a strong northerly current against them.

Captain Henry Pease described the passage he made on the *Champion* in 1870. His experience was by no means unique.

On the 4th of October we put away for the straits, in company with the Seneca, John Howland and John Wells—a gale from northeast, and snowing. On the evening of the 7th it blew almost a hurricane; hove the ship to south of Point Hope, with main-topsail furled; lost starboard bow boat, with davits—ship covered with ice and oil. On the 10th entered the straits in a heavy gale; when about 8 miles south of the Diomedes, had to heave to under bare poles, blowing furiously, and the heaviest sea I ever saw; ship making bad weather of it; we had about 125 barrels of oil on deck, and all our fresh water; our blubber between decks in horse-pieces, and going from the forecastle to the mainmast every time she pitched, and impossible to stop it; ship covered with ice and oil; could only muster four men in a watch, decks flooded with water all the time; no fire to cook with or to warm by, made it the most anxious and miserable time I ever experienced in all my sea-service. During the night shipped a heavy sea, which took off bow and waist boats, davits, side-boards, and everything attached, staving about 20 barrels of oil. At daylight on the second day we found ourselves in 17 fathoms of water, and about 6 miles from the center cape of St. Lawrence Island. Fortunately the gale moderated a little, so that we got two close-reefed topsails and reefed courses on her, and by sundown were clear of the west end of the island. Had it not moderated as soon as it did, we should, by 10 a.m., have been shaking hands with our departed friends.

Getting through the Straits was only part of the problem: they still had to traverse the Bering Sea—justifiably known as the cradle of storms—in one of its nastiest months. The gales there were doubly dangerous because they "blew half way around the compass," coming first out of the southeast and then from the northwest, giving the mariner two lee shores to fear. To make matters worse, in late autumn one good day, called a "weather breeder," usually introduced five days of gales.

There was little the men could do to protect themselves from those storms except to batten and lash everything down and to put the boats on their upper cranes and turn them on their sides to save them from being burst by tons of water. If they were approaching the Aleutians when the storm hit, they turned back into the Bering Sea to avoid being on a lee shore when it swung into the northwest.

William F. Williams described tacking ship in the Bering Sea to claw off a lee shore:

> With all hands on deck, port watch forward, starboard watch aft, spray flying the whole length of the ship and scuppers underwater [the captain] would stand on the lee side of the quarter deck with one hand holding the mizzen shrouds, looking forward under the lee bow at the line of white foam that marked the breakers, until it seemed sometimes that we could not possibly miss going ashore. Then, as the ship started to rise to a sea, would come the order like the crack of a rifle, "Hard a-lee," and in a thunder of slatting sails and rigging, with decks covered with water, the men executing orders as they came through the roar of the wind, the good old ship swung slowly but surely to the offshore tack.

In extremis they hung canvas oil bags over the sides. The oil seeping from the bags and spreading over the water calmed the seas somewhat. But even these precautions rarely saved the ships from extensive damage if they ran into a bad storm, as the mate of the *Orca*, a particularly sturdy and seaworthy steam bark, recorded:

> Had a very heavy gale at SW, it blowing with great violence until 6 PM. About 12 hours with horrible rugged sea. Shipping great quantities of water. We lost 4 boats off the davits & one spare boat & the steam launch off the upper deck with engine room sky lights, rails & stanchions etc etc [from] the cabin deck.

Ships that were too lightly ballasted risked capsize in heavy weather. As the autumn wore on, if a vessel had caught few whales, the men ballasted the ship by filling the empty casks with salt water for their passage south. On their way south they also took down the stoves in the crew's quarters, cleaned the decks, scraped the tryworks' soot from the masts, repaired and tarred the rigging, scoured out the tryworks, and pumped the "gurry" (oily sludge) overboard, or if the ship were full and bound home directly, the tryworks was broken up and, saving the pots, thrown overboard.

The boats also had to be prepared for a passage at sea. Whalecraft, except the oars, was removed from the boats and stored below. Any boats worn by the ice were broken up for firewood or traded to the natives. The bow and waist boats, being more vulnerable to the seas, were often lashed on top of the boat racks and those that were carried outboard were put on the upper cranes.

The cutting tackles were taken down, the lines coiled and stowed below. The cutting stage was taken in. The crow's nest came down and the spars and sails that had been taken down in the spring were sent up. Once past the Aleutians, the anchors were unshackled, lashed inboard, and the chains stowed. The captain often towed a "patent log" over the taffrail to determine the distance run.

Whether they were leaving with a full ship or not, departing from the Arctic, from its hardships and dangers, from the chance of shipwreck and lonely death so far from home, was cause for heartfelt thanksgiving. The log books and journals are eloquent and succinct: *Litherland:* "Took our departure from Bherings Strait this morning at 10 AM. Thank God." *Robin Hood:* "Passed the Diomedes and Fairway Rock, and now thank God we are out of the Arctic Ocean." *Constitution:* "Steered to the south thro' the straits. Strong gales northward so that we are leaving the Arctic Ocean astern very fast and without the least regret either." *Nassau:* "Bound to the south and god forbid that we ever come back again."

As they worked south into the Pacific, on the first warm day they threw overboard their worn and dirty fur clothes and broke out garments that had been packed away in their sea chests for six months. On drawing near Hawaii's green mountains, forgetting that they would make one or two more northern cruises before their voyage would end, thoughts of shore leave took over. Celebrating the return, William Abbe wrote the following song about Maui ("Mohee"):

Once more we are waft by the northern gales
Bounding over the main
And now the hills of the tropic isles
We soon shall see again
Five sluggish moons have waxed and waned
Since from the shore sailed we
Now we are bound from the Arctic ground
Rolling down to old Mohee
Now we are bound from the Arctic ground
Rolling down to old Mohee

Through many a blow of frost and snow
And bitter squalls of hail
Our spars were bent and our canvas rent
As we braved the northern gale
The horrid isles of ice cut tiles
That deck the Arctic sea
Are many many leagues astern
As we sail to old Mohee
Are many many leagues astern
As we sail to old Mohee

Through many a gale of snow and hail
Our good ship bore away
And in the midst of the moonbeam's kiss
We slept in St. Lawrence Bay

And many a day we whiled away
In the bold Kamchatka Sea
And we'll think of that as we laugh and chat
With the girls of old Mohee
And we'll think of that as we laugh and chat
With the girls of old Mohee . . .

Ted Pedersen, cabin boy, at the helm of the Nanuk *in the North Pacific, bound to San Francisco at the end of an Arctic cruise, 1921* Private collection

4 / *Pursuit and Capture*

When a ship reached the Bering Sea and its boats were rigged, the real business of whaling began. It was surprisingly difficult to catch a bowhead. On the average, a ship caught only one bowhead every three weeks in the Bering, Chukchi, and Beaufort seas. The final delivery of oil and whalebone to the wharf depended on the successful conclusion of a long chain of events.

First, the captain had to place the ship where whales could be found. The ability to do so was the result of years of whaling experience and reflection on whale behavior.

The whalemen then had to find whales. Except when fog, gales, or darkness prevented them from searching, an officer, a boatsteerer, and usually a foremast hand were in the crow's nest scanning the water. From that vantage, in periods of exceptional clarity, the men could see a whale breeching or spouting as many as eight miles from the ship. In similar conditions they could hear the whale's great whooshing exhalation, a long PAHHHHHHHHH, from as far away as two miles.

But at great distances it was often difficult, even for an experienced lookout, to differentiate bowheads from less desirable species—humpbacks, finbacks, blues, or gray whales. Green hands were often confused by two phenomena: the slick, rounded back of a diving walrus can look like a sounding bowhead, and the sea swell, exploding upward through a hole in an ice floe, can be remarkably like a whale's blow.

When the cry, "Bloooooooows," came from the crow's nest, the captain immediately started aloft with his spy glass to confirm that the lookout had, in fact, seen the V-shaped and slightly forward-leaning spout of a bowhead. If the whales were bowheads, and if the wind were blowing less than half a gale, the captain usually immediately ordered three or more boats lowered to go after them while he stayed in the crow's nest keeping an eye on the whales.

When in the troughs of the seas or when working through patches of ice, the men in the boats often lost sight of their quarry. The captain signaled the boats by setting certain combinations of sails and flags to indicate whether the whales were on the surface or below, ahead or astern, to windward or to leeward. The signals varied from ship to ship and each ship's crew kept the meaning of the signals to itself to prevent another ship's boats from reaching the whales first.

When the captain gave the order to lower, the boatsteerer and mate jumped into the boat and lowered away. The seamen slid down the falls to the boat as soon as it was in the water. Once clear of the ship, the men "lined the oars" by taking the

Boat signals for the Northern Light

Whaleboats cruising in the Arctic American Museum of Natural History

whaleline from the after-tub, passing it around the loggerhead, then leading it forward over the oars and under the kicking strap and through the bow chock, and then aft where they fastened it to the harpoons.

If the whales were to leeward of the ship, the men immediately raised the mast and sailed down toward them, but if the whales were to windward, they had to use the oars to reach a position from which they could use the sail to approach the whales quietly.

Silence was important. A bowhead's hearing is very good. The dull rattle of oars in the muffled oarlocks was usually enough to "gally" the whale and drive it under. Sailing was the best way to approach the animals, but if there was no wind, paddles had to be used, and without the noise of waves on the water, most bowheads heard the boats well before they came within striking range. The foreign sounds of a turning propeller or of a boat striking even a small piece of ice were enough to drive them off. A bowhead swimming fast could easily outrun a whaleboat in light airs and if it were blowing any stronger, no boat could catch a whale going directly upwind. The men were frequently frustrated in these conditions. "Lowered for the spirit of a bowhead whale going like hell to the NW," wrote John Peabody. "I hope he will fetch up against a cake of ice and break his damned neck."

Even in good conditions if a number of ships were together and there were only a few whales, the number of boats in the chase would virtually insure that, one way or another, one of the boats would gally the whales.

There was an informal agreement among whalemen that if a lone whale was sighted, the first ship's boats to get into the water would be the ones to chase it. But this arrangement was impossible to adjudicate or enforce; the poorer the luck, the more it was ignored. Likewise, when a ship had its boats down chasing whales, another captain often lowered his whether he could see the whales or not. In such cases, the captain of the first ship often sent one or two of his boats off in another direction to draw off the interlopers. Another ruse was also tried—unsuccessfully—as a whaleman aboard the *Montreal* smugly reported: "The Lydia hailed us when she saw our boats going in the ice, shouting at the top of their lungs, 'it's a fin back.' . . . Wonder what they thought of the fin back when they saw the boats towing a bowhead to the ship." At times the captains of steam auxiliary vessels, seeing another's boats chasing, would allow their propellers to turn slowly, driving the whales under.

When a boat did strike, the bowhead apparently uttered a cry, inaudible to whalemen, that gallied all the whales for miles around. "Saw a lot of bowheads," wrote a whaleman aboard the *John P. West*. "Finally struck and it was a sight to behold to see the whales make for the ice." Once the whales had been gallied, they remained for a time very "shy" and extremely wary of approaching boats.

One other impediment to catching whales the whalemen brought upon themselves. Almost until the Civil War a number of masters strictly observed the sabbath and would allow no whaling on that day. This practice often built terrible strains between the master and the rest of the crew. In fact aboard the *Tiger* in Bering Strait, even the captain's wife felt the urge to hunt whales, contrary to her husband's wishes: "What a time for remembering it was the sabbath," Mary Brewster wrote.

> Perhaps [the crew] kept it better than I did, for my only trouble and thought was—I was afraid they would not have the whale. . . . I cannot see how a whaling master can, in the midst of whale[s], have sufficient firmness to resist the temptations of trying to get them. I cordially believe there is not a man in the business (how much he may feel disposed to keep the sabbath) if amongst whale[s] wishes the day would pass so Monday would come, and during the day his mind is more taken up with what he is losing than in hoping to keep the day according to God's requirements.

The procedure for actually fastening to and killing a whale was changing when Thomas Roys discovered the Bering Strait grounds.

Whaling irons underwent basically a three-part change. Beginning in the 1830s blacksmiths began to experiment with new designs in whaling irons to improve their holding power. The two-flued whaling iron, which had been in use for more than five hundred years in Europe, had a tendency to cut its way out of a whale when strain was put on the whaleline. The first major innovation that gained wide acceptance was the single-flued iron, which came into general use by around 1840. When strain came on the whaleline, the single-flued barb would put an asymmetrical force on the soft iron shank and allow the head to bend in the wound, creating a greater surface area than size of the entry wound.

The most important development in whaling irons came in 1848 with the development of the toggle iron, a device in which the entire head was mounted on an axle at the end of the shank. Being asymmetrical, the head could turn at right angles in the wound. If it was seated properly, it proved extremely difficult to withdraw.

In the nineteenth century whaleships setting out for the Pacific usually carried 150 to 200 whaling irons. The toggle iron quickly captured much of the market. Its acceptance was stimulated, no doubt, by the testimonials of men such as Captain Asa Tobey, whose report appeared in the *Whalemen's Shipping List* in 1853.

> When I sailed from New Bedford July 1st, 1850, I had thirty-five toggle harpoons. With these I have struck 23 polar whales, 10 sperm and 8 right whales, and not one of them broke or drew, with one exception where the whale rolled upon the bottom, and bent the head so much that both ends were within three inches of each other. There has been but a single instance in which they have not entered, and then I believe the line was broken before. They had heavy strains on them, with five or six hundred fathoms of line out and held on. Out of seven right whales struck with them in the South Atlantic five had the scars of other harpoons on them.

Despite the efficiency of the toggle iron, it did not help kill the whale quickly. Some still escaped. When a whale was heading for the ice, the best the men could do was to haul their way up the whaleline, hand over hand, and "heave in a drug iron," which was an iron towing a drogue, a square piece of wood tied to the line to slow and tire a running whale. Occasionally the men also bent on "pokes," inflated sealskin floats, an Eskimo invention to tire the whale.

Around 1850 a more practical device, the bomb lance, was developed. The bomb lance was a small metal cylinder that was fired from a shoulder gun. It was filled with an explosive and fitted with a time-delay fuse that allowed it to explode a few seconds later inside the whale. The basic design was modified by a number of whalecraft makers. In one form the gun was made of solid brass weighing twenty-seven pounds. It gave a brutal kick—enough to throw the man over backwards. It was known to kill at one end and wound at the other. According to Captain Hartson Bodfish, it "was a marvelous creation, almost as deadly at the breech as at the muzzle." The guns were dangerous and occasionally exploded. Several men accidentally shot themselves with them. At other times the bombs failed to ignite or the projectile charge was a dud.

On the whole, however, the bomb lance was effective, having an accurate range of about sixty feet. They were most often used by the mate. He kept his gun in a

A toggle iron twisted by a whale

A whaleboat cruising in scattered ice

wooden box in the bows and usually tied the gun to the boat with a lanyard because the recoil could make the gun hard to hold.

The bomb lance was used in the western Arctic as early as 1851—a few may have been used before that. It was widely used in the North because it allowed the men to try to kill the whale before it escaped into the ice, and before the men could pull the whaleboat close enough to use the lance. Nevertheless, right to the end of the fishery the standard equipment of a whaleboat usually included three hand lances for use when all the bomb lances had been shot off or, in the interest of economy, for giving the *coup de grace*.

Although the shoulder gun and bomb lance proved a more effective means of retaining a wounded whale, even a badly wounded whale could escape into nearby ice. The peculiar conditions of the Arctic demanded a way to fasten to and kill a whale simultaneously.

The solution was the darting gun. It was simple and effective. The darting gun was mounted at the end of the harpoon shaft alongside the toggle iron, which was detachable and mounted in gudgeons. When the toggle iron had penetrated far enough into a whale, a trigger rod hit the whale's skin and fired the gun, shooting a bomb lance into the whale, where it would explode several seconds later. The recoil from the first explosion jumped the gun free of the toggle iron. The boatsteerer then hauled the gun inboard by a nine-fathom "gun warp" while the whaleline remained fast to a large loop on the shank of the toggle iron.

Captain Ebenezer Pierce invented the darting gun in 1865 and the following year he brought a few guns with him to the Pacific aboard the *James Allen*. He distributed the guns among his friends for evaluation and so remarkable was their success that by 1867 news of his invention had reached the pages of Honolulu's *Pacific Commercial Advertiser*. The darting gun was well adapted to Arctic whaling. It spurred a number of imitations and modifications. Both the darting gun and the shoulder gun remain in use today, essentially as they were invented, among the Eskimos of northern Alaska.

With or without explosive weapons, when the whaleboat approached close to a bowhead, the officer steering the boat told the boatsteerer (the harpooner) to rise. He stood, jammed his thigh into the clumsy cleat for balance, and picked up the first harpoon. The officer tried to keep the boat in the whale's blind spot, approaching the whale from either directly behind or straight ahead. At the last moment he would swing the boat off to one side or the other, depending on whether the boatsteerer was left or right handed.

The officer tried to "go on" the whale just as it rose, for only when it was blowing was its acute hearing temporarily reduced. When the time was right the officer barked, "Give it to him!" and, depending on the distance, the boatsteerer darted or pitched his iron. The whale's most vulnerable spot was its neck—ten or twenty feet aft of its nose. Before the development of the darting gun, it did not much matter where the boatsteerer struck the whale—apart from its upper jaw, blow holes, and the "small" of its tail, places where there was so little flesh that the iron would not hold. But with the advent of the darting gun, and hence the potential for an immediately killing shot, the boatsteerer tried to strike the animal between the two shiny black humps that projected from the water and thus to sever its spinal cord with the bomb lance.

A boatsteerer could throw his iron accurately up to thirty feet. As soon as one iron was fast, he immediately threw the second, if there was time. If not, he threw the second iron overboard to prevent it from slashing around in the boat as the whaleline raced out. As quickly as he could, he then picked up the shoulder gun and tried to bomb the whale before it sounded.

Because of their weight and poor balance, darting guns could be thrown accurately only a very short distance. An iron was used only with the first darting gun. The boatsteerer, as soon as he was fast to the whale, then tried to strike it with the second darting gun.

It was possible, but extremely rare, to kill a whale with the irons alone. Even with

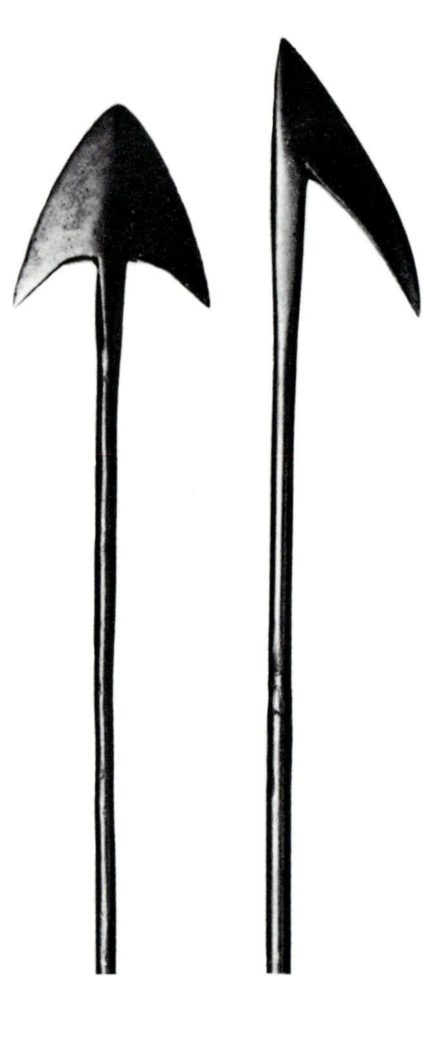

Whaling irons: two flued iron, single flued iron,

the use of the shoulder gun and darting guns together, only about one whale in twenty was killed instantly. Some whales required a dozen bombs or more to kill them. There were, consequently, many ways to lose a whale after it had been struck the whalemen, to their disgust, lost between 15 and 25 percent of their whales. Some whales offered a boatsteerer only "slack blubber" by hollowing their backs, leaving a slack, wrinkled section where the skin did not offer enough resistance for the iron to penetrate.

A wounded whale could turn ugly and try to demolish the boat with its tail, although this behavior was more common with gray whales or right whales. Most often bowheads ran or sounded. If the whale sounded fast, it might take all the boat's whaleline before the men could "bend on" to another boat's line. Even if they succeeded in attaching more line, the whale could easily take six hundred fathoms or more. The whale could stay below for as long as an hour and simply outwait the men if the seas were rough or darkness were approaching. If the whale reached the bottom, it could also "roll the irons out," as the tortuous shapes of many used toggle irons attest. If the whale ran on the surface, it could tow the boats three miles or more in making for the safety of the ice. If it reached the ice before the men could get close enough to kill it, and if the whaleline did not part, the men would be forced to cut loose from the whale to avoid having their boat dashed to pieces in the ice.

The men lost whales through human or mechanical mishaps as well. When darting, the boatsteerer occasionally missed the whale entirely, or he might pitch his irons in too flat an arc so that they bounced off the whale's back, or dart into its head where the iron would not toggle, or only dart one iron and dart it too lightly so that the whale was "hay hooked" and the iron easily drew free. If the boatsteerer missed a good chance he rarely got another. The captain usually immediately "broke him," sent him forward to the forecastle, and "fleeted up" the "preventer boatsteerer" or another likely crewman.

A soft pine shear pin held the toggle head in place until force was put on the line; if the pin was missing, the head could toggle in mid-air and strike the whale broadside, or if a hardwood pin was used mistakenly, as it occasionally was, the head would not toggle at all and would draw from the wound. If too much ice fouled the box warp it would spend the harpoon's momentum and hay hook the whale. If the first iron drew, the force on the second could part the short warp; or the line could part from weakness, by abrasion from the ice, or because the second iron, if not fast, thrashed about and cut it. A poorly aimed bomb lance could cut the line. The line could also come adrift from the iron, or the iron could break.

On some days it seemed as if it was not worth even trying to catch a whale. The following events were recorded aboard the *Sophia Thornton*: ". . . lowered and fastened. The first iron drawed. The second broke. Lost the whale. At 4 PM saw one whale, lowered and fastened. Got foul line and boat stove and line parted. Lost the whale, 3 irons, spade, and some line."

Freezing weather could play havoc with the guns, immobilizing the hammers and triggers. But the guns could also be a problem in their own right. Not only could the

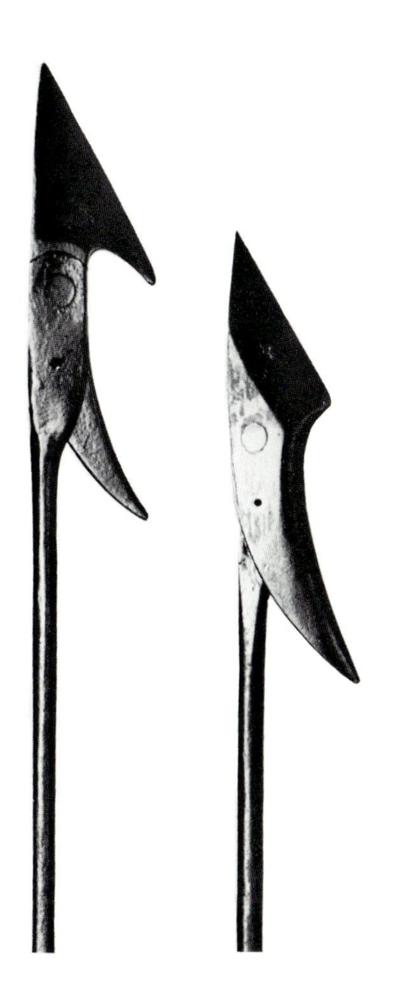

early toggle iron, improved toggle iron

explosion of a poorly aimed bomb lance blow the iron loose, but more than once an accidentally discharged darting gun or shoulder gun blew the bow off the whaleboat—and there are several reports of guns wounding or killing whalemen in the Arctic.

Although these accidental deaths were sufficiently rare to be dismissed by the whalemen as a remote possibility, they genuinely feared two equally rare occurrences that might happen when they were fast to a whale. The chance of being towed far beyond the ship and disappearing into the fog was a palpable danger. When a boat did disappear the men aboard the ship immediately began firing guns and blowing horns to help them find their way back. Ships always took up the lost boats of another; they could well imagine themselves in the same situation.

But the most terrifying way of losing both whale and life was to be "taken down" by the whaleline. If a whale was sounding fast and the whaleline fouled on any part of the boat, it could be taken under in a flash. At the very least it would be capsized. Death came in minutes in those icy waters. If a man got a turn of the whaleline or lance warp around him, he could be jerked out of the boat with such violence that he often knocked several men out with him, breaking their bones as well as his. If he were taken under, it was a near miracle for him to survive, even if the men in the boat cut from the whale instantly.

It took skill to kill a "fast" whale. Bowheads often dive deep first, then, after waiting, rise and run near the surface to windward or toward the ice if it is near. If the whale sounded, all the ship's boats tried to gather around the spot where they judged it would rise to bomb or lance it as it surfaced. If the whale ran, then the men in the "fast boat" had to endure a "Nantucket sleigh ride," the boat streaking through the water behind the whale, hobby-horsing over the swells and smashing into the seas while sheets of spray soaked the men.

While the men in the boat tried to pull themselves close enough to bomb or lance the whale, the other boats tried to anticipate where the whale was heading. Because the whale often rose to blow just before diving under the ice, they could occasionally get a killing shot if they happened to be in the right position. Other boats nearby would follow the whale's turbulence (the "suds"), hoping for a shot if it surfaced unexpectedly.

Before shoulder guns gained universal acceptance, the men tried to stop a running whale by the dangerous practice of "spading flukes," that is, by using the whaleboat's sharp cutting spade to cut the tendons in the whale's small, thus hamstringing it and preventing it from diving. In 1887 James T. Brown noted that veteran whalemen believed that a certain amount of excitement had gone out of the business of killing a whale because of the advent of the bomb lance. The elderly whalemen felt that to have "fought under the flukes of a whale" was a mark of skill and daring.

If the whale began to tire before reaching safety, the men pulled the boat hand over hand up the whaleline toward the animal while the boatsteerer and officer changed places, the boatsteerer taking the steering oar, the officer going to the bow for the kill. When the whale was tired and wounded, "spouting thin blood," the

Toggle iron

WHALEMEN ATTENTION!

Bomb Guns and Lances.

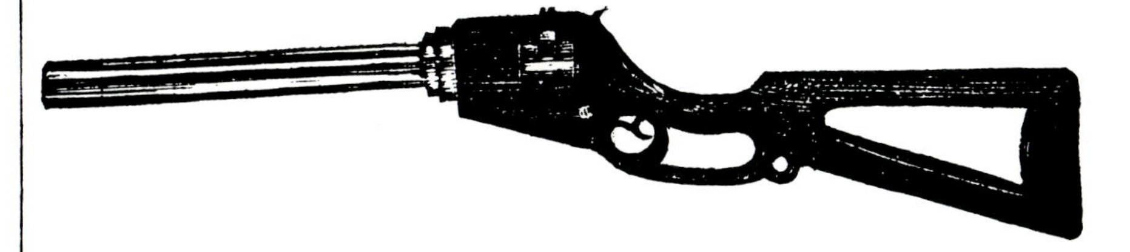

THE UNDERSIGNED HAS THE AGENCY AND OFFERS FOR SALE

CUNNINGHAM & COGAN'S CELEBRATED PATENT BREECH-LOADING BOMB GUNS,

—ALSO—

BOMB LANCES AND CARTRIDGES COMBINED.

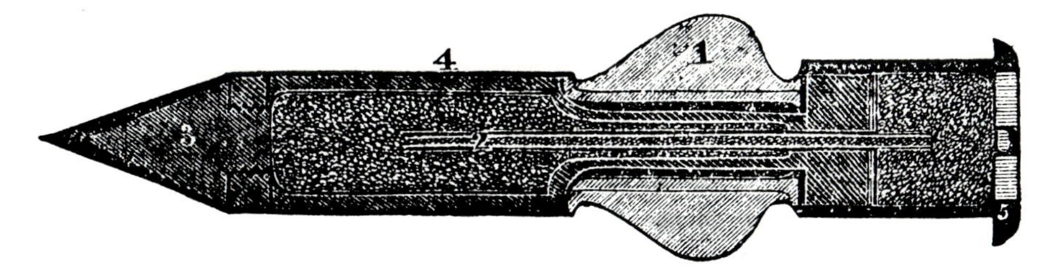

These Guns and Lances are pronounced by all who have seen and examined them, to be SUPERIOR to all other kinds, and are recommended in the STRONGEST terms.

The Superiority of these Guns over all others, is that they can be Loaded and Discharged

TEN TIMES A MINUTE.

You do not have to carry any Powder, Caps, &c., in the boats. Can be fired as wel under water as above, as water has no effect upon either Gun or Lance.
For particulars and prices apply to

WILLIAM LEWIS, Agent,

No. 4 Taber's Wharf, - - - - New Bedford, Mass.

ap4'76 tf

Advertisement in the Whalemen's Shipping List

An exploded bomb lance

officer would often try to kill it with the lance without further bombing. The men began by "bowing on" the whale, bringing the boat up very close to the whale and cleating the whaleline to the gunwale. When the boat was in this position, "wood to blackskin," the officer thrust or pitched the lance into the whale, hauling it out by its warp and repeating the action and sometimes "churning" it (pumping it in and out). He was aiming "just aft the foreshoulder," trying for the whale's "life," the heart and lungs.

When the whale began to "spout thick blood," the dark globs of congealed blood that signalled it was mortally wounded, the men quickly backed the boat down, away from the whale, lest in its death flurry it lash out and stove the boat. When the whale "finned out," rolled on its side with its fin weakly moving in the air, its life was fast ebbing. The men waited until it ceased moving, then approached cautiously and the officer poked its eye with the lance to make sure that it was dead, not merely comatose.

When the whale died, and if there were other whales nearby to chase, the men merely put a "waif" on the carcass. The waif, a small flag mounted on a pole with a pointed shaft, helped the ship to spot the animal. The whaleboat was then free to set off again in pursuit of the other whales as the men re-coiled the line in the tubs.

The captain watched the chase from the crow's nest. As soon as he saw the waif he worked the ship toward the carcass, which could be several miles away. Usually the boat's crew had to do at least some towing to get the whale to the ship for processing. Many officers simply put a line around the whale's flukes and towed it tail first, but some, particularly those who had endured long tows, felt that it was easier to tow a bowhead head first, with a line through a hole in its lip. Many considered it essential to tie the fins together. The fins in death stick out at right angles, making the tow a slower job.

The vast majority of bowhead carcasses floated on the surface, but occasionally one sank that had died with little air in its lungs. Most bowheads were captured on the continental shelf, however, where the depth was rarely greater than thirty fathoms. The men would haul on the whaleline until the boat was directly over the sunken carcass, then loosely attach a "messenger iron," a huge toggle iron weighted with a heavy object such as a boat anchor or chain, so that it would slide down the whaleline, plunge into the carcass, and fasten itself securely. Often they used more than one messenger iron. Once they judged they had a firm hold on the carcass, they took all the lines to the ship, rove them through a block at the masthead, and tried to haul up the carcass with the windlass. One ship which carried a steam winch—as most did in the later years of the fishery—hauled a whale off the bottom in 2,400 feet of water near Banks Island. Despite precautions the lines sometimes parted or the irons drew. If the captain thought this was likely to happen, he would simply anchor near the whale and wait for it to "blast"—for the gases that are generated in putrefaction to lighten the whale and bring it to the surface.

The whalemen were always on the lookout for a dead whale floating free. These whales were easy to spot; once they blasted, they floated high in the water and had masses of feeding sea birds wheeling about them. Occasionally the men found one that killer whales had killed and more rarely one that had been killed by the natives, but usually it was a whale that had been struck and lost by another ship.

The men often found an iron in the carcass. The initials of a ship were always punched on the head of the iron. A boat from the ship that had struck and lost the whale could claim the carcass before the iron was withdrawn, but like the agreement as to the priority in chasing a lone whale, the whalemen usually bent this custom to their best advantage. Some crews withdrew other ship's irons and replaced them with their own. One crew is reported to have gone through the motions of striking and bombing a dead whale to confuse the rightful owner nearby. Disputes over the ownership of a whale were occasionally brought to court in the United States.

The oil and baleen from these "stinkers," as the whalemen called them, added to the ship's cargo with a minimum of effort, but if the whale had been dead for more than a few days, its odor was so strong that their good fortune was a mixed blessing. If the carcass was too rotten, the blubber made foul oil and the baleen had usually fallen out of the head.

Special tactics were required when the ships or the whales were deep in an ice field. This situation usually occurred during May and June when the ships were pressing their way north through the ice near Cape Navarin. The bowheads, traveling through the pack ice, came up to blow in the small ponds of water between the ice floes. The men learned that when open water was severely restricted, the whales behaved strangely and surfaced close to the boats and ships. Each spring a

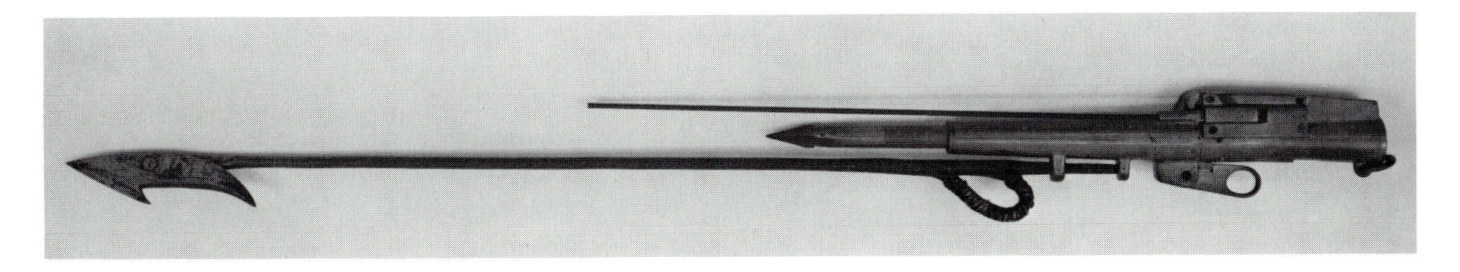

Patrick Cunningham's patented darting gun, shown with mounted darting iron and bomb lance

Whaleboat going on a whale in the Arctic

number of whales were bombed directly from the ships. At other times, when the whales were farther from the ships, the captain sent groups of men across the ice on foot to wait at holes in the ice. If the ponds were large enough, they lowered the boats onto the ice and dragged them to the water. In this specialized type of whaling it was essential to fasten an iron in the whale at the same time it was bombed; even a mortally wounded whale could move a small distance under the ice before dying, vastly complicating the problem of retrieving any oil or bone, if, indeed, the carcass could be found at all.

When the boats were cruising in relatively open conditons, and a "fast" whale ran into the ice, the problem was the same. The boats came together. As one ran out of line another would bend theirs on to it. One whale took 9,000 feet of whaleline out of Captain Hartson Bodfish's boats. Bodfish and others found that when a whale went into the ice it was best not to put any retarding force onto the whaleline but merely to put a "poke," a sealskin float, on the end when they ran out of line: "I

Righting an overturned whaleboat

discovered that when a whale swims under the ice with such a drag behind him, he will stop and rest, then start again and feeling the drag continue will often swing about and come out near where he went under. In this way we saved many a whale that went under the ice when first struck."

When a "fast" whale died deep in the ice, the men, if they could, took the whaleline to the ship and used the windlass to haul the carcass out. At times it was impossible to get the line to the ship and the only alternative was to have the boats tow the whale out. A tow was slow and tiring work: it could take twelve hours to move the whale through a mile of packed ice, with the men constantly forcing the floes apart, poling the boats through, and keeping the line and the whale clear of the thousands of projecting points of ice that could snag them and bring them to a halt. Sometimes it took two or three days to reach the ship. If a whale were too far in the ice to haul out, the only alternative was the even more difficult process of cutting in the whale where it lay and hauling the blubber and baleen back to the ship.

A dead bowhead

If they could not find bowheads, the whalemen went after any other animal that yielded oil, although each of the several species they took was far less desirable than the slow, fat, and docile bowheads. After bowheads, walrus were the most sought. A mature walrus gave only one-half to two-thirds of a barrel of oil—and per barrel of oil they required a greater amount of work to butcher and try out—but the whalemen could count on finding them in vast herds at the edge of the retreating ice. The ships took nearly 150,000 walruses in the lull between the spring and fall seasons, when the bowheads were safe in the unexplored reaches of the Beaufort Sea.

During the summer, or "'tween seasons'" lull, the whalemen also turned their attention to gray whales. They called these mottled and somewhat scabrous-looking creatures, among other names, "scrag whales," referring to their meager yield: only twenty-five to thirty barrels of yellowish or reddish oil compared to the bowhead's hundred barrels of white oil. Orson F. Shattuck, aboard the *Frances*, described them best: "Struck and killed [a gray whale] but he sunk and we had to raise him up. They are a very poor whale, not worth taking. The blubber is not more than 3 or 4 inches thick and yields very little oil and that very inferior in quality, and their bone is from 8 to 10 inches long, white and brittle, being of no use whatever."

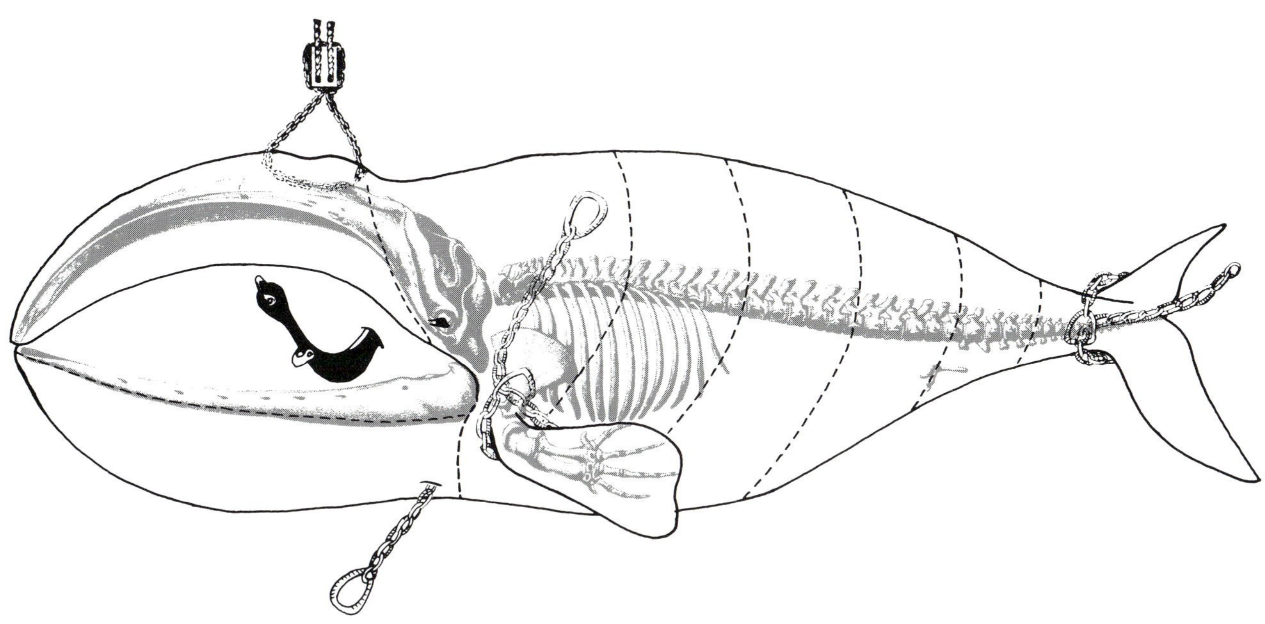

Cutting-in diagram. Whale is secured to the ship with a fluke chain. The cutting begins at the left fin, which is hoisted with a fin chain. The lip is usually raised with a blubber hook and the headbone is taken up with a head strap. When the whale is supine a throat toggle is used to raise the tongue and throat.

Because these whales had a well-deserved reputation for being vicious fighters, another term the whalemen used for them was "devil fish." "Chasing devil fish," the mate of the *George Howland* wrote, "the waist boat struck and got the head of the boat knocked off." The mate of the *Oregon* wrote: "Lowered for scrags, struck and had a boat stove all to pieces."

Some captains resorted to using a Greener gun for fastening to gray whales. A British invention, the Greener gun was a large swivel-mounted smooth bore fixed to a post on the clumsy cleat of the whaleboat. It could shoot a whaling iron thirty or forty yards. The Americans often used the Greener gun in the California gray whale shore fishery so that they could fasten to the whale without getting within range of its flukes. In the calmer waters of California the gun could be aimed with some precision, but in the boisterous waters of the Arctic it was too inaccurate to gain general acceptance. Throughout the history of the fishery the whalemen only took about five hundred gray whales, and may have killed and lost another three hundred in the Bering and Chukchi seas.

The whalemen also went after humpback whales when no bowheads were available, but like the gray whale, the humpback's bone was so short it was essentially worthless and a humpback only yielded about thirty-five barrels of oil. Thomas Roys found the humpback "difficult to approach on account of its shyness and when fastened to [,] . . . difficult to kill as it runs with railroad speed." Dead humpbacks often sank. If they could be cut in, their blubber was so soft that the gear frequently tore through it.

In an exceptionally poor year, the whalemen abandoned Bering Strait entirely and went to Bristol Bay, south of Saint Matthew Island to hunt for right whales. The right whale was, as Thomas Roys put it, "a most furious and ungoverned character." It was certainly the most dangerous and pugnacious of baleen whales. When struck, it swept its flukes from side to side ("eye to eye") like a scythe. By the time the bowhead fishery began to have sparse hunting patches during the season, the right whale population had already been severely reduced by the whalemen, giving the Bristol Bay fishery a very low return for effort.

The Arctic whalemen also took small numbers of virtually every other whale they encountered. They occasionally chased finback whales, but the finback's speed, which was even greater than the humpback's, often caused the irons to draw. They only yielded about thirty-five barrels of oil and no commercial whalebone. The men also frequently found numbers of killer whales around the carcasses of the whales, walruses, or seals that the school had killed. Occasionally the men lowered a boat to retrieve what remained of the carcass and to try to take a killer whale or two, but the killer whale only yielded about five barrels of oil at the most. Killers were hunted more to keep the crew active than for any serious commercial purpose. For the same reason the whalemen also occasionally took porpoises, beluga whales, and even the giant bottlenose dolphins.

The process of cutting in—"the act of removing [the whale's] oleaginous blanket and transferring it to the vessel"—began when the dead whale reached the ship. The men chained the whale tail-first along the starboard side of the ship. They passed a fluke chain around its small, ran the chain through the "fluke chain pipe" (a small port in the bulwarks) and secured it to the fluke chain bitt, a sturdy post on deck. The boats might have been chasing and towing for twelve hours or more, and the captain usually sent the men below for a meal before beginning to cut in the whale.

In the meantime others lowered the cutting stage over the whale and got the cutting tools out on deck where the carpenter or cooper began sharpening the cutting spades, a process that would continue until they had finished stripping the blubber from the whale.

Some headway was necessary to keep the whale close alongside. The ship was set on a course that put the whale to windward. This way the pressure of the wind on the sails would counterbalance the enormous weight of the whale when hauling up the blubber. If the seas were very rugged, they would cut in to leeward, but this was done only as a last resort as it substantially increased the heel of the vessel.

When the men came back on deck, a couple of the officers, often the first and second mates, went out onto the cutting stage. The stage was essentially three heavy planks, bolted together and suspended by heavy tackle over the whale. The main plank was usually about twenty-two feet long and the legs were each ten feet. The legs held the main plank far enough away from the ship's side to give the men a better angle for cutting in. A fin chain was secured around the whale's left fin.

The first cut: hoisting the fin

Shackled to the chain was one of the cutting tackles, a massive two-sheave block, sixteen inches wide, rove with cutting falls, rope five and a quarter inches in circumference. The end of this line was led forward to the windlass drum where ten men or more began heaving on the windlass, taking up the slack to put a strain on the fin chain.

Heaving on the windlass for hours at a time was hard, tiring work and the men usually sang songs to make the job lighter: "Blow the Man Down," "Rolling Rio," "John Brown's Body," "Blow, Boys, Blow," and "Dixie," among others. "They would heave away with a will and make the pawls clank and clatter as they roared out the chorus," wrote Walter Burns, who sailed to the Arctic before the mast near the turn of the century. "The day laborer on land has no idea how work at sea is lightened by these songs. Their favorite was 'Whiskey for the Johnnies.'"

It had a fine rousing chorus and we liked to sing it not only for its stirring melody but because we always harbored a hope—which, I may add, was never realized—that the captain would be touched by the words and send forward a drop of liquor with which to wet our whistles. Gabriel would begin in this way:

"O whiskey is the life of man."

And the sailors as they heaved would chorus:

"O whiskey, O Johnny.
O whiskey is the life of man,
Whiskey for the Johnnies."

Then Gabriel would sing:

"Whiskey killed my poor old dad,
Whiskey drove my mother mad,
Whiskey caused me much abuse,
Whiskey put me in the calaboose,
Whiskey fills a man with care,
Whiskey makes a man a bear."

And the men would come through with the refrain:

"Whiskey, Johnny.
I drink whiskey when I can.
O whiskey for the Johnnies."

Once the men put a strain on the fin chain, the whale rolled over on its side with its left fin in the air. They then secured that cutting fall to the bitt. The whale's left lip was now exposed, and the officers on the cutting stage used their razor-sharp spades on eighteen-foot poles to cut a hole through the base of the whale's left lip and to make a deep incision all along the jaw bone at the base of the lip. The men on deck then shackled the blubber hook—weighing 75 to 150 pounds—to the second cutting tackle and sent a man down onto the whale to hook it through the hole in the whale's lip. The second cutting tackle's fall was taken to the windlass, and the men put a strain on this line. Meanwhile the officers were cutting along the base of the whale's lip with their spades. Gradually the lip rose as the men heaved on the windlass and the officers hacked through the blubber, pausing frequently to get sharp spades.

When the lip was free it was hoisted up and swung inboard and then lowered through the main hatch into the blubber room. The officers with the cutting spades now moved slightly forward on the stage to work on the fin. The officers cut through the skin and blubber to the "lean," first cutting a scarf line around the fin. Then they scarfed parallel cuts diagonally across the whale's body. These parallel cuts were about six feet apart and determined the width of the "blanket piece" of blubber that would be removed from the whale.

The men on the windlass then put strain on the line again, and while the officers cut around the fin, the tackle tore the whale's humerus from the shoulder socket,

Hoisting the lip with a blubber hook

disjointing the "knuckle bone." When the knuckle bone was free, the officers continued to cut between the blubber and the lean as the men on the windlass slowly hauled the fin higher, beginning the blanket piece.

When the blanket piece had risen high enough to rotate the whale's body about eighty degrees from its starting position, the whale now floated nearly in a prone position, close to its posture when swimming but still slightly over to the right side. The officers on the cutting stage then lowered one of the boatsteerers onto the whale to begin the process of removing the baleen. The boatsteerer had a "monkey rope" around his waist to keep him from falling into the icy water that swirled around his legs. He first cut a hole beneath the spout holes and inserted the "head strap," a looped piece of heavy chain with a large wooden toggle at the end. He then shackled the head strap to the other cutting tackle (which had previously been freed of the lip) and, taking an axe, cut half-way through the upper jaw just behind the spout holes.

Chopping through the rostrum to free the headbone.
At the right is the blanket piece

Raising the headbone

The job as "monkey" was hardly sought after, but it usually was the responsibility of the boatsteerer who had first struck the whale. It was extremely unpleasant to stand aboard the slippery mass, leaning against two crossed blubber spades for balance, while trying to chop through two feet of bone with the ship pitching and rolling. With the seas surging around him, the monkey was sometimes up to his neck in the water.

When the monkey had finished his cut on the left side of the upper jaw, the men hove the blanket piece still higher and rolled the carcass through about twenty degrees to expose the other side of the upper jaw. They also put some strain on the head strap at the same time. The boatsteerer then continued his chopping, and with

Lowering the headbone on deck University of Alaska

Split headbone California Academy of Sciences

a cracking noise the headbone—the upper jaw with the baleen attached—swung free and was hoisted inboard.

The men lowered the headbone until it was just clear of the deck so that others, armed with broad "bone spades" could move around it, cutting at the base of the baleen until the two great sections of baleen fell on the deck. Each section held about three hundred slabs and weighed one thousand pounds or more. The two sections were usually then lashed on deck until the men finished cutting in the whale. They then returned to the baleen, chopped it into pieces of five or six plates apiece, and stored it below decks to await a less busy time for scraping the gum from each plate.

When they found time, the crew returned to the whalebone, bringing the bunches on deck, and splitting the plates apart from one another. The three hundred plates in each section were held together in one mass by the "oyster," the whitish gum.

Preparing to chop out the whalebone California Academy of Sciences

It was very important to clean the plates well. Whalemen scraped the tenacious white gum from the plates with pieces of sharpened coconut shell or with a variety of special draw knives. The Eskimos enjoyed visiting the ship at this time because the gum was a delicacy to them. Herbert Aldrich tried it and declared, "It was as I should imagine a decayed raw peanut would taste."

When they had finished this preliminary scraping, the men put the plates' butts into water to soften the remaining bits of gum. On some ships the plates were soaked in hot water in the trypots; on others the men merely filled a tub or dinghy with cold water, or even simply plugged the scuppers and flooded the deck.

After the bone had soaked for fifteen to twenty hours it was removed from the water and the men "washed" it—they scrubbed any remaining gum from the plates with sand and water and then polished them until they were shiny. If the baleen was not carefully cleaned, it developed an "ancient and fish-like smell" that was highly disagreeable and impossible to eradicate, lowering its value.

Scraping the oyster from the root of the whalebone Kendall Whaling Museum

Once the men had finished cleaning the whalebone, they spread the slabs out on deck or in the rigging and allowed them to dry slowly. Baleen usually lost 5 percent of its weight between the kill and wholesaler, but if the bone was not dried properly, it could suffer a shrinkage of at least 10 percent of its weight on the passage home.

When the bone had dried sufficiently, the crew lashed the plates into bundles of about eighty pounds apiece and stored them below. Once the bundles were below decks they were inspected regularly for signs of rat-gnaw. If it appeared that rats were eating the whalebone, the men moved it back on deck and waited for a favorable time to "smoke ship"—to kill the rats by setting charcoal fires below decks, sealing all cracks with paper and molasses paste, and keeping the crew above decks for twelve hours or more.

Preparing to soak the whalebone in a dinghy

Washing whalebone. In the background strips of whalebone are soaking in the trypots

While the raw headbone was being freed of its baleen, the blanket piece was slowly rising as the men heaved and the officers cut it away from the carcass. The ship, heeling over ten or fifteen degrees, groaned from the enormous weight. When the upper and lower cutting tackles had nearly met, the captain gave the order to "board" the blanket piece. The third mate, stationed in the waist of the ship, then took a boarding knife, a sort of long dagger mounted on a handle like a garden spade's, and cut a circular hole through the blanket piece near the deck. He then took the second tackle (which had been disconnected from the head strap) and forced the loop at the bottom of the block strap through the hole, plugging the loop from the other side with a massive wooden "blubber toggle." Later in the fishery the

Whalebone drying in the Bowhead's *rigging, San Francisco, 1883*
Peabody Museum of Salem

third mate cut two holes in the blanket piece and, instead of using the toggle, rove a chain through the holes.

The men then heaved on the windlass to raise the second tackle somewhat, and the third mate sliced through the blanket piece horizontally with his boarding knife. This operation required some care. The blanket piece was twelve to eighteen feet long, six feet wide, and twelve to eighteen inches thick, and weighed tons. Once the piece on the upper tackle was free, it could swing across the deck with devastating effect as it did aboard the ship *Ontario* in the Arctic in 1853, crushing the captain.

The blanket piece was maneuvered through the main hatch into the blubber room where two men cut it into "horse pieces," about four feet by six inches. The blubber room was fitted with "shifting boards" that divided it into several pens, to prevent the slippery blubber from sliding about as the ship rolled. The men divided the horse pieces in the pens to keep the ship on an even keel.

The second blanket piece was now rising, further rotating the whale's carcass. When the carcass had turned 180 degrees from its starting position, the right lip was exposed and cut off. The men continued hauling up the blanket piece, and when the whale was in a belly-up position they began to work on the tongue and throat. They cut a small hole through the tongue, pushed the "throat chain toggle" through it, and attached the cutting tackle to its ring. As the windlass raised the tongue, the officers cut along its edges, slicing it free of the lower jaw bones.

At the same time work continued on the blanket piece, with the first mate usually "leaning," cutting the blubber away from the muscle, and the second mate "scarfing," determining the width of the blanket piece, until the whale had been stripped of its blubber. The process was like peeling an apple.

Only the flukes then remained to be taken. The men cut a hole through one of the flukes, attached a chain, and as strain was put on the tackle, one of the officers used a broad-headed "bone spade" to cut right through the back bone near the small. Once they had severed the vertebrae, the flukes were hoisted aboard and the carcass was cut adrift. About five- or six-hundred pounds of the tenderloin in the tail section, considered very good meat, was saved.

The flukes were then cut from the backbone and the remaining blubber trimmed from the head and tail sections. The bones were thrown overboard. Once all the blubber was aboard, the crew, hoping for another whale yielding at least forty-five barrels, gave the cheer used since the earliest days of the American whale fishery, "Hurrah for five and forty more."

The cutting in could take anywhere from about two and a half hours to more than two and a half days, depending on the size of the whale, the skill of the crew, and, most important, the weather. Once the men had the fluke chain on the whale, they wasted no time in cutting in because the threat of heavy weather was always present in the Arctic. If the seas were too rugged, the combination of the ship's rolling and the great weight of the whale could part even the strongest head strap or fluke chain or bend open the blubber hook or force the straps, hooks, and chains to rip right through the blubber, causing part or all of the whale to be lost. If a ship happened to be on a lee shore when a gale rose, the carcass might have to be cut adrift to allow the ship to claw off the coast.

Even if the blubber was safely aboard, heavy weather could be a problem. Horace P. Smith of the *Narwhal* wrote, "I was obliged [to run into Herschel Island harbor] or lose the oil from the blubber in the between decks as the rolling of the ship is working it out fast. As it is I shall at the very least lose 50 bbls if not more."

Raising the blanket piece

The captain usually gave orders to begin boiling, trying out the blubber, as soon as they had finished cutting in the whale. The blubber-room men had cut the blanket piece into horse pieces. They began "leaning" them, using sharp knives to trim off the bits of muscle still clinging to the blubber. If the flesh were heated with the blubber in the trypots, the oil was discolored, reducing its value.

Boarding the flukes University of Alaska

The blubber-room men then pitched the horse pieces back up on deck, where others put them in tubs and dragged them to the "mincing horse." The mincing horse was a bench designed to hold the horse piece, skin down, while one of the men drew a two-handled mincing knife back and forth across it, cutting down to the skin, making "leaves." This slicing exposed more surface area, hastening the rendering process. The minced pieces, often called books or bibles, fell off the end of the horse into a mincing tub at the man's feet. Some ships carried a heavy hand-cranked machine to mince the blubber, but this invention never gained wide acceptance.

A man then dragged the mincing tubs to the tryworks. The fires under the trypots had been started using waste wood. A boatsteerer, in charge of the works, pitchforked the books into the pots. The oil was stirred continuously to prevent the blubber from burning and sticking to the bottom of the pots. Burning also discolored the oil and reduced its value.

When the blubber had melted away from the skins and these scraps of skin had become brown and twisted like hard lumps of cooked bacon rind, the boatsteerer scooped the scraps out of the pots with a skimmer. He then dropped the scraps into a strainer called a "scrap hopper" where the oil drained off and was saved. Some ships carried a scrap press for squeezing oil from the scrap. Most of the drained scrap was used to feed the fires, but some was put into a cresset, or "bug light," a wire basket that was set afire and hung between the tryworks smoke stacks to allow the men to continue their work after dark.

The boiling went on night and day. At night the fire from the bug light and the flames leaping from beneath the trypots and from the tryworks chimneys lit up the sails and rigging with an eerie glow that was visible for miles. By day the ship was equally visible by its plume of greasy black smoke. The sticky soot would have to be scraped from the masts at the end of the cruise.

The men kept the oil in the trypots very hot—so hot that the solder on the tools sometimes melted. It was important not only to melt the blubber but also to boil off all the water in the blubber. If any water remained in the oil, the oil would turn rancid in the casks. The tryworks were usually covered with a housing to prevent water or spray from getting into the hot oil. The water could explode into steam and burn the men with scalding vaporized oil droplets. In rugged weather, when the vessel shipped a big sea, the trypots threw up a huge cloud of steam and bursting oil droplets. If the hot oil spilled from the pots and reached the firepits, it exploded into flames, pouring over the tryworks and across the decks. A cracked trypot discovered too late produced the same effect.

Although the oil was kept very hot, it was not allowed to boil. To make sure that the pots did not boil over the men tested the temperature by spitting into the oil, an action which Charles Stevenson described as "producing a peculiar crepitating noise when the blubber has been sufficiently cooked." When the pots were full the boatsteerer bailed away part of the oil into a copper cooling tank on the starboard side of the tryworks. They always left some oil in the pots to help speed the heating of the next batch of blubber.

The deck pot, full of minced blubber, is chained to the tryworks. Behind it is the scrap hopper and mincing horse. In front of the tryworks is a pile of scrap that will be used to feed the fires California Academy of Sciences

The men found that the blubber from bowheads was not all the same. They most disliked taking a "dry skin," a whale whose blubber was "hard, compact, and tenacious" and gave off a "thin and watery" oil. Captain Charles Scammon, a nineteenth-century scientist and whaleman, believed that dry skins were very old whales. Once the men found they had a dry skin, they often simply threw the blubber overboard. Very young whales, by contrast, were full of oil, but often their blubber, when fresh, was as hard to boil as a dry skin's; the men left it in the blubber room for a few days to allow it to "ripen" before boiling. A bowhead's tongue also presented problems in boiling when it was "green," and it, too, was allowed to ripen in the blubber room. Nor did all whale oil boil out to the same color. The preferred oil had a white tint but some was pink or yellow, and the rotten blubber from a stinker could yield "black" oil.

Once the oil in the copper cooling tank was cool enough, the men pumped or drew it off from the tank to a large deck cask or to an intermediate tank between decks, where it was left until it had reached the air temperature.

The hot oil could not be pumped directly into the casks in the hold because it would shrink the barrel staves, causing the casks to leak. The casks in the hold were hosed down at regular intervals to keep them tight.

Boiling out a bowhead took from thirty-six to forty-eight hours, depending on its size and the sea conditions. The watches alternated the work while the cooper "set up shooks"—made casks from the staves, heads, and hoops that had been sent out, knocked down, in other casks. Meanwhile, as the tryworks went full blast, the thick, greasy smoke coated everything downwind. If whales were sighted the watch that was off duty was sent after them.

If the men were lucky, the ship became "blubber logged"—the blubber-room filled with horsepieces and no room to stow more. It was dangerous for a ship to be blubber logged. The blubber from an average-sized bowhead (one hundred barrels) weighed more than fifteen tons. With the weight of more than one bowhead stowed so high in the vessel, the ship became very crank (top heavy) and prone to capsize should a storm blow up. It was imperative, therefore, to get the oil into the casks in the lower tiers as quickly as possible.

Still, being blubber logged was not the worst fate the men could imagine. To have the blubber room full and a whale alongside meant they were well on their way toward catching the twenty or so bowheads they needed to fill the ship. After his eighth whale, when the bark *Coral* was blubber logged in the Arctic, Captain Frederick A. Barker wrote, no doubt only half-annoyed: "Ship's cabin stinks like a Chinese market." Nevertheless, as J. F. Beane noted aboard the bark *Java* in the Okhotsk Sea in 1866, "the blubber room was full . . . , but there wasn't a man on board who would not have given up his stateroom, berth and sea-chest, if necessary to stow the stuff in."

Even better was to have a full ship and be bound home. It is best described by the arctic whalemen themselves. *Pioneer,* 1850: "Hove in the throat and lips of last whale. Gave three cheers for home." *Java II,* 1858: "Threw about 4 cords of wood overboard so as to make room for stowing oil. We get duff every day now." *Shepherdess,* 1850: "Stowed down the remainder of the oil, hove the tryworks overboard, cleared up the decks, made all sail. Hurrah for home."

5 / *Development of the Fishery, 1849-64*

Roys's discovery of new whaling grounds set off an oil rush to Bering Strait. In 1849 there were more than one hundred and fifty whaleships in the North Pacific fleet, and fifty of them—forty-six American, two French, and two German vessels—went north into unknown waters on the strength of Roys's news.

Not only had Roys filled the *Superior* in less than two months, but "during the entire period of his cruise there," wrote the editor of *The Friend*, "no ice was seen and the weather was ordinarily pleasant." Furthermore, the report continued, Roys found that the waters were shallow, with good holding ground everywhere, allowing him to anchor at will. Roys had taken his first whale at midnight—and the summer light of those high latitudes allowed him to hunt the "quite tame" whales around the clock. Most important, he took what he perceived to be three different species of whales, the largest of which yielded an average of two hundred barrels of oil. Incredibly enough Roys also reported a fourth species, a whale so large that he thought it would have yielded more than three hundred barrels. It was so big that he believed the *Superior* could not have borne the strain of cutting it in. "It was not through fear for themselves, but the 'whaling gear' of the vessel, that they allowed the King of the Arctic Ocean quietly to hold his own way!" the editor added.

This news could not have come at a better time for the world's whaling fleets. By 1845 the industry's expansion had nearly come to a halt. One after another, known whaling grounds had been depleted. Antarctic waters would remain inviolate until advanced technology allowed their penetration in the twentieth century. With a few exceptions, only the waters of the extreme northern Pacific rim remained to be exploited by whalemen and by 1845 the yield of one northern Pacific ground—the Gulf of Alaska—had begun to deteriorate markedly.

Fortunately for the whalemen, in 1845 Captain Mercator Cooper of the Sag Harbor whaleship *Manhattan* found a new right whaling ground near the Kurile Islands in the Okhotsk Sea. In 1848 other ships pushed farther into the Okhotsk Sea, beyond fifty-five degrees north on the west coast of Kamchatka to discover the stock of bowheads in those waters, but they failed to understand the significance of their discovery. It would be half a dozen years before the Okhotsk fishery enjoyed its brief but intense exploitation, an exploitation that was accompanied by an equally brief period of gray whaling by the same whaleships during the winter months in the lagoons of Baja California.

But these were minor discoveries. It was the incredible richness of the western Arctic that prolonged the life of the American whaling industry, which otherwise would certainly have been moribund by 1870. Roys's discovery marked the industry's high point—its final flash of glorious success.

The period from 1846 to 1851 was the zenith of the American fleet's vigor. An average of 638 vessels, worth close to $20,000,000, sailed from thirty-nine ports on the East Coast of the United States. Another $70,000,000 was invested in the industry that employed in aggregate 70,000 persons. (The combined $90,000,000 is equal to approximately $997,200,000 in 1982 dollars). In 1849 one fifth of the world's whaling vessels were milling about in the North Pacific—without prospects. It is not surprising, then, that one third of the North Pacific fleet would have pushed one thousand miles beyond their usual grounds into poorly charted waters—all on the report of only one ship.

Despite their numbers and eagerness to reach Roys's whaling grounds, the ships moved north gingerly in the spring of 1849, not knowing what to expect of the seas before them. Once past the Aleutians all were nervously on the look out for Roys's vast herds of whales. At first the whalemen must have wondered whether they had been hoaxed; they saw only large numbers of ducks and gulls. "No fish of any kind," one whaleman noted. "All looks as barren as Buzzards Bay." As they worked north and began to see the ice, the lack of whales and fear of the unknown set some to grumbling. Aboard the *Tiger,* Mary Brewster, who would soon become the first western woman to pass Bering Strait, wrote, "I pray next week may be a little more greasy so these scowling brows may look smooth."

But soon enough, as the ships neared Saint Lawrence Island, the men found that Roys had not exaggerated. Their success was phenomenal. From late June until early September the ships sailed amid shoals of bowheads; first they found them southwest of Saint Lawrence Island; then in "the Straits"; and finally in the southern Chukchi Sea, part of which became known as the "cow yard" because of the big, slow-swimming whales they found there.

The entries in the *Ocmulgee*'s 1849 logbook are eloquent, displaying at first boredom, then doubt, then joy. Joseph Dias, Jr., the mate of the *Ocmulgee,* began to realize his good fortune in early June.

June 7. Saw a great many whales, and the right kind. Steeple tops [bowheads], but it was too rugged to lower.
June 8. First and middle part strong gales and whales without number.

Later he recorded a whaleman's nirvana.

July 25. Plenty of whales in sight but all hands too busy [boiling] even to look at them.
July 26. Blubber logged and plenty of whales in sight.
July 27. The same.
July 28. Blubber logged but were obliged to cool down for 6 hours for want of casks. Whales a plenty.
July 29. Blubber logged and whales in every direction.

The most successful ship was the *William Hamilton* with 4,200 barrels of oil. When the fleet returned to Hawaii they found that the average catch per ship had been about 1,324 barrels and that the whales had averaged nearly 150 barrels. One whaleman reported that with the twenty-four-hour daylight they had been in such a "hurry of whaling" that it had been difficult to keep track of the date. The whaling had been so good near Bering Strait that only a few ships had gone as far as the seventieth parallel and all had left the Arctic by early September.

Although Roys had confused several ages and sexes of bowheads for four separate species, he had not exaggerated when he described his "King of the Arctic Ocean" as one that would have yielded more than 300 barrels of oil. A number of the bowheads taken in the early years of the fishery are known to have exceeded 300 barrels and one is recorded as having yielded 375. The cows were found to be much bigger than the bulls. By contrast, an average sperm whale gave 45 barrels and an average gray whale about 30 barrels.

The baleen from these "noble whales" was equally impressive. The *Sarah Sheaf* probably took the longest baleen on record, fourteen feet. The longest slabs from a number of other whales reached more than twelve feet. Whales of this size would easily have yielded more than 3,000 pounds of whalebone.

With the return of the fleet to Hawaii in the autumn, the news of the Arctic season of 1849 confirmed Roys's report. The story ricocheted around the world from one maritime newspaper to another. But the ships would never see the same success again. The news lured more than 130 ships to the north the following year, drawing vessels not only from all the American whaling ports but also from Bremen, Le Havre, Nantes, and Hobart, Tasmania. The 1850 season was highly successful for all the ships, but the catch per whaling day did not equal 1849's amazing returns.

The ships had great success near Cape Navarin and in Anadyr Gulf and the weather, like that of 1848 and 1849, was comparatively mild, prompting a fit of hubris in the following report in Honolulu's *Polynesian:*

> We doubt if so much oil was ever taken in the same period, by the same number of ships and attended with so few casualties. In fact, a cruise in the Arctic Ocean has got to be but a summer pastime, as is proved by the fact that the wives of some half dozen of the captains accompanied their husbands in the last cruise, with the same willingness that they would have gone to Saratoga or Newport.

But nemesis follows hubris. Spurred by this bright outlook, more than 170 ships went north in 1851 to find the season a disaster. The average catch per whaling day was only one-third that of 1849. It was also a comparatively icy year. For the first time the masters forced their ships into the pack ice, rather than skirting it as they had in the two previous years. Finding the whales near Cape Navarin and in Anadyr Gulf both "scarse and unusually wild," they kept pressing forward toward the Arctic Ocean and the result was the loss of seven ships in the Bering Strait region. By the end of July ships began leaving the grounds. By mid-August many were bound south, tired of "plenty of fog, ice, and broken down hopes." Captain Bailey of the

Bundles of whalebone in New Bedford, ca. 1860

Champion of New Bedford sourly summarized his season: "I have done nothing but get my ship stove most to pieces by ice. The Anadir Sea has been a dead failure. Ships have not averaged a whale."

The 1852 season was almost as bad. Believing that the results of the 1851 season had not reflected the true value of the Arctic grounds, more than 220 ships went north. More vessels were criss-crossing the Bering Strait region that year than have been there in any year since. The catches were about half of the 1849 level, and they were erratic: some ships did well, others came home clean.

A few captains quickly caught the drift and left the grounds before the middle of June to try their luck in the Okhotsk Sea. With so many ships cruising about, the whales were extremely wary and hard to catch. "A whale has scarcely to draw the natural breath of life before a score of boats are after him," wrote Edwin Pulver, noting that the whales had learned to head for the safety of the ice when danger threatened. In such crowded conditions, the ships that remained in the Arctic were forced to take risks to get their whales and the result was the loss of five ships.

The 1853 season was scarcely better. More than 160 ships went to the Bering Strait region. Although they tried to increase their catches by arriving earlier in the spring and staying later in the autumn, their average catch was less than 600 barrels per ship. Several vessels were lost as well. However, the average catch in the Okhotsk that summer was more than 1,600 barrels per ship.

By 1854 most of the captains had learned their lesson; about 160 ships headed to the Okhotsk, while only 45 chose Bering Strait. Those that went to the Okhotsk averaged 1,000 barrels per ship; those that went to Bering Strait found the worst season yet. "Almost half of the Arctic fleet obtained no more oil than was necessary to keep their own binnacle lights burning." The average was less than two whales per ship, and fifteen came home clean. The whales were so scarce that year that when one whaleman aboard the *Roman* spotted a whale, he wrote, "Saw a bowhead. Joy to the world." Some ships simply gave up early and headed south to try for right whales on Kodiak, Kamchatka, or in Bristol Bay.

With whales so scarce, those ships that remained on the grounds were forced to push toward the limits of navigation throughout the Chukchi Sea. By the mid-summer of 1854 the ships had thoroughly scoured this basin—with the exceptions of the northeastern and northwestern corners, which they considered to be both barren and dangerous.

Until 1854 they feared the passage toward Point Barrow in the northeastern corner of the Sea because of the lack of a safe harbor and the narrowness of the lane of open water between the land and the polar pack ice. By chance, however, an officer of HMS *Plover,* which was part of the fleet searching for the lost Arctic expedition of Sir John Franklin, told a few whaling masters that year that Elson Lagoon behind Point Barrow had provided the *Plover* with safe quarters for two winters. Then, in mid-August, a few whaleships met HMS *Enterprise* in Port Clarence, and Captain Richard Collinson, one of the ablest of the Royal Navy's Arctic navigators, told the whaling masters that he had just returned from Point Barrow, where he had sailed in relatively ice-free seas and had seen whales.

Whaling in Bering Strait. Detail of lithograph by Benjamin Russell

Armed with the knowledge that there was at least one port of refuge on the American coast north of Bering Strait, the *Franklin, Gideon Howland, Hobomok, Rousseau,* and *William Thompson* headed to Point Barrow and became the first pelagic vessels to hunt whales in the Beaufort Sea. They took a few small whales there, but the reward and the risk were unacceptable, and the fleet returned to Hawaii secure in the belief that the Arctic had been "fished out."

The following year, 1855, the Arctic was almost deserted. More than 130 ships went to the Okhotsk Sea. It is likely that only eight vessels passed north of the sixtieth parallel in the Bering Sea; of these, most spent only a few unsuccessful days in Anadyr Gulf before returning to the right whaling grounds in Bristol Bay.

Although the 1855 season in the Okhotsk Sea was only fair, nearly 150 ships returned there in 1856. The Bering Strait whaling grounds were nearly deserted. It is unlikely that more than one or two whaleships passed north of the sixtieth parallel and those that did probably devoted more time to trading with the natives than to whaling.

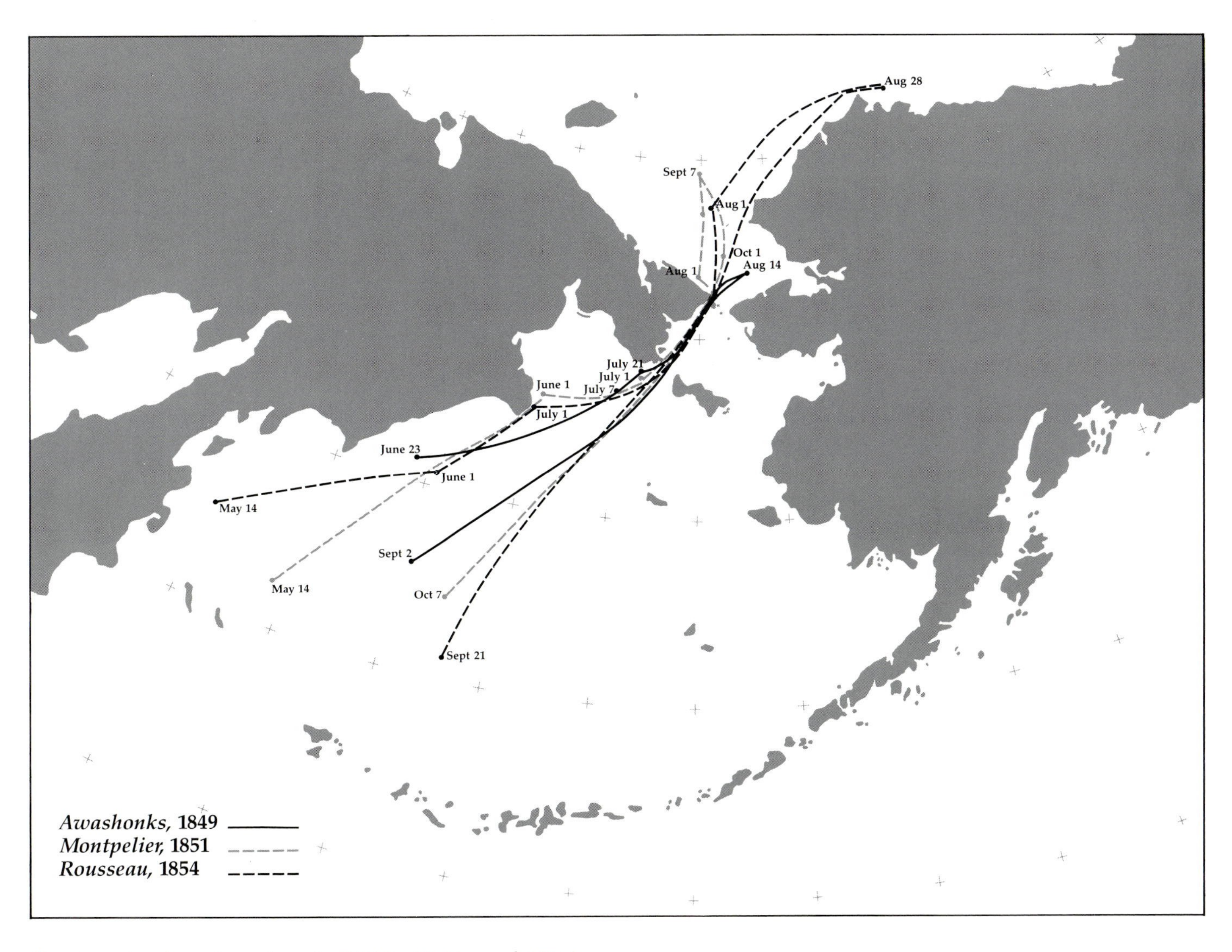

Routes of selected cruises in 1849, 1851, and 1854

By the autumn of 1856 the vast majority of whalemen believed that the western Arctic had been largely swept clean of whales. As early as 1850, at least a few whalemen believed they were doing irrevocable damage to the bowhead population there. A rare flash of humor appeared in Honolulu's *Friend* on October 15, 1850.

We are somewhat surprised that a member of the whale-family should condescend to make his appeal through our columns, inasmuch as we have ever aimed to direct whalemen to the best cruising grounds. We feel honored by the compliment and shall feel bound, on no consideration to betray the confidence thus reposed in us: —

A POLAR WHALE'S APPEAL
Anadir Sea, North Pacific
The second Year of Trouble

Mr. Editor,—In behalf of my species, allow an inhabitant of this sea, to make an appeal through your columns to the friends of the whale in general. A few of the knowing old inhabitants of this sea have recently held a meeting to consult respecting our safety, and in some way or other, if possible, to avert the doom that seems to await all of the whale Genus throughout the world, including the Sperm, Right, and Polar whales. Although our situation, and that of our neighbors in the Arctic, is remote from our enemy's country, yet we have been knowing to the progress of affairs in the Japan and Ochotsk seas, the Atlantic and Indian oceans, and all the other "whaling grounds." We have imagined that we were safe in these cold regions; but no; within these last two years a furious attack has been made upon us, an attack more deadly and bloody, than any of our race ever experienced in any part of the world. I scorn to speak of the cruelty that has been practised by our blood-thirsty enemies, armed with harpoon and lance; no age or sex has been spared. Multitudes of our species (the Polar), have been murdered in "cold" blood. Our enemies have wondered at our mild and inoffensive conduct; we have heard them cry, "there she blows," and our hearts have quailed as we saw their glittering steel reflecting the sun beams, and realized that in a few moments our life-blood oozing out, would discolor the briny deep in which we have gambolled for scores of years. We have never been trained to contend with a race of warriors, who sail in large three-masted vessels, on the sterns of which we have read "New Bedford," "Sag Harbor," "New London." Our battles have hitherto been with simple Indians in their skin canoes. We have heard of the desperate encounters between these whale-killing monsters and our brethren the Right whales on the North-west coast. Some from that quarter have taken shelter in the quiet bays of our sea, others of the spermaciti species from Japan, have also visited us and reported their battles and disasters; they have told us it is no use to contend with the Nortons, the Tabers, the Coffins, the Coxs, the Smiths, the Halseys, and the other families of whale-killers. We Polar whales are a quiet inoffensive race, desirous of life and peace, but, alas, we fear our doom is sealed; we have heard the threat that in one season more we shall all be "cut up," and "tried out." Is there no redress? I write in behalf of my butchered and dying species. I appeal to the friends of the whole race of whales. Must we all be murdered in cold blood? Must our race become extinct? Will no friends and allies arise and revenge our wrongs? Will our foes be allowed to prey upon us another year? We have heard of the power of the "Press;" pray give these few lines a place in your columns, and let them go forth to the world. I am known among our enemies as the "Bow-head," but I belong to the Old Greenland family.

Yours till death,
POLAR WHALE.

P. S. I send this by the __________ of __________ Don't publish the name of the vessel. P. W.

The author understood that at the current pace of killing, the bowheads would soon be exterminated. The whalemen were, however, held captive by the same forces that drive all boomers to deplete a resource: without the universal acceptance of a common goal in preserving the yield of a renewable resource, those who show forbearance are merely those who profit less.

Nevertheless, we know that the richness of the region's bowhead stock sustained the industry well into the twentieth century. Why were catches so poor from 1851 to 1856? By 1856 the ships had killed only about 7,000 bowheads out of the nearly 20,000 they would kill from 1849 to 1914. It is likely that the size of the bowhead population in 1848 was more than 20,000 whales, and most likely about 30,000. Clearly then, by 1856, there were still plenty of bowheads left.

There were probably several reasons for the precipitous decline in the catches. Certainly, however, one of the principal causes of the decline was the whale's adaptation to their new threat. They learned that there was safety deep in the pack ice where the whaleboats could not go.

In 1852 Captain Asa Tobey remarked that the bowhead "is no longer the slow and sluggish beast we at first found him." The whalemen soon found that where they made a large catch in one year, it was unusual to have good whaling again for three years or more. One captain wrote in 1851, "where I whaled last voyage now looks like a deserted village." A whaleman aboard the *Bowditch* summed it up in 1850 when he wrote, "The whales appear very shy. They don't like cold iron."

The whales also learned a number of tricks to avoid capture. They developed the ability to distinguish between sounds, recognizing the threatening ones: "If a bowhead is near an ice pack, and a chunk of melting berg suddenly drops into the sea with a loud splash, it does not disturb the whale," wrote Captain Foley of the *Monterey.* "But let a whaleboat, proceeding ever so cautiously toward [the whale], strike a floating piece of ice in his neighborhood, and . . . [it] vanishes. . . . It would be useless to look for the whale for several hours." Occasionally the whalemen also noticed a whale's body stiffen slightly at the sound of an approaching boat. Often, then, the animal would jump backward, roll sideways, or sink like a stone.

Whales that had escaped after being struck were particularly wary. Jim Allen, an accomplished whaleman, told of capturing the "whistling whale." Famous for a blow that sounded "like a steamboat whistling," this whale had been recognized and hunted for many years. When Allen finally killed it in 1899, the men found an iron and a hole in its side where it had been struck "a great many years before." The whistling whale's baleen was eleven feet long, an indication of great age. Allen believed that the whale had become exceptionally canny because of its wounding. "Whales that have once been struck are always hard to get," said Allen. Allen may well have been right; in 1890 the steam bark *Beluga* took a whale with an iron in it from the ship *Montezuma.* The *Montezuma*'s last Arctic voyage had been in 1853.

The bowheads' responses to their new threat were at first very effective. When the ships were nearby the whales kept near the ice for much of the season in the Bering and Chukchi Seas. Moreover, many of the bowheads normally migrated into the Beaufort Sea in the summer. Whaleships did not regularly cruise in those waters until the 1870s and the whales consequently enjoyed three months of safety before returning to the Bering and Chukchi seas. In the 1850s the whalemen still approached the ice with extreme caution and had not learned that in the late autumn the whales congregated to feed near Herald Island and Herald Shoal in waters that by then were ice free. Understandably chary about challenging the violent weather of late autumn, very few ships passed north of the seventieth parallel and most cleared Bering Strait, bound south, by mid-September.

Another possible cause for the dramatic drop in the whalemen's catches is that the western Arctic bowhead stock may have comprised two or more subpopulations. In the years immediately following Roys's discovery, a number of bowheads were taken in the height of summer from fifty-three to fifty-five degrees north near the

Commander Islands and the coast of Kamchatka, in the southwest corner of the Bering Sea. Within a few years after Roys's discovery, however, no bowheads were found there. It seems that at least one subpopulation did not travel north in the summer. Lacking the safety of the pack ice in the height of summer, these bowheads were quickly captured—as were most of the bowheads in the Okhotsk Sea where the ice cover is gone by the end of July. Similarly, after the first twenty years of the fishery whales were only infrequently taken during the height of summer in the ice-free waters of Anadyr Gulf, in "the Straits," or in the southern Chukchi Sea. Today, bowheads are rarely, if ever, seen south of the sixtieth parallel and are found almost exclusively in the Beaufort Sea during the height of summer.

In 1857 the Kodiak grounds were on the decline and with nowhere else to go, more than one hundred whaleships headed for the Okhotsk Sea. The only thing that kept more ships from hunting for the Okhotsk's bowheads was the price of sperm oil. It had begun to rise again and a number of captains headed to the sperm whale grounds. Nevertheless, a handful were willing to risk a return to the Bering Strait. To their surprise they found plenty of whales, although the harsh weather and rough seas limited the catch.

In 1858 nearly one hundred ships headed to Bering Strait because there was nowhere else to go. They found an exceptionally cold and icy spring. Well into July ice plugged Bering Strait and, worse, the men saw only a few whales. By the middle of the summer a number of ships had left again for the Okhotsk or Bristol Bay, "fully satisfied that the Arctic ground is a humbug." But those who stayed north until the end of the season found good whaling near Cape Lisburne from late August until mid-September.

The rest of the Pacific was so barren that in 1859 more than eighty ships went north to Cape Lisburne. They spent as much time as possible cruising near that headland, but, of course, the whales had deserted the area and the ships took very little oil. The 1860 season was much the same.

At this point the outlook for the industry was not good. Many investors in the industry took their money elsewhere, but for those who remained a pattern had clearly developed: the western Arctic would be the mainstay worldwide for half of each year. To achieve moderate success the ships now had to arrive in the Bering Sea earlier in the spring and to stay in the Chukchi Sea later in the autumn. The whales would never be found in the same place in successive years and the fleet would now be successful only in those years when the ice and the weather favored the ships. After one or two poor seasons the size of the Arctic fleet would decline; then, the year following a good catch, it would increase—only to have the pattern repeated again.

6 / *Civil War in Bering Strait: The Cruise of CSS* Shenandoah

The sun rose shortly after 1:00 A.M.at Bering Strait on June 28, 1865. The great surging rollers that usually travel north out of the Pacific Ocean, steepening and shortening as they reach the Strait, were still. The wind that normally howls through the narrow gut was quiet, the sky an even light-grey. Only the ice floes, drifting north with the current, disturbed the smooth waters of the Strait, marbling the calm dark surface with a brilliant white.

The massive black headlands of East Cape, Siberia, and Cape Prince of Wales, Alaska, framed the sun as it floated in a rosy haze on the northern horizon. Between them lay the two steep-sided, flat-topped Diomede Islands, as black as the headlands but streaked by vertical slashes of unmelted snow in the islands' steep ravines.

It was a sparkling dawn—the sort that happens only once or twice a year.

Under the bight of East Cape in the mirror-calm and quiet, nine whaleships stood at anchor around another that lay listing severely. More than a dozen whaleboats were moored to the listing whaler, the *Brunswick* of New Bedford. Aboard her the captains of the nine ships were auctioning her gear and cargo.

Only the day before, in a strong breeze, the *Brunswick* had smashed into a large cake of ice and stove a hole in her hull below the waterline. By heeling her over, Captain Alden Potter kept the gash above the water, but there was no hope of repairing her more than a thousand miles from even the most rudimentary shipyard. Following the customary procedure, Potter hoisted the Stars and Stripes upside down to signal that his vessel was in mortal jeopardy. The other ships closed in to offer assistance. Potter called their masters aboard and they confirmed his opinion, unanimously agreeing that there was nothing that could be done to save the vessel. Before abandoning his ship, Potter auctioned her movable gear and cargo to raise what money he could for the ship's owners and insurers.

Two heads of baleen had already been put aboard the *James Maury* and the auction was proceeding for the remaining oil and baleen, running rigging, whalecraft, boats, and other gear, when a steamer hove into sight from the south. The masters were baffled by its presence. No auxiliary vessel had ever been seen in those waters before.

The steamer was flying the American flag. Some thought it was a supply vessel for the Western Union Telegraph Expedition, which was attempting to lay a cable across Bering Strait to link Paris and New York. A few thought it might be the Confederate raider that had been in Australia in February, but they were quickly overruled. In the spring when the fleet had left San Francisco and Honolulu, the Civil War was all but over and the Confederate Armies were disintegrating. It had been clear then that all armed resistance would soon cease.

But, to the shock of the whalemen, five armed boats put off from the steamer carrying uniformed officers of the Confederate Navy. The officers announced to the crews of the becalmed ships that they were the prizes of CSS *Shenandoah*. They ordered the men to abandon their vessels and come aboard the *Shenandoah* as prisoners of war or, the Confederates threatened, they would be blown out of the water.

The whalemen panicked. The captains raced to their ships. Some slipped their cables and tried to reach within one marine league of the Siberian shore, where, theoretically, they would be under the protection of the Russian government. But it was no use. In the flat calm the ships could move nowhere. The whalemen began abandoning their ships and rowing to the steamer while Confederate prize crews boarded the vessels, axed open oil casks and set them afire.

The Southerners released two ships to take the prisoners south, but eight ships were burned in what was the last offensive action of the American Civil War. Occurring as it did two and a half months after Lee's surrender at Appomattox and one and a half months after the capture of Jefferson Davis, the event had severe repercussions. Among other things, it soured relations between the United States and Great Britain for another seven years.

Four years before the *Shenandoah*'s fiery cruise, the South had no navy. The Confederacy's leaders realized that to survive they had to keep their supply lines open to Europe and, at the same time, to disrupt the Union's shipping. They needed ships, fast. But Britain, the logical source for these vessels, issued a proclamation of neutrality that prohibited the South from buying ships of war in England.

The proclamation fell short of recognizing the Confederacy as an independent nation; it did, however, assign the South the status of a belligerent, allowing Confederate ships to visit British ports—a point that was hotly contested by Charles Francis Adams, the United State's Minister to the Court of St. James.

The South was desperate for fighting ships. The only way to get them was to buy them in Europe. The Southern leaders immediately dispatched an agent, James D. Bulloch, to begin secretly buying ships and supplies for the Confederacy. Bulloch, a former officer of the U.S. Navy, reached England in June 1861. His mission was, as he wrote, "to destroy the enemy's commerce, and thus to increase the burden of war upon a large and influential class at the North." This would weaken the Union's blockade of the South by drawing U.S. naval ships away in pursuit of the Confederate commerce raiders.

The first raider Bulloch built was the *Florida*. Shortly thereafter he started work on the *Alabama*, which had a spectacular career sinking Northern shipping until she herself was sunk by the USS *Kearsarge* in 1864. He was frequently blocked by Charles

Francis Adams and others who forced the British to honor their Proclamation, but Bulloch was successful in cutting back the American carrying trade. His cruisers sank many Northern ships and by 1863 three-quarters of the importations to New York were carried in foreign hulls, whereas in 1861 two-thirds had come in American bottoms. Northern shipowners, protecting their investments, transferred 750 vessels to foreign registry between 1861 and 1864.

The South was clearly losing the war when Bulloch unleashed his second phenomenally successful raider, the *Shenandoah.* Sent out in response to the desperate order given by the South's secretary of the navy, Stephen P. Mallory, to increase the attacks against the North's "coasting trade and fisheries," this raider would cause the most anxiety among the American whaling industry.

Bulloch knew that it was useless to have a vessel built in England. Charles Francis Adams had an effective network of informers and he had already forced the British to seize two of the South's ram vessels while they were still in the stocks. Bulloch decided to convert ships that he could buy. He reasoned further that because Mallory now wanted cruisers to range worldwide after the Northern whaling fleets, most of the propeller steamers available for sale would be unsuitable as "they were constructed almost exclusively for steaming, had no arrangement for lifting their screws [to add speed under sail], and were masted only with the purpose to furnish auxiliary sailing-power. . . ."

Bulloch needed a vessel that was fast under both sail and steam. In the autumn of 1863 he found her. The *Sea King,* then about to set out for Bombay on her maiden voyage, was, as Bulloch noted, "well suited for conversion into a vessel of war." At more than 1,100 gross tons, with a length of 222 feet, a 35-foot beam, and clean, raking lines, she was big and fast. "She carried a cloud of canvas, having cross-jack, royal studding sails, jib topsail, and all the high flyers," and she could steam comfortably at nine knots. She was well built: Lloyd's gave her a rating of A-1, good for fourteen years, and she had an iron frame and knees and was planked with six inches of East India teak. A full-rigged ship, her lower and topmasts and bowsprit were of iron and hollow; she carried roller reefing topsails, steel yards, wire standing rigging, and an engine that delivered 850 horsepower. She proved her potential on her maiden voyage when she covered 330 miles under sail in twenty-four hours.

But the *Sea King* had one important disadvantage: she looked remarkably similar to the *Alabama.* As Bulloch noted, "her fitness for conversion into a cruiser would be manifest at a glance. . . . It was absolutely necessary to permit no one having the faintest odor of 'rebellion' about him to go near her." Bulloch was on the mark. Rumors had already begun to circulate that the South intended to outfit the *Sea King* as a raider.

When the *Sea King* returned, Bulloch sent a British subject, Richard Wright, to buy the vessel. Bulloch planned to send her out again as if on another merchant voyage, but to secretly outfit her at sea as a Confederate ship of war.

To do this he needed a tender to rendezvous with the *Sea King* to provide her crew, guns, and supplies. Bulloch bought an iron propeller steamer, the *Laurel,* and put her "in the hands of a shipping agent in Liverpool, who advertised her for a voyage to Havana to take freight and a limited number of passengers."

Meanwhile the *Sea King* was moved through several British ports to give the appearance of the preparations for a merchant voyage, and Richard Wright gave his power of attorney to the *Sea King*'s captain, G. H. Corbett, allowing him to sell the vessel "at any time within six months for the sum of not less tha £45,000 sterling." On the eighth of October in London, Captain Corbett weighed anchor "on a voyage which, to all appearances, was precisely similar to the former one." Word of the departure reached Liverpool the same day, and many of the officers and some of the men destined for the *Sea King* slipped aboard the *Laurel* in the evening. The *Laurel* was then taken into the Mersey to begin her voyage.

Both vessels headed for a rendezvous at Madeira. First, however, the *Sea King*, armed only with her two twelve-pound "pop guns" (the normal armament for an East Indiaman), had to slip past USS *Niagara* and USS *Sacramento*. The *Sea King* ran the gantlet without difficulty and arrived at Funchal, Madeira, on October 18, three days after the *Laurel*. The next day both vessels stood around to Desertas, an uninhabited island, and frantically began transferring the crew and cargo, fearful that they would be discovered by a U.S. naval ship while they were unarmed.

After thirty-six hours, when the work was completed and the supplies and equipment lay in a wild jumble in the hold and on the *Sea King*'s deck, Captain Corbett, as ordered by Bulloch, announced to the crew that he had sold the ship to the Confederacy. Immediately Captain James I. Waddell, C.S.N., who had secretly come out aboard the *Laurel*, appeared on deck in the uniform of the Confederate Navy and announced that the *Sea King* was now commissioned as CSS *Shenandoah*, a raider outfitted to destroy federal shipping. He asked the crew of the *Sea King* to remain aboard. Only four of the fifty-one elected to stay, and five more men from the *Laurel* joined the thirty-four men and officers who had slipped aboard her in Liverpool, giving *Shenandoah* a crew of only forty-three, a third or less of the number needed to make her an efficient fighting ship. They were so shorthanded that the crew alone could not raise the anchor and the officers were forced to pitch in alongside. The vessels then separated. The *Shenandoah*'s cruise had begun.

Bulloch's orders to Waddell were succinct:

> You are about to proceed upon a cruise in the far-distant Pacific, into seas and among the islands frequented by the great American whaling fleet, a source of abundant wealth to our enemies and a nursery for their seamen. It is hoped that you may be able to greatly damage and disperse that fleet, even if you do not succeed in utterly destroying it.

To avoid running afoul of Britain's proclamation of neutrality, Waddell was told to allow Corbett to return to England with the crew and to cancel the *Sea King*'s register. Waddell was then to begin operations, going to Australia via the Cape of Good Hope. From there he was to head north, traveling through the various whaling grounds to the Caroline Islands and then on into the North Pacific. There he was to position himself 60 to 130 miles north of Oahu in November to intercept

Captain James I. Waddell, C.S.N., master of CSS Shenandoah U.S. Navy

the whaling fleets returning from their summer cruises in the Gulf of Alaska, the Okhotsk Sea, and Arctic Ocean.

Bulloch added, however, that in all these directives Waddell was to be allowed the broadest discretion because he would be so far from the direct control of his government. Bulloch could not have known at that time just how far from direct control Waddell would eventually be.

At the outset Waddell was ill-equipped to carry out these charges. Although the *Shenandoah* had aboard four eight-inch smooth bores, two thirty-two-pound Whitworth rifles, and her original two twelve-pounders, only the twelve-pounders were rigged. With that armament, as she lay off Desertas on October 20, 1864, she could have made little impression on the whaleships, let alone federal ships of war.

Once underway Waddell faced the problem of converting the *Shenandoah* from a merchantman into a fighting ship, a job that would normally require a dock and ship riggers. Even before they could begin, they had to clear the deck of its chaotic mess and sort out the supplies. Once that was done, they mounted the guns on carriages and cut gunports. They were ready to rig the gun tackles when they discovered that there were none aboard the ship. Without the tackles the guns would be impossible to steady in a sea; if fired, the recoil would send them careering across the deck and smashing into the far rail. Waddell knew that he could not jury-rig the tackles. He would have to seize them from his first capture. Until then he would have to bluff his way through any encounter with his tiny twelve-pounders.

Bluff or not, his presence at sea was a real advantage for the South. The news was out almost at once. On October 7, 1864, the day before the *Shenandoah* and the *Laurel* sailed, Thomas H. Dudley, the capable and industrious U.S. consul in Liverpool, wrote to Charles Francis Adams that the *Laurel* was taking on a cargo of guns and gun carriages and twenty-one more men that would be required for a merchant voyage. At first Dudley thought that the *Laurel* herself was being outfitted as a privateer. By October 12, however, he had begun to suspect that she was a supply vessel for another ship. On the eighteenth he learned of the *Sea King*'s departure and suspected the *Laurel*'s true purpose.

All was confirmed by November 12. Dudley interviewed some of the sailors who had returned from the *Sea King*. Adams put the word out and protested the outfitting of the *Shenandoah* to Earl Russell, the British Foreign Secretary. By then the *Shenandoah* was at sea and at large.

On October 30, due west of Dakar, *Shenandoah* claimed her first prize: the bark *Alina* of Searsport, Maine, on her maiden voyage, carrying railroad iron to Argentina. Overhauling her, they fired a blank cartridge from one of the twelve-pounders and forced her to heave to. They stripped her of her stores and took enough blocks for their gun tackle. After taking the *Alina*'s crew prisoner aboard the *Shenandoah*, they scuttled the bark.

They sank three more ships in the South Atlantic before capturing the clipper ship *Kate Prince* of Portsmouth, New Hampshire. Because Waddell wanted to be rid of his prisoners from the four previous captures, he "ransomed" the *Kate Prince* by forcing the captain to sign a bond, agreeing both to pay the Confederacy $40,000 ($252,000 in 1982 dollars) within six months of the end of the war and to carry the prisoners directly to Recife (Pernambuco), Brazil.

The next day the Southerners burned the little schooner *Lizzie M. Stacey*. She was carrying barrel staves and other cargo to the whaling fleet at Hawaii. On the fourth of December they burned the old whaling bark *Edward* of New Bedford. Waddell then headed for Tristan da Cunha, a group of tiny volcanic islands midway between Cape Horn and the Cape of Good Hope. He put twenty-eight prisoners ashore at the small settlement there. Fortunately, the men did not have to wait long for passage off that infrequently visited rock. On December 28, USS *Iroquois* arrived in pursuit of the *Shenandoah*.

On leaving Tristan, Waddell had decided to steer directly for Australia. But to confuse the prisoners ashore, who, he felt, would soon be in contact with the Union ships that must now be after him, he steered on a bearing for Cape Town until he was below the horizon, then swung to the more southerly course.

To his dismay, shortly after leaving Tristan, one of the crew discovered a crack in the propeller shaft coupling. Should it give way, the ship would be without auxiliary power. The nearest dry docks were at the Cape and Australia, but Waddell did not want to get caught dry-docked at Cape Town, so he decided to continue on for Melbourne. The *Shenandoah*'s passage through the roaring forties was difficult and stormy, the crew's life made miserable by increasingly leaky decks. After the capture of another freighter and a brief stop at St. Paul (3,800 miles south of India), the *Shenandoah* dropped anchor in Hobson's Bay off Melbourne on the twenty-fifth of January, 1865.

She was immediately surrounded by a fleet of harbor craft that carried a large and friendly crowd, sympathetic to the South's cause. Waddell wrote to Sir Charles Darling, the governor, requesting permission to make repairs on his ship and to buy coal and provisions there. Darling granted the permission the following day but also firmly reminded Waddell of Great Britain's neutrality.

Although Waddell had reached a friendly port, he had not outrun the Union's network. Within twenty-four hours of the *Shenandoah*'s arrival, William Blanchard, the U.S. consul, had sent a letter to Charles Francis Adams giving details of the *Shenandoah* down to the serial numbers on the guns, information he had probably received from the prisoners who had been captured in the Indian Ocean and released in Melbourne.

Blanchard also immediately protested to Darling that the *Shenandoah* had not touched at any Confederate port during her cruise. Thus, he claimed, the ship had not been properly naturalized by the Confederacy, that she was in fact a pirate, not a ship of a belligerent power, and that she should consequently be seized.

Darling rejected this rather weak argument as well as the subsequent blizzard of protests and affidavits filed by the U.S. consul, but the presence of the *Shenandoah* in port clearly made the governor uneasy. Through his commissioner of trade and industry, he kept pressure on Waddell to leave as soon as possible.

Blanchard was on far firmer ground when he claimed that Waddell was recruiting in Melbourne for his undermanned ship—in direct violation of Britain's Foreign Enlistment Act. Neither Blanchard nor anyone knew at the time that his protest was to have far-reaching implications seven years later. On February 10 he produced testimony that more than ten Melbourne recruits were aboard the *Shenandoah* (later it was claimed that the repairs were drawn out for two weeks longer than necessary while more than forty men were recruited). Presented with such direct testimony, the local authorities demanded to search the ship, but they were refused permission by the officer of the day, and apparently the search was not pressed. The *Shenandoah* cleared for sea on the seventeenth of February and sailed the following morning.

As soon as they dropped the pilot outside Melbourne's heads, forty-two men who had been smuggled aboard (the Southerners claimed they were stowaways) came out of hiding: fourteen from the hollow iron bowsprit where they had nearly

The Shenandoah *dry docked in Melbourne, Australia* U.S. Navy

suffocated; twenty from some empty water tanks; the rest from the lower hold. The *Shenandoah*, now refitted and with her crew doubled, sailed into the western Pacific where no federal ships were cruising at that time. "A fast ship in an era of slow communications, the *Shenandoah* had every advantage of surprise."

Waddell at once headed north. He assumed that his long stay in Melbourne had allowed Blanchard to put the word out ahead of him and that a federal ship of war might be after him, so he continued on past the whaling grounds of New Zealand. On March 30, near the equator he came upon the little Hawaiian trading schooner *Pfiel*, a vessel that frequently spent its summers in the Arctic, trading whiskey for furs and ivory, and its winters in the Pacific, trading for tortoise shell. Claiming his was the British ship *Miami*, Waddell talked with the *Pfiel*'s captain and learned that several whaleships were in the harbor at Ascension Island (Ponape) in the eastern Carolines. He arrived there two days later to find the ships *Hector* of New Bedford and *Edward Carey* of San Francisco and the barks *Pearl* of New London and *Harvest* of Honolulu.

Waddell captured them without a shot and locked the crews below decks on the *Shenandoah* for fourteen days while he stripped the ships of gear and provisions, then ran the vessels aground and burned them. The *Harvest* was registered in Hawaii, which was at that date a sovereign nation. Waddell noticed, however, that her registry had been transferred there from New Bedford only three years before. He concluded that the transfer was a mere canard to escape the Southern cruisers. He put her to the torch as well.

Waddell sent the 130 prisoners ashore with a few provisions and sailed away. The men lived there for five months, cared for by the natives and eating mostly yams and bananas, until they were rescued by the Hawaiian whaling bark *Kamehameha V* and taken to Honolulu.

Waddell then steered north, armed with more than his guns: among the items he had confiscated at Ascension were whalers' charts of the Okhotsk and Bering Seas. On the charts were the captains' notes of the places where whales had been taken in previous seasons and where, therefore, whaleships were likely to be found in 1865.

After a few days of unsuccessful hunting in the sea lanes between Hong Kong and San Francisco, Waddell turned the *Shenandoah* north toward the Okhotsk Sea. The Southerners sighted the Kurile Islands on May 20, passed through them, and ran up the west shore of the Kamchatka Peninsula. On May 27 they captured the fifty-five-year-old bark *Abigail* of New Bedford.

The *Abigail*'s captain, Ebenezer Nye, had run toward the *Shenandoah* thinking she was a Russian vessel, and was astonished when the Russian flag was replaced by the Confederate colors. Nye was incredulous at finding a raider so far north in the Pacific. Joshua Minor, one of the *Shenandoah*'s officers replied, "Why the fact of the business is, Captain, . . . we have entered into a treaty offensive and defensive with the whales, and we are up here by special agreement to disperse their mortal enemies."

Nye, who had caught few whales on the *Abigail*'s three-year-long voyage, answered, "The whales needn't owe me much of a grudge, for the Lord knows I haven't disturbed them this voyage, though I've done my part at blubber hunting in years gone by."

It took the *Shenandoah*'s crew two days to transfer the *Abigail*'s supplies. The work was complicated by the discovery of several dozen barrels of corn whiskey (labeled "to be used in case of sickness") that Nye planned to use in fur trading on the Siberian shore. The first party sent to the *Abigail* got roaring drunk on the stuff; then a second was dispatched from the *Shenandoah* to rescue the first, but they succumbed to the same temptation; then a third. "There was not a dozen sober men on board the ship except the prisoners," wrote Cornelius Hunt, the master's mate. "We never captured a prize that created so much excitement as this, and we never captured one of so little value."

But one valuable article they captured was the person of Thomas Manning, the *Abigail*'s second mate, who agreed to join the *Shenandoah*. A native of Baltimore, Manning had served on several whaling vessels before joining the *Abigail* in San Francisco. He was, apparently, thoroughly disliked by the *Shenandoah*'s officers—as much for being a liar as a turncoat—but he was extremely useful to them, for he agreed to guide the raiders to the Northern whaling fleet.

Initially, however, Manning's aid bore no fruit. The *Shenandoah* was quickly caught in the ice while steering for the whaling grounds at Jonas Island in the center of the Okhotsk Sea. To the Southerners, the pack ice was a fearsome thing: "A woman's temper," Hunt wrote, "is not more capricious than the movements of ice in these northern seas." While lying in their bunks the men could hear "the huge blocks thundering and chafing against the side of the ship as though [they] would dash her to pieces." And as soon as the pack eased, to the crew's relief, Waddell headed the ship out of the Okhotsk.

The *Shenandoah* passed again through the Kuriles, then headed north along the east coast of Kamchatka, and on June 16, 1865, entered the Bering Sea. She was now poised at the beginning of the most destructive fortnight of her career.

With Manning's knowledge and the captured whale charts, Waddell had a good idea of where to find the Arctic whaling fleet. Hunt, however, tried to increase their stock of information; under the guise of an interest in the techniques of Arctic navigation, he frequently chatted with Ebenezer Nye, attempting to draw more details of the fleet's whereabouts from him. He found himself charmed by Nye's vaguely larcenous demeanor and by his schemes to add to the profits of his voyage by trading cheap whiskey and secondhand clothing to the natives, but he could not have foreseen that Nye's grit would cost the *Shenandoah* a number of prizes.

As the *Shenandoah* headed north toward the ice and the fleet, she was enveloped by the cold and murky fogs that are part of summer in the Bering Sea. The Southerners must have learned from the *Abigail*'s men that every spring the whaling fleet worked its way north along the Asian side of the Bering Sea, through the ice, toward the three areas where bowheads could usually be counted on to congregate briefly during their northern migration. The waters near Capes Navarin and Thaddeus, Indian Point, and Bering Strait usually proved good hunting for the whalers, and Waddell hoped they would be good hunting for the *Shenandoah* as well.

After five days in the clammy fog, the *Shenandoah* broke into clearer weather and the men found themselves within sight of Cape Navarin. Almost immediately they began seeing scraps of blubber and meat floating in the water, indicating that a whaleship was cutting in nearby. Judging from the direction the current was setting, Waddell correctly concluded that the whaleship was southwest of him. He steered in that direction, and in the northern twilight shortly after midnight, he saw a smudge on the horizon, the smoke from a whaler's tryworks. At 9:00 A.M. the men spotted two ships to port, at 10:00 they could smell the smoke from their fires, at 11:00 Waddell sent a boarding party to the largest vessel in the fleet, the ship *William Thompson* of New Bedford, and an hour later he seized the *Euphrates*, also of New Bedford.

As soon as they had the *Euphrates* the lookouts reported another sail, and Waddell went after her. It was the bark *Robert Towns* of Sydney, Australia, and it proved to be an unfortunate encounter for the *Shenandoah*. As the raider passed close to the *Robert Towns*, her captain, Frederick A. Barker, a Nantucketer, asked the ship's name. Waddell, who had run up the Russian flag, replied it was the ship *Petropavlovski*. Barker smelled a rat. He had been in Australia when the *Shenandoah* was there, and he no doubt had a description of her. As the ships parted Barker set out at once to warn the fleet farther to the north, while Waddell, unable to detain a neutral vessel, returned to his two prizes. After taking the crews aboard the *Shenandoah*, he set fire to the vessels, then "steamed away to the northward in search of more Yankees."

Almost at once the raider was enclosed in the fog again, and it was not until late the next day that she emerged from it to find eight sails in sight. Again, unfortunately for Waddell, two of the ships were flying foreign colors.

The first vessel they overhauled was the *Milo* of New Bedford. As the *Shenandoah* approached, the officers onboard the *Milo* assumed she was one of the Western Union Telegraph Expedition ships. The *Shenandoah* passed close under her stern and Waddell asked Captain Jonathan C. Hawes to come aboard with his ship's papers. Captain Hawes, a "fine looking old veteran, standing over six feet two, and straight as an arrow," came over the rail and announced that he believed the war was over. Waddell asked for proof, but Hawes could produce none. Because he would be rapidly overflowing with prisoners, in lieu of burning *Milo*, Waddell forced Hawes to sign a bond for $40,000—payable to the Confederate government at the end of the war—and agree to go directly to San Francisco with the men.

Captain Hawes sensibly agreed, and of course the bond was never paid. Cornelius Hunt could not resist the dig two years later when he cheekily reported:

> this and a number of similar vouchers taken by us during our cruise, have not yet been paid, and if they ever intend to take up these obligations, no better time than the present will ever offer. To be sure the war terminated disastrously to our cause, but we are, therefore, so much the more in need of any trifling sums that may be owing us. The above amounts, therefore, may be sent to me, care of my publisher, who is hereby authorized to receipt for the same.

In the meantime, the two vessels nearest the *Milo* had figured out what was going on and were making all sail to get into a nearby ice field. But with her auxiliary power the *Shenandoah* quickly overhauled the bark *Sophia Thornton* and brought her to with a warning shot. The other, the *Jireh Swift* was a fast bark and under a rising wind made every effort to escape. It took the *Shenandoah* three hours to catch her. She was immediately put to the torch, and the crew was sent in boats to the *Milo* while the *Shenandoah* turned to chase a third ship. This one, however, had a running start and made it to the safety of the ice.

Abandoning that chase, Waddell returned to the other whaleships nearby but found that they were now protected by a vast field of drift ice that had moved in on them, floes that the *Shenandoah*'s men, still shy from their brush in the Okhotsk, were in no mood to challenge. Hunt himself saw the irony in this—the ice, which had taken such a toll of whaleships, had now become their protector.

Cornelius Hunt, the Shenandoah's *master's mate* U.S. Navy

So Waddell turned back to the *Sophia Thornton*. Anxious to be off after other prizes, he told the *Thornton*'s men that they must go to the *Milo* in their boats, that they were free to take what provisions they wanted with them, but that they must set fire to the ship as they left. As a precaution that they would not merely sail away once the *Shenandoah* was out of sight, he ordered his boarding party to cut down her masts. "As we glided seaward," Hunt wrote,

we could see the disabled vessel, with her masts dragging alongside, and the paroled prisoners in their whale boats, transferring from her to the Milo whatever suited their fancy. I have no doubt that craft was thoroughly ransacked, but ere the sun made its brief disappearance below the horizon, a bright tongue of flame shot heavenward, telling us that the prisoners had performed their distasteful task. A more unpleasant duty, I trust, will never be assigned to any of them. It is hard enough to see the oaken cradle in which one has rocked for so many weeks and months destroyed by the incendiary torch, but when necessity compels a sailor to light with his own hand the fire that is to consume the ship he has learned to love, he has good grounds for complaint against the fates.

Later the same day, the twenty-third, they came upon a small trading brig, the *Susan Abigail* of San Francisco, carrying a cargo of "guns, pistols, needles, calico, twine and Yankee notions in general" for fur trading. To the dismay of the *Shenandoah*'s men the brig carried newspapers reporting Grant's capture of Richmond, the Confederate capital, on April 3 and Lee's surrender at Appomattox on April 6, as well as Jefferson Davis's flight to Danville and his proclamation that the South would fight on. Papers reporting Lincoln's death on April 15 were also circulating about in the fleet. The more whalemen he captured, the more he heard from them that the outcome of the war was a foregone conclusion.

At this point, Waddell must have begun to have doubts about the consequences of his actions. He at least had to consider the possibility that his side had already lost the war (as it had); that he and his ship were without a country; that his actions could be considered piracy (as they were by many).

The following note in Waddell's hand, signed by him and dated June 27, is affixed to one of the pages in the *Shenandoah*'s log:

The brig 'Susan Abigail' sailed from 'San Francisco' about the 19th of April bringing [newspapers] to the 17th of April. I read from one of the April papers dispatches for surrender by Gen'l Lee to Gen'l Grant and an announcement of a proclamation issued by President Davis at 'Danville' to the people of the South declaring the war would be carried on with renewed vigor.

In view of the havoc he was causing, and would cause, Waddell wanted to emphasize that he was operating as an officer of the Confederate Navy. As such, he would continue to assume that the Confederacy existed. But the fact that he chose to state this in the record suggests that he was in some doubt about the outcome of the war—or possibly that he knew more details of its progress than he was admitting to.

And so Waddell burned the *Susan Abigail* and moved on toward Bering Strait seeking other prizes. But as he did this, word was going out among the fleet that a raider was loose in the Arctic. Frederick Barker had immediately taken the *Robert Towns* north after his encounter with the *"Petropavlovski"* on June 23. He had headed for Cape Bering and Indian Point where he knew he would find other ships. While the *Shenandoah* burned the *Jireh Swift* and *Sophia Thornton*, the *Gustave* of Havre had lain two miles away. She was lead gray in color, and warped to her and painted the same color was the bark *William Gifford* of New Bedford. At a distance they looked like one vessel. As soon as the *Shenandoah* moved on, the ships separated; the *William Gifford* heading south and Captain Vaulpré taking the *Gustave* north in search of ships he might warn. Vaulpré was repaying an old debt: during the Crimean War he had been warned by an American whaleship that a Russian vessel was after him. The Hawaiian vessel *Kohola* under Captain Barney Cogan, an Irish-American, and *Hae Hawaii* under Captain John Heppingstone, an Australian-American, had also seen the action on the twenty-third and sailed ahead, warning the fleet.

Certainly the most selfless act was Ebenezer Nye's. As soon as the *Shenandoah* was out of sight of the *Milo*, he and his first mate took two whaleboats and set out for Cape Bering to warn the ships there, 200 miles away. Not only was it a risky passage for open boats through ice and open water, but it was a great risk for the men themselves. Because they had been forced to give their parole that they would go directly to San Francisco, had Waddell caught them, the consequences would have been severe.

On June 25 the New Bedford bark *Mercury* was chasing whales in a fresh gale off Cape Bering. "Saw a boat standing in toward the ship from seaward," the log keeper wrote.

> Ran down for her. It proved to be Capt. Nye, late of the ship Abigail. He stated that his ship had been burnt by an English pirate in the Okhotsk Sea & had been a prisoner on board of the pirate Shenandoah ever since till [two] days previous. Capt. Nye stated that the ship was within a few miles of us & said if we went any further east we should be in danger. . . . We hauled our wind & spoke the Florida . . . Corinthian, and Peru & we all put to the westward hoping to keep out of sight.

The *Minerva* under Captain Edward Penniman was near Indian Point when the *Gustave* reached him. Captain Vaulpré came aboard and announced he was direct from Cape Thaddeus where the *Shenandoah* was burning ships.

> What to do I did not know. All of my boats were off, and there was such a light air that it was no use to set a signal. Then I thought of an old cannon between decks. We got it up on deck and lashed it down on the gangway. Mrs. Penniman by this time had gone down to breakfast and did not know what we were doing.
>
> . . . I thought I might call back my boats by firing the old thing. . . . When the gun went off it rose right up in the air and, coming down on deck, made a great hole on the planking. Everybody was scared half out of their wits, while the concussion was so great that it broke all the glass in the skylight to the cabin. Mrs. Penniman didn't know what had happened.

But not everyone was as receptive to the *Gustave*'s news. Captain Penniman started back to the west to get out of the *Shenandoah*'s path. "While standing on my course I fell in with the *Governor Troup*, Captain E. R. Ashley," he recalled. "I pulled over to him (in one of my boats) and explained what had been told me by the Frenchman. 'I don't believe a damned word of it,' said Ashley. 'It is a damned French trick to get north of us and best us in whaling,' was his angry reply. . . . and he parted with me saying that he was going up the strait."

The next day the *Minerva* was off the west end of St. Lawrence Island and the men could see smoke to the northeast. "The man at the mast head reported a ship coming down on us with all sail set. . . . It proved to be the *Governor Troup,* and Ashley, overhauling us, yelled out: 'Come on, Penniman, let's get out of here; that was no French trick; I saw six ships burning after I left you.'"

On the twenty-fifth in clear weather and light airs Waddell got up steam to go hunting. He knew the whaleships couldn't move fast in so little wind. He soon sighted two sails and went after them, but they proved to be an Hawaiian ship, possibly the brig *Victoria*, and a French ship, probably the *Winslow* of Havre, and they no doubt went on to the north warning others of the threat. Finally, near St. Lawrence Island, Waddell came upon the ship *General Williams* of New London cutting in a whale. While *Shenandoah*'s men were busy removing prisoners and setting fires (they usually lit fires in the cabin, forecastle, and main hold and then drove cutting spades through the bottoms of the boats left on the cranes), a group of twenty-five or thirty boats of Eskimos swarmed over the raider trying to trade for rum with their furs and ivory.

At one o'clock the following morning in beautiful clear weather and a near calm, the *Shenandoah* steamed up to three others, helpless in the light airs. As the men were burning the *Nimrod, William C. Nye,* and *Catharine* of New Bedford, the lookouts sighted five more ships to the north. Now overcrowded with prisoners, and worried that with their numbers they might take over his ship, Waddell simply towed the men along behind the *Shenandoah* in twelve whaleboats. Cornelius Hunt described the scene: "It was a singular scene upon which we now looked out. Behind us were three blazing ships, wildly drifting amid gigantic fragments of ice; close astern were the twelve whale-boats with their living freight; and ahead of us the five other vessels, now evidently aware of their danger, but seeing no avenue of escape."

Waddell gave a wide berth to one of the five, the bark *Benjamin Cummings* of New Bedford, because he had learned they had smallpox on board. But he captured three of the remaining four, burning the *Isabella* and *Gypsey*, and sending the *General Pike* to San Francisco with the prisoners aboard and under a $30,000 bond. The other ship was probably E. R. Ashley's *Governor Troup.*

Captain Hudson Winslow of the *Isabella* had the temerity to complain to Waddell that with two hundred men on board the *General Pike* they were likely to run out of food before they reached port. Referring to the large number of Hawaiians among the prisoners, Waddell replied, "Eat the Kanakas if you get hungry." The men aboard the *General Pike* never had to resort to cannibalism, but with so many men on board they did have to cook in their trypots on their way south.

By now Waddell had realized that his cat could not catch many pigeons if they were on the wing. A stiff north breeze had come up, nullifying the *Shenandoah*'s advantage of auxiliary power. So, after finishing with the *Isabella* and *Gypsey*, he let the ship's fires go out, hoisted her propeller for speed under sail, lowered her stack, and beat to the north under canvas to avoid alarming the whalers.

On the afternoon of the twenty-seventh they raised several sails to the north, but in that breeze Waddell knew he couldn't take them all, so he kept well away from the ships, hoping that the wind would die. Early on the twenty-eighth the wind fell calm, and, as the light improved, they spotted sails all around them. They got up steam and set off on what was to be their last and most destructive day's action and the last offensive action of the Civil War.

At 6:30 A.M. they were twelve miles south southwest of the Diomede Islands and they decided to go for the southernmost ship. At 10:00 they caught the bark *Waverly* of New Bedford, took thirty-three prisoners aboard, set fire to her and were off again at 11:00. Approaching East Cape at 1:30 P.M., they spotted ten ships clustered together. Nine stood by around the crippled *Brunswick*.

Soon Waddell's boats had fanned out among the fleet, and there was nothing the whalemen could do. The only resistance came from Thomas Young, the feisty old captain of the *Favorite* of Fairhaven. As part owner of the ship he was loath to see her taken by a "pirate." Young, nearly seventy years old, loaded his bomb gun, armed his crew with muskets, and climbed onto the dog-house roof as the *Shenandoah*'s boat approached.

The boat retired to the *Shenandoah* and Waddell then brought the raider around so that her guns were trained on the *Favorite*. By this time, the crew, realizing that this puny resistance was useless, secretly removed the cap from Young's bomb gun and then, as the boat returned from *Shenandoah*, made preparations to leave the whaleship. Young, who had apparently liberally fortified himself with liquor, stood his ground.

The officer in the boat ordered Young to strike his colors.

"I'll see you damned first," was the reply.

"If you don't, I'll shoot you."

"Shoot and be damned."

The officer then ordered his men to board the *Favorite*, and Young raised his gun and pulled the trigger, but the hammer only clicked. That done, he realized resistance was useless and surrendered.

Alexander Starbuck shared the whaleman's outrage when, twelve years later, he wrote: "His inhuman captors, who were unable to appreciate bravery, put him in irons in the topgallant forecastle, and robbed him of his money, his watch, and even of his shirt-studs."

But, in fact, Young's captors did appreciate his bravery, although several thought that at least a measure of his courage was drawn from a bottle: "When he came on board, it was evident he had been seeking spirituous consolation, indeed to be plain about it, he was at least three sheets in the wind, but by general consent he was voted to be the bravest and most resolute man we captured during our cruise."

That day Waddell burned the *Brunswick, Hillman, Nassau, Isaac Howland, Martha 2nd, Favorite, Covington* and *Congress 2nd*—all of New Bedford, save the *Favorite* from

The Shenandoah *burning whaleships off Cape Thaddeus, Siberia, June 23, 1865. Left to right:* Susan Abigail, Euphrates, Shenandoah, Jireh Swift, William Thompson, Sophia Thornton, *and* Milo, *with prisoners aboard. Watercolor by Benjamin Russell*

Fairhaven, which lies on the other side of New Bedford harbor. The *Nile* he bonded for $41,000 (about $250,000 in 1982 dollars) and sent to San Francisco with 190 men in addition to her crew. He found that the captain of the *James Maury* had died, leaving his widow and two small children aboard, so he bonded that vessel as well and sent her to Honolulu with 222 persons in all, plus the deceased captain whose remains had been preserved in a barrel of whiskey.

Also at East Cape that day the *Shenandoah* apparently passed close to the brig *Kohola* of Honolulu. Captain Barney Cogan hailed the ship:

"Have you heard the news?"

"No, what is it?"

The photograph's caption reads: "Captains of the whaleships captured and burned in the North Pacific on the 21st and 22nd of June, 1865, by the rebel cruiser Shenandoah; the last act of expiring insolence."

"Lee has surrendered, the North has beaten the South, the war is over, and you are done for."

"You go to Hell," said Waddell.

Waddell moved on north in search of other prizes, but the word was by now well ahead of him. The efforts of Ebenezer Nye, the *Robert Towns*, the *Gustave*, and the Hawaiian vessels had been magnified tenfold as one vessel told another, and by June 29 most of the fleet had been warned and had taken evasive action. In fact, well after the *Shenandoah* had left the Arctic the word was still spreading.

The remaining sixty or so ships scattered like pigeons, heading in all directions for safety. Some went northeast to Cape Lisburne, where the *Gratitude* was stove and

The Shenandoah *towing prisoners in their whaleboats, June 25, 1865. From Cornelius Hunt's* My Last Cruise

sank, trying to get the protection of the ice. Others went northwest to hide behind Burney (Kolyuchin) Island. In the Anadyr some went to Holy Cross Bay and Mercury Harbor. One went east into Norton Sound and holed up in Golovnin Bay, and many headed south to St. Paul Island, Bristol Bay, the Aleutians, or simply abandoned their cruises and went to Hawaii.

The bark *Martha,* warned by the *Robert Towns,* headed with the *Cornelius Howland, Eliza Adams, Oliver Crocker,* and *Louisiana* for the inner reaches of Kotzebue Sound and anchored at Chamisso Island for three weeks. On their way in, *Louisiana* ran aground and was wrecked, and thus she and the *Gratitude* indirectly became the twenty-first and twenty-second victims of the *Shenandoah*'s arctic cruise.

A secondary effect of the *Shenandoah*'s cruise was not lost on the whalemen: valuable summer whaling time was slipping away from them while they stayed out of sight. The logbook keeper on the *Oliver Crocker* saw yet another chance for profits to slip away: "Expect to lay here until the 25th, if we do not get burnt before," he wrote on July 16. "A fine chance to trade for furs if only we had something to trade with, such as guns, powder, caps and rum."

A measure of the *Shenandoah's* threatening presence can be seen in the fact that few ships returned to the whaling grounds before late July, although, unkown to them, the raider had spent only a few hours north of Bering Strait on June 29 before heading directly south to the Pacific.

Waddell's abrupt departure is difficult to explain from his own account. If he believed that as many as sixty whaleships had passed north of Bering Strait, as Hunt says he did (the actual number was probably closer to half that figure), then why did he spend less than twenty-four hours there?

He complained that the ice was too threatening to allow them to operate safely. "In consequence of [*Shenandoah's*] great length," he wrote, "the immensity of the ice and floes, the danger of being shut up in the Arctic Ocean for several months, I was obliged to turn her prow southward." But a few pages later he gave another reason: "If intelligence of my movements had reached any of the enemy's cruisers . . . or specially from the ransomed ships which were sent from Behring's Sea, it would have been easy for them to blockade the *Shenandoah* or force her into action."

Neither of these statements is convincing. If a large number of whaleships were north of Bering Strait in the course of their normal cruises, then presumably the *Shenandoah*—the only vessel with the advantage of auxiliary power—could have worked the ice safely. Also there were only a handful of Union ships in the Pacific at the time, and it was well known that they were fully engaged in guarding San Francisco and her supply lines to the Panama Railroad. Similarly, it is unlikely that the news of the *Shenandoah's* position—only six days after the release of the first ship bonded in the Bering Sea—could have reached anyone quickly; nor would it have been a simple matter to locate the *Shenandoah* even if word of the raider's whereabouts had reached a Northern warship.

A clue to Waddell's behavior can be found in Midshipman Francis Chew's diary. He mentions that they had captured newspapers that brought them up to the events of April 22. This date is a week later than the date that Waddell claimed to have known of, and April 22 was only four days before Johnston's surrender of the Southern Army to General Sherman. The news by that date provided incontrovertible evidence of the swift disintegration of the Confederacy. "The perusal of these papers has deeply effected us all," Chew wrote, "making due allowances we know that great reverses have befallen us, and that our country is in a very perilous condition, if not over run. Under these circumstances our anxiety may be imagined."

Unconvincingly, Waddell claimed that once in the Pacific, he planned to go to San Francisco and blockade that city. It is likely, however, that the course of the war suddenly became real to him, and his abrupt disengagement from action in the Arctic, when many prizes still remained there, was due to his hunch that the Confederacy had collapsed. His move to the waters off San Francisco was probably to gather information on the progress of the war.

All was confirmed on August 2 when they overhauled the British bark *Barracouta*, thirteen days out of San Francisco. The Southerners asked about the progress of the war. The reply was, "What war?" Francis Chew wrote: "From her we received the

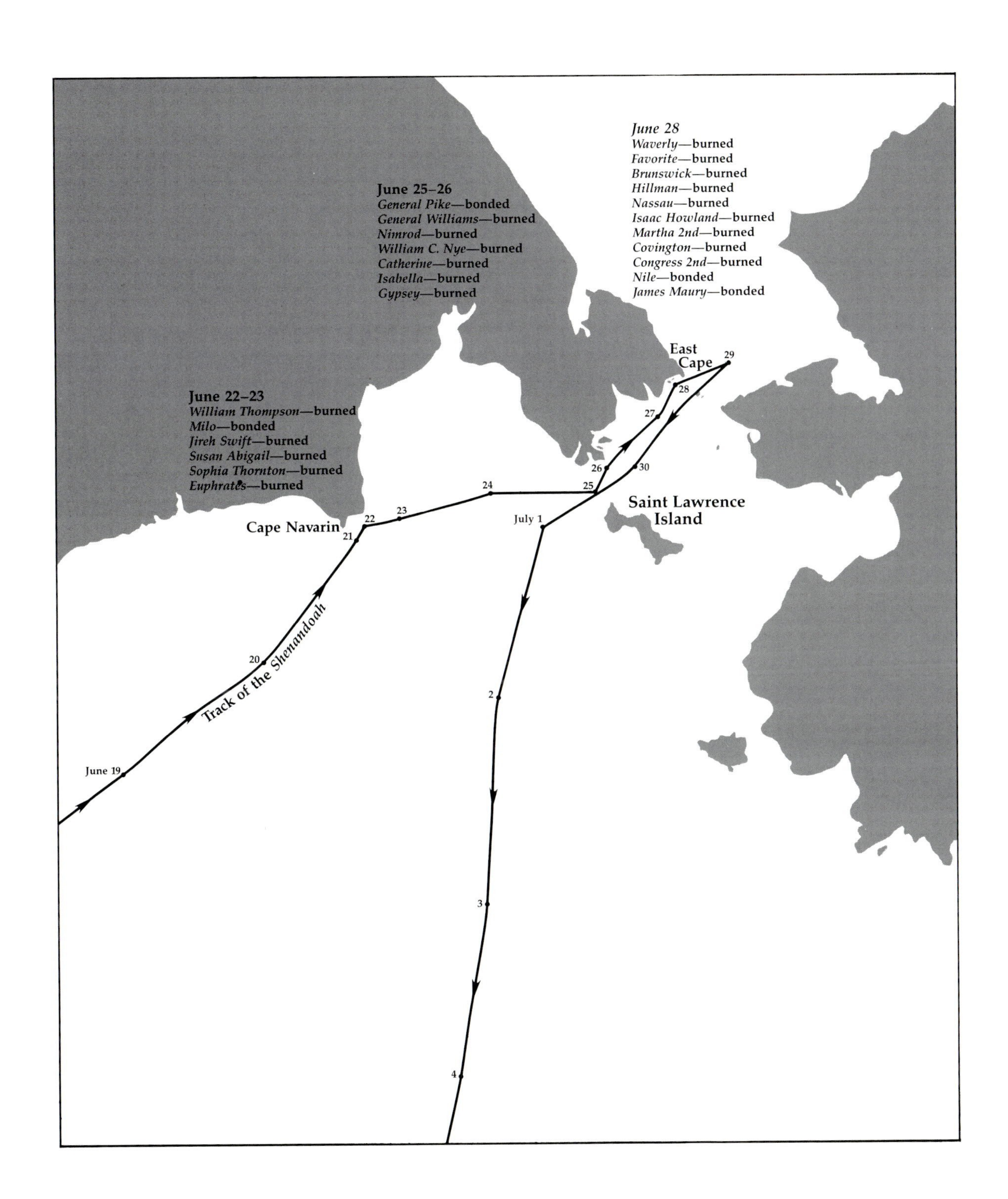

The Shenandoah's *Arctic cruise*

sad intelligence of the overthrow of the Confederate Government! . . . We are exiles! Our situation now is . . . very dangerous. The powers which extended to us the 'Rights of Belligerents' have withdrawn them and excluded us from their ports. We are . . . now called 'pirates' and will be picked up by any man-of-war that happens to see us."

With that news Waddell ordered his ship disarmed. The guns were dismounted and put in the hold, along with all firearms, and a note was made in the log that all offensive actions had ceased against vessels of the United States.

With foreign ports closed to him, Waddell at first set a course for Australia but soon changed his mind and steered for England, despite the possibility of being intercepted by a U.S. naval vessel on the way. The men hoisted her screw, lowered the stack and whitewashed it, boarded up the gun ports, and generally set about disguising the ship as best they could. Knowing full well that things would not go well for them should they be captured by the U.S. Navy, they resolved to avoid all contact with ships until they returned to the nation from which they had set out, a nation which could not really reject them but which surely would not welcome their return.

The fact is that on June 19, before the *Shenandoah* reached Bering Strait, James D. Bulloch, not knowing the ship's whereabouts, had written to Waddell in care of the British consuls at all ports where he might touch, ordering him to cease offensive actions. Bulloch tried to inform Waddell that President Andrew Johnson had declared the war to be over, that European nations had withdrawn their belligerent rights and had forbidden Confederate vessels to enter their ports, and that the president's amnesty had excluded most of the *Shenandoah*'s officers.

Simultaneously the news of the *Shenandoah*'s destructive cruise began to travel round the world. On June 24 Honolulu's *Pacific Commercial Advertiser* reported that the *Pfiel*, recently arrived from the Carolines, had been overhauled by a ship that likely was the *Shenandoah*. On July 20 the *Milo* reached San Francisco and newspaper headlines across the nation screamed:

TERRIBLE HAVOC BY PIRATE SHENANDOAH!

EIGHT WHALESHIPS DESTROYED!

WHALESHIP MILO BONDED!

Editorials throughout the land branded Waddell and his men as the vilest sort of criminals. "It harmonizes with the policy of a set of men," wrote the editor of Honolulu's *Friend*, "who did not blush to publish to the world that they were founding a Government with 'Negro chattel slavery as the corner stone.'"

Admiral George Pearson, commander of the U.S. Navy's Pacific Squadron, received the news from San Francisco on July 29 and immediately sent USS *Saranac* and USS *Swanee* out in search of the *Shenandoah*. And, on October 1, the British, clearly worried lest their pigeon come home to roost, ordered their naval vessels to cooperate with the U.S. Navy and detain the *Shenandoah* in a port or seize her at sea should she be offensively equipped.

But Waddell evaded all attempts to catch him, rounded Cape Horn safely and reached Liverpool on November 5, dropping anchor for the first time in more than eight months, having come 17,000 miles directly from the Arctic Ocean. Not only had the *Shenandoah* become the only Confederate vessel to circumnavigate the world, sailing 58,000 miles, but it had also flown the Confederate colors more than six months after the end of the Civil War. She had taken 1,053 prisoners and destroyed 34 vessels valued at $1,361,983 ($8,580,600 in 1982 dollars) and caused the loss of two more.

The *Shenandoah*'s return to England, thirteen months after she had set out from there, was extremely awkward for the British government. As early as May 1865, Britain had tried to head off any possible American claims over shipping losses to the raiders. It was embarrassing to have the ship return from her cruise after her unimpeded departures from both England and Australia and after her sinking of a score of vessels well after the end of hostilities.

The London *Times* summarized the problem:

> The reappearance of the Shenandoah in British waters is an untoward and unwelcome event. When we last heard of this notorious cruiser she was engaged in a pitiless raid upon American whalers in the North Pacific. . . . It is much to be regretted . . . that no federal man-of-war succeeded in capturing the Shenandoah before she cast herself, as it were, upon our mercy.

The presses of the United States cried out that England should try Waddell for piracy or hand him over to U.S. authorities, but although the *Shenandoah* was turned over to the U.S. consul, Waddell and his men, after a short detainment, were allowed to go free.

The American press was not going to take this lying down, coming as it did after what many Northerners viewed as the most thinly veiled British collusion with the South during the Civil War. San Francisco's *Alta California* announced:

> The frozen regions of the Arctic seas are illuminated by the bright lights of British neutrality, and covered with the charred remains of American whalers, the unresisting victims of British perfidy.

Although Britain asserted that it had observed the letter of its law in its neutrality during the Civil War, America demanded payments of reparations for the activities of those Confederate raiders that were built in British shipyards: the *Alabama*, the *Florida*, and the *Shenandoah*. The United States hinted that in the future event of Britain being at war and the United States neutral, the United States might reciprocate in kind. Because of political changes and strong feelings on both sides of the Atlantic, it was not until the signing of the Treaty of Washington in 1871 that an agreement could be reached over articles of arbitration on claims arising from the three ships' depredations. These claims were referred to as "the Alabama Claims."

The Tribunal of Arbitration was convened at Geneva on December 15, 1871. By mutual consent it was composed of five members: British and American

The Shenandoah *in the Mersey, November 6, 1865.*
From the Illustrated London News

representatives, a Swiss, a Brazilian, and an Italian. Both sides presented their cases and then the tribunal adjourned to consider their action. The United States claimed that the total cost of the ships destroyed by the raiders would be $14,000,000, but that the personal claims would substantially raise the amount for recompense.

The Tribunal reached its decision on September 2, 1872. The group agreed that Britain was to be held responsible for the damages caused by the *Florida* and the *Alabama* and by the *Shenandoah* after she left Melbourne. The amount of the award to the United States was $15,500,000 ($125,000,000 in 1982 dollars) in gold.

For the injured parties, however, this was just the beginning. The government immediately invested the sum in bonds at 5 percent interest, but it was not until 1874 that Congress created the "Court of Commissioners of Alabama Claims" to give judgment on the claims for damages arising directly from the actions of the three cruisers. This class of claimant included the owners and insurers of the vessels sunk directly by the three cruisers and the officers and seamen who had lost personal effects and wages. The treasury paid, with interest, a total of $9,316,120.25. The remainder of the award, as of March 31, 1877, amounted to $9,553,800 because of the interest. However, it was not until 1882 that Congress acted to dispense the remainder. It established a new act to cover two new classes of claimants; principally those who had suffered direct losses from the "exculpated cruisers" (those Confederate ships that had been acquitted from Britain's culpability by the Geneva Tribunal) and those who had paid increased war risk insurance as the result of the cruisers' actions. Well into the mid-1880s these claims were still being settled.

It has been maintained that the Confederate raiders, along with the Arctic disasters of the 1870s, mortally wounded the American whaling industry. This is unlikely. From a fleet of 186 whaleships in 1861, 46 were lost to Confederate vessels and 40 of the fleet's oldest hulls were bought by the government and sunk in Charleston and Savannah harbors in an unsuccessful attempt to blockade those ports. The Civil War, however, did hasten the industry's demise by stimulating the growth of the nascent petroleum industry, which soon undercut the price of whale oil.

In fact, the activities of the raiders and of the *Shenandoah* in particular can best be described as a case of too little, too late. The strategy for outfitting them was, as we have seen, "to destroy the enemy's commerce, and thus to increase the burden of the war upon a large and influential class at the North." Although the raiders did force a significant decline in the number of American hulls carrying cargoes to Northern ports, the fact that the cargoes then came to the same ports in foreign hulls did not seriously hurt Northern merchants. Had all the raiders been outfitted at once in the early days of the war, they might have contributed to the Southern effort by blockading Northern ports or by drawing the U.S. Navy away from its blockade of the Southern ports. Thus it appears that the plan was both strategically and tactically flawed. As it was, the *Shenandoah* had no effect on the outcome of the war. Even if she had been successful in "utterly destroying" the Arctic fleet, as Bulloch wished her to, it is unlikely that the act would have caused more than hardship. By that time, the whaling industry was not vital to the North's existence.

Regardless of this, reminders of the *Shenandoah*'s destructive cruise were visible for many years. Initially the Arctic whalemen noticed that the natives of Bering Strait were equipped with irons and lances from the burned ships, and the wreck of the *Louisiana* was still visible on shore in Kotzebue Sound for more than a decade. Most of the whalemen involved in the action continued in their trade, and the columns in the New Bedford papers well after the beginning of this century record their stories of the *Shenandoah*'s cruise, stories told to grandchildren and great-grandchildren—with growing hyperbole as the years passed.

James D. Bulloch, who had outfitted the *Shenandoah*, never returned to the United States. He remained in England for the rest of his life. He was visited there by his nephew, Theodore Roosevelt.

The *Shenandoah* herself was eventually sold to the sultan of Zanzibar and was lost in the Indian Ocean in 1879. Waddell remained in England for a decade and finally was hired as captain of one of the Pacific Mail Steamship Company's vessels. Despite the passage of years, the company had to obtain a pardon for Waddell from the Hawaiian government for his destruction of the Hawaiian ship, the *Harvest*. And, when this news reached San Francisco, the following column appeared in the *Alta California*:

It has been reported along the city front that a number of whalemen who suffered so terribly ten years ago, and who are serving on board the whaling fleet now recruiting at this port, have arranged to abduct Capt. Waddell and give him a remarkably lively cruise in the Arctic before the mast of one of those self-same and much-abused blubber hunters.

After several years with the company Waddell moved to Maryland and briefly took charge of a state boat, suppressing the oyster pirates who were raiding the beds in Chesapeake Bay. He died in Annapolis in 1881.

Perhaps it is fitting to close with Waddell's words, for despite the destruction of ships, which he carried out so efficiently, he was, elementally, a man of the sea. And like all true sailors, it hurt him to see a ship die. Concomitant with his desire to avenge the South he clearly had distaste for his duty:

When we see a great vessel rolling lonely at sea, her masts gone, her gear loose and adrift, and sheets of foaming sea pouring in and out of her helpless sides . . . practical and unimaginative people may say, what difference does it make to the ship, but no sailor will listen to that.

There is life in the craft from the time she leaves the ways into the tide, to the hour when her timbers are laid on the sand or rocks, or, the saddest of all, in the ship-breaker's yard.

7 / Walrusing

"Passing through thousands of walruses. The water was alive with them for a mile as far as you could see," wrote Samuel Broadbent aboard the whaleship *South Boston* west of Point Hope in 1852.

When the first whalemen reached the northern Bering Sea they were amazed by the vast herds of walruses in the waters there. It was then, as it is today, an awesome sight to gaze across Bering Strait in June and everywhere see groups of up to several hundred walruses lying asleep, jumbled on top of one another, aboard the floes drifting slowly northward. Grunting and bellowing, or shifting slowly in their tight agglomerations, occasionally waving flippers in the air to cool themselves, or rearing briefly to display their tusks in small territorial conflicts, these mounds of wrinkled brown skin seem to be part of a constantly fermenting mass. At a distance of half a mile you are hit by the strong odor that is peculiar to the walrus. Closer, you see the magnificent old bulls that weigh nearly four thousand pounds and carry tusks almost a yard long; the younger bulls; the cows with their thinner, more delicately curved tusks; and the pudgy calves, lying next to or on top of their mothers or woofing and scrambling along after them.

The whalers first regarded these animals with the same mixture of curiosity, affection, and fear that strikes a traveler in Bering Strait today. The walruses' gleaming white tusks, broad muzzles that sprout heavy bristles, their small, bulbous, bloodshot eyes, and their huge, thick necks and warty, scarred skin mark them as one of nature's curiosities. Up close, their smell, size, and surprising speed over short distances on stumpy flippers, and—most impressive—their sheer mass and momentum when a frightened herd stampedes pell-mell into the water, all testify to the strength and exotic charm of these wonderful animals.

And angry young bulls or mothers charging a boat, hooking their tusks over its gunwale or swimming under it and punching their tusks up through the hull, can be dangerous opponents, as the whalemen quickly learned.

When the first whaleships reached Bering Strait the walrus population there stood at a high level, certainly in excess of 200,000 animals. This Pacific walrus population could be found throughout the northern Bering Sea and Chukchi Sea and near shore in both the southern Beaufort Sea and parts of the East Siberian Sea. Their range also extended into the southern Bering Sea—to the west, halfway down the Kamchatka Peninsula and, to the east, all the way to the Alaska Peninsula.

Now, as then, most of the animals spend their winters amid the southern reaches of the pack ice and then migrate north in the summer on the retreating and melting floes.

The population is divided into two groups; one group winters primarily southwest of Saint Lawrence Island and the other winters in the southeastern Bering Sea near Bristol Bay. The western group is the first to migrate north in the spring and the last to return in the autumn. It passes west of Saint Lawrence Island and through Bering Strait in May and then divides with the currents, some of its members going along the coast to the waters near Point Barrow and the others going northwest toward Wrangel Island. The eastern group migrates to and from Bering Strait east of Saint Lawrence Island. These animals probably travel northwest in the Chukchi Sea, toward Wrangel Island.

Walruses feed primarily on the clams and other invertebrates that grow in especially great abundance on the shallow silty bottoms of the Chukchi and northern Bering seas. They locate the invertebrates by brushing their muzzles along the sea floor. When a walrus finds mollusks, it grasps a shell betweeen its lips, then quickly withdraws its strong, fat tongue, creating a vacuum that rips the soft part of the mollusk from the shell. To this day Eskimo hunters often eat the clean clam meat that they find in the stomachs of freshly killed walruses.

The natives of these regions have always depended on the walrus for their livelihoods. The meat and blubber were used for food, the skin was used for boat and house coverings and rope, and the tusks and the dense jaw bones and bacula (penis bones) were used in a variety of tools, particularly harpoons. Because of their vast numbers these animals could be counted on as a constant and bounteous resource. From the middle of the seventeenth century onward there was also a small commercial market for the tusks, which were traded by the natives to obtain iron and tea and other manufactured goods.

When the whaleships first arrived in Bering Strait, the herds had not been disturbed by pelagic hunting. A walrus "showed no fear and came close to the ship to look at it," wrote Mary Brewster, wife of the captain of the ship *Tiger* of Stonington in 1849. "William got an iron to fasten to it, but it looked so innocent he left it and was not sorry."

Although the whalemen were preoccupied with chasing the bowheads in those opening years of the fishery, during lulls in the whaling activity several ships caught one or two walruses, more for curiosity than for profit. On the whole, these early encounters were chaotic or at best disorganized. Washington Fosdick, steward aboard the New Bedford ship *Montreal* described one hunt in 1852: "The mate's boat pulled up to a large cake of ice on which were about a dozen of them. As soon as the boat struck the ice, the crew sprang upon it and started for them on the full run, when such a rolling, tumbling, wallowing, and scratching as there was among them would seldom be witnessed . . . and as large and clumsy as they were, they all succeeded in rolling into the water but one large bull which they succeeded in killing with lances and iron poles which they plied vigorously over his head."

The reason for this messy and unproductive approach to the hunt was, no doubt, that the whalemen were using tactics devised for hunting elephant seals in the Antarctic and on the coast of the Californias. These enormous, sluggish animals, some weighing 8,000 pounds, could be stalked while they slept, or herded into cul-de-sacs, and then killed with clubs and lances.

The walruses, however, were half the size of elephant seals, and the whalemen quickly discovered that they were far more difficult to capture. One participant described a skirmish as "a pitched battle with an army of walruses." The men learned that walruses have dense, thick skulls, and very tough hides, especially the males. Their surprising speed over short distances, combined with their ability to flop quickly off the ice into the water, made clubbing mostly ineffective and lancing very difficult, requiring careful aim and some strength.

A fighting walrus could be an ugly opponent. "The mate tackled two of them with lances when they both pitched off the ice together, one on each side of his boat, the tusk of one of them hooking to the gunwale on one side and the tusk of the other going through the boat on the other side," Fosdick reported. "It was a narrow escape for had they gone into the boat instead of on either side their immense weight would have dashed her into fragments and perhaps proved fatal to the entire crew."

Apart from such sporadic brushes, during the first dozen years of the Bering Strait fishery, the whalers paid little attention to the walruses. But their success in capturing the whales was already beginning to tell on the numbers that were left to catch. In fact, the bowhead population was in steep decline: by 1852 the whalemen had already killed a third of the entire number of bowheads that they would catch from 1848 to 1914 and by 1869 they had already taken two thirds of the total, leaving the remaining third to be captured over the next four and a half decades.

The whalers understood—if only implicitly—that they were hunting a finite whale stock and that they were swiftly reducing its numbers. They realized that the great successes of the first few years of the Arctic fishery had not been reached again, nor had significant new whaling grounds been discovered in other parts of the world; consequently it was logical for them to search for another source of oil. The *Cleone* of New Bedford was one of the first vessels to begin walrusing in earnest. In August 1859, after nearly two months of catching nothing on the Kodiak grounds, Bristol Bay, the Anadyr, or the Arctic, she began taking walruses in the Chukchi Sea. The men, who previously had not considered walruses "worthy of their steel," found that a mature walrus yielded between two-thirds and three-quarters of a barrel of oil, and thus, in the absence of any whales to catch, the walrus oil and tusks would at least provide some income. Every year more ships took up walrusing and by the 1870s virtually the entire fleet was hunting walruses.

The reasons for this are clear: the Civil War had stimulated the market for oils. After the end of the war the price held relatively firm, and walrus oil generally could be sold for two or three cents more per gallon than whale oil because it was easier to refine. Furthermore, the whalers' summer voyages were becoming divided into three parts: "early," "middle," and "late" seasons. During the early and late seasons they could catch bowheads. But in the middle season—from late June to early August—the migrating bowheads were far ahead of the ships and amid the safety of the ice in the Chukchi and Beaufort seas. Because the costs of a whaling voyage were fixed, regardless of whether or not the ship was catching whales, it made sense to derive some revenue from this otherwise unproductive period of time. One whaleman wrote in his journal aboard the *Trident* in July 1870 that walrusing was "hard work and slow getting oil, but it is all that can be done this month."

One may wonder why the whalers did not simply concentrate on the gray whales that were often more accessible to them during the middle season. A few did, in the absence of other game, but, on the whole, several factors militated against this. The grays had no baleen that was commercially valuable. Also the gray whales that were taken in the Arctic yielded only about twenty-five barrels of oil, which sold for about five cents a gallon less than "whale" oil. They required somewhat more skill and speed to strike than a bowhead, and some of these "devil fish," when wounded, could be exceptionally ugly fighters, demolishing boats with their tails. But most important was the fact that by the end of the Civil War the gray whale population of the eastern Pacific (the one that annually migrates between the lagoons of Baja California and the Chukchi Sea) had itself been severely reduced. By 1874 the intense lagoon fishery that took up the winters of many of the Arctic whalemen in Baja California had suppressed the gray whale population to less than 15 percent of its former size.

Once the whalers made the decision to hunt walruses seriously, they quickly learned that the tactics developed for elephant sealing were not effective for walrusing. They required efficient procedures for killing and butchering the animals and for trying out their blubber. These methods evolved quickly and typically followed a pattern.

As the ships worked their way north through Anadyr Gulf in June, once the men were reasonably sure that the bowheads had passed them and gone into the ice north of Bering Strait, they began preparations for walrusing. They took the whaling gear out of most of their whaleboats and replaced it with walrus equipment. They had learned that the old technique of using lances and harpoons and clubs to capture the walruses was not sufficiently productive for the small amount of oil that each animal would yield. They knew the walruses had to be taken *en masse.*

The blacksmith would already have begun making knives, hooks, and other gear. Also needed was an umiak or light fishing dory, and if the carpenter had not built one, the captain might buy a skin boat from the natives. Some captains put their oldest whaleboats on the cranes and took in the good ones. The *Helen Mar*'s logbook records why: "went into the ice with three boats and got six walrus and three stoven boats."

Once the hunt began, the ship cruised near the ice in all weathers in constant readiness to lower to herds of walrus. William Fish Williams, who was aboard the *Florence* in 1874, remarked that this procedure was particularly nerve-wracking for the captain and officers because of the constant danger in thick weather of collisions with ships or ice. When a herd was spotted, a walrus boat and a whaleboat were lowered together to go after them. The walrus boat carried only three men, an officer and two crewmen, as well as two large-calibre buffalo guns (often Sharps or Henry 45-70s) and a box of ammunition.

If the weather was clear, the ship would often stay three miles away while the whaleboat towed the walrus boat toward the animals. The approach was always made from downwind because the walrus has a good sense of smell. Although a walrus's eyesight is poor, the walrus boat was usually painted a light color and the men wore white cloth shirts and trousers over their clothes to avoid spooking the herd. If they happened to see any polar bears while they were on their way to the herds, they would try to kill them. The scent of a bear would frighten the walruses, quickly driving them off the ice and into the water.

The whaleboat, (or, in later years, sometimes a steam launch) would stop towing about a half mile from the animals, and the sailors would then row the walrus boat very quietly toward the herd; sometimes it took two or three hours to get within close range if the herd appeared nervous and "easily gallied." Once the boat reached the piece of ice where the herd was sleeping, the officer either crawled out onto the ice to take up a good position for shooting, or he simply shot directly from the boat.

He took great care to kill the first animal with one shot, usually in the temple, because as Calvin Leighton Hooper noted, "At the first sound of the rifle they all raise their heads, and if one has been wounded and goes into the water the rest all follow; but if the shot has been effective, they soon drop their heads and go to sleep again. This is repeated a few times, until they become so accustomed to the firing that they take no notice of it. Then they are approached within a few feet and dispatched as fast as the guns can be loaded and fired." At times the officer fired so rapidly that the rifle became too hot to hold. He then would drop it into the water on a lanyard to cool while he went on shooting with another. Hartson Bodfish estimated that he once saw Captain Leander Owen kill 250 walruses on one piece of ice.

Once this bloody business was done and the shooting had stopped, the prodigiously tiring work of cutting in the carcasses and towing the blubber to the ship began. Anyone who has hunted walruses with the Eskimos in Bering Strait can appreciate the back-breaking work involved in butchering. Each boat's walrus equipment was now ready for use: it carried a small block and tackle to help heave the blubber around or haul the whaleboat out on the ice should it need repairs; six gaffs and six pikes for the crew to help roll the cumbersome carcasses over during butchering; six sheath knives for cutting the blubber off; a whetstone, a file, and several steels for sharpening the cutting equipment that dulled quickly against the tough hide; ice hooks for tethering the boats; a hand lance for dispatching any wounded walruses; three "rippers," blades from straight razors set in wooden handles for cutting through the hide; six hand hooks to pull away the hide as they cut beneath it with the knives; an axe for cutting off the head or chopping out the tusks; line for warping the blubber rafts together; and several boat waifs, small flags on poles, to indicate where the blubber was temporarily cached.

First a boatsteerer worked on a carcass with his ripper, sectioning the hide into

WALRUS TEETH!

WALRUS TEETH.

WALRUS TEETH.

THE HIGHEST MARKET PRICE PAID
for this article by IRWIN & CO.
177-3m Custom House Brokers, Kaahumanu st.

Advertisement from the Pacific Commercial Advertiser, *Honolulu, late 1860s*

horse pieces about a foot square. Then the boat's crew followed with hand hooks and knives, cutting the horse pieces of hide and blubber away from the carcass. When the horse pieces had been taken off from one side, the men used the pikes and gaffs to roll the animal over and expose the other side. They then chopped the tusks out of the upper jaw and frequently saved the tongue, heart, liver, and some meat as well. The heart and liver made good eating and the meat, which has large coarse fibers, was ground into sausages. The tongue was often pickled.

If only eight or ten walruses were taken in one place, the blubber, meat and tusks could be loaded into one of the whaleboats and taken to the ship. But if more were killed, as happened frequently, the captain worked the ship as close as possible and the men rafted the horse pieces over, either in the water or by towing them on a small piece of ice.

Once the blubber was aboard, the next step was to remove the hide from it. The men set up a cutting table on top of a couple of casks and got to work with their knives. When the hide was off, they minced the blubber so that it would try out quickly. Originally they did the mincing in big tubs which the cooper had fitted with false bottoms so that the blubber could be chopped up with spades, but they found that the wood splinters made the oil blackish when the blubber was tried out; so instead, they minced the blubber on the cutting table or put it through a mincing machine.

They then pitched the blubber below to allow it to age for a day or two so that it would not be "green" and consequently difficult to try out. At the beginning of the walrus season, the crew would have set up shifting boards in the blubber room to hold the blubber in bins until they were ready to boil it. If they took 700 walruses in forty-eight hours as Leander Owen did in 1877, space below decks would be at a premium and the ship would be in trouble should the cargo shift.

When they found time to begin boiling, the blubber had to be pitched back up on deck, fed into the pots, and tried out like whale blubber. But the men at the tryworks had to take care to cook it slowly and to stir it well—if the blubber stuck to the sides or bottom of the pot it would burn and discolor the oil.

Later on, the crew scraped and washed the blood and pulp from the tusks and dried them. During peak walrusing years (roughly 1868 to 1883), ivory sold for between fifteen cents and $1.25 per pound on the docks. Tusks from females were more valuable because they have less "core" of marbled dentine, which in walrus ivory has a granular appearance. In Honolulu and San Francisco the whalemen often used the tusks for what A. Howard Clark called "spare change" to pay for provisions and repairs to the ships. The ivory was then sold to markets in New York, China, Japan, and London, where it was carved into the same articles as elephant ivory, although, because of its granular core, it was generally considered to be inferior.

Toward the beginning of August, when the captain judged that the bowheads would be moving out of the Beaufort Sea, he stopped walrusing and rigged the boats for whaling once again. It would often take the carpenter a week to repair the damage they had suffered in the ice.

Once they had begun walrusing in earnest, the whalers found they could make their first big catches of the season in June in the waters between Saint Lawrence Island and Bering Strait. But their best hunting was in July on the "western grounds"—once the ice had allowed them to reach Cape Serge (Cape Serdze Kamen) on the north shore of the Chukchi Peninsula. There the northerly current from Bering Strait meets the southeasterly one that flows along the Peninsula, creating an eddy that traps the ice and its passengers, the great nursery herds of walruses. Later in July the whalemen often moved to the northeastern Chukchi Sea and made good catches there on the "eastern grounds" in another segment of the population.

It is likely that the whalers captured a total of about 150,000 walruses, and 85 percent of these were taken from 1869 to 1878.

As appalling as the size of this catch was, the damage to the population was almost certainly greater. The total kill was probably more than twice the size of the catch. Mortally wounded animals often escaped from the whalemen and, at times, the warm blood flowing from the slaughter broke up the ice floes, resulting in the loss of part or all of a particular harvest. The native harvest must have increased, too, if only briefly, because of the market that the whalers and traders created by buying ivory from them in return for manufactured goods or alcohol. Also, the small trading schooners that sailed annually from Honolulu and San Francisco to trade for furs and ivory in Bering Strait occasionally went walrusing after their trading was done. The ability of the population to reproduce itself must have been severely suppressed because the bulk of this harvesting occurred among the nursery herds north of the Chukchi Peninsula.

After the period of intense slaughter in the 1870s, the walruses became so "scarce and shy" that some ships stopped hunting them. No longer could ships take as many as 1,600 animals in a month. As John Murdoch reported from Point Barrow in the early 1880s:

> The whalemen complain very much of the increasing scarcity of walrus on their usual walrus-hunting grounds, the ice-field just north of Bering Strait. Where they were formerly accustomed to get a hundred walrus a day by shooting on the ice; they now consider eighteen a good day's work. Not only have the walruses been killed off by the indiscriminate slaughter which has been the custom, but they have grown cautious, and have learned to withdraw to inaccessible parts of the ice fields, where they cannot be reached with a boat. This habit will go a good way towards preserving the species from utter extinction.

It was the natives who suffered the most from this gross slaughter. Many of the native groups of the Bering Strait region depended on the walrus herds, not the whales, as the most reliable food resource available to them. This was particularly true for the inhabitants of Diomede, King, and Saint Lawrence islands, who had no significant terrestrial food resources to fall back on. In the years following the end of the Civil War, these peoples watched as the whalemen drove the walrus and whale populations toward extinction.

Not all whalemen were insensitive to this problem. It was forcefully driven home to Captain Frederick A. Barker after his vessel, the *Japan* of Melbourne, Australia, was wrecked near East Cape in a punishing autumn gale in 1870. The Eskimos there had taken him and the rest of the survivors into their huts and cared for them as best they could until they were picked up by whaleships the following summer.

Dependent all winter upon the largess of the natives, Barker, an intelligent man and a sensitive observer of their lifeways, saw how narrow the margin of their existence was and how dependent they were upon the walrus herds. Barker wrote to the New Bedford *Republican Standard,*

> Should I ever come to the Arctic Ocean to cruise again, I will never catch another walrus, for these poor people along the coast have nothing else to live upon. . . . I felt like a guilty culprit while eating their food with them, that I have been taking bread out of their mouths, yet although they knew that the whaleships are doing this, they still were ready to share all they had with us.

Barker's letter was reprinted in the *Whalemen's Shipping List* and many whalemen agreed with him. No doubt they realized that they too could easily be shipwrecked in the Arctic and be forced to rely on the kindness of the same people to whom their walrusing was causing such hardship. One captain who replied to Barker wrote in the *Republican Standard,*

> The worst feature of the business is that the natives of the entire Arctic shores, from Cape Thaddeus, in the Anadir Sea, to the farthest point North, a shore line of more than a thousand miles on the west coast, with the large island of St. Lawrence, the smaller one of Diomede and King's Islands, all thickly inhabited, and our own coast of northern Alaska are now almost entirely dependent on the walrus for their food, clothing, boats, and dwellings. Twenty years ago whales were plenty and easily caught, but the whales have been destroyed and driven north, so that now the natives seldom get a whale. . . . Several captains lately arrived home have told me that they saw the natives thirty or forty miles from land, on the ice, trying to catch a walrus to eat, and were living on the carcasses of those the whalemen had killed.

Barker's concern was strongly echoed in Honolulu's newspapers as well, but it did little or no good, and the fleet continued its destruction of the herds.

With such gluttonous waste, a disaster was inevitable. The natives in Bering Strait had been reported "very poor" after the winter of 1877-78—they had caught few whales or walruses—but the disaster occurred the following winter, 1878-79.

The first news reached the outside world when the trading schooner *General Harney* arrived at Honolulu with the report that many natives in the Arctic had died in the previous winter as the "result of the killing off of the walrus by whalers." Soon thereafter, another trading vessel, the schooner *Pauline Collins* anchored at Unalaska and Captain J. J. Nye told the officers aboard the revenue cutter *Richard Rush* that he found all the people dead at three settlements on Saint Lawrence Island. "From information he received at another settlement, he learned that they had all died from starvation during the winter and early spring on account of their inability to get seal, walrus, and whale-meat, the ice having broken up early, and a continuance of southerly winds having kept it packed in such quantities against the island, and for so long a time that they were unable to get any food."

Then Ebenezer Nye's letter arrived. Captain Nye, aboard the bark *Mount Wollaston* at Cape Lisburne, wrote the following to the editor of the New Bedford *Standard* on August 2, 1879:

> About half the fleet are in this vicinity; the other half are all over to Cape Serge and the western ice walrusing, destroying them by the thousands; about 11,000 have been taken and 30,000 or 40,000 destroyed this year. Another year or perhaps two years will finish them,—there will hardly be one left, and I advise all Natural History societies and museums to get a specimen while they can. They are such stupid animals, lying still to be killed by the hundreds, that I see no help for them if the whalemen continue to take them as they have in past years; they, like the dodo, will be extinct; then God help the natives of this whole northern country; all must perish,—there is really no help for them, as the seal is not plenty and the whales are so wild and shy that they cannot be taken.
>
> The past winter has been a severe one; an unusual amount of snow has fallen and little or

no game could be had. The shore people depend almost entirely on the water for their food, clothing, boats and dwelling. The seal and walrus have furnished them with all they needed to make them comfortable. In 1863-4-5 I was here and then the people were numerous, had good houses, excellent boats and plenty of clothing; they could supply the large fleet of that day with all they wanted of seal skin clothing; now nothing can be got excepting boots, and they are very poor; they say they cannot get skins enough to clothe themselves.

Fully one-third of the population south of St. Lawrence Bay perished the past Winter for want of food, and half the natives of St. Lawrence Island died, one village of 200 inhabitants all died excepting one man [emphasis added]. Mothers took their starving children to the burying grounds, stripped the clothing from their little emaciated bodies, and then strangled them or let the intense cold end their misery. It is sad and heart-rending to hear them tell you how they suffered.

"Cornelius," an intelligent and educated native at Plover Bay and well known by some of our New Bedford people, told Capt. ["Barney"] Cogan how they suffered. He has a bright little girl of four years that he took from her mother, who was taking her to the graveyard to leave her to die.

Capt. Cogan is, and always has been, a great friend to the natives of Behring Strait and vicinity. Last year he took down fifty tons of food to them, and Capt. [Leander] Owen and others less amounts. Had they not been partially fed by the whaleships many more must have died. The four white men that stopped at Plover Bay [for the winter] also had a large stock of provisions, which they gave to the natives freely, thus saving many lives.

What they will do in the future God only knows. One intelligent man who is on board the Frances Palmer now and has a wife and three children at Plover Bay, told me that he got a little trade of the white men, then went a long journey with his dogs and bought two reindeer, which he gave to his family in this manner (marking on his finger), a small piece for himself and one for his wife each day, two small pieces for his boys, night and morning, and small pieces for his baby, three times a day. He said, "Suppose I no do this all die."

Many natives are now on board the ships and schooners for the purpose of getting some whale or walrus meat to take to their families in October, and they have a story current amongst them that a Russian man-of-war will be here this Summer and burn the whaleships for destroying the walrus; no doubt their wishes and hopes go with the story.

Capt. Cogan has taken very few walrus; he says that for every one hundred walrus taken a family is starved, and I concur in his opinion. *I don't want or need money bad enough to go for the walrus* [emphasis added]. I saw fifty or sixty dead ones lying upon a cake of ice a few days ago (just killed) with the young calves sucking and crawling over their dead mothers; it was a cruel sight, as the young all die. I also saw female walrus in the water with several calves that have lost their mothers swimming about them.

I should like to see a stop put to this business of killing the walrus and so would most of those engaged in it. Almost every one says that it is starving the natives, and if one of our whalers should be so unfortunate as to be wrecked on the coast in the Fall, the crew must perish. The people have eaten their walrus skin houses and walrus skin boats; this old skin poisoned them and made them sick, and many died from that. They also eat about all of their dogs, and there are but three boats and three dogs left at what was once the largest settlement of Plover Bay.

The tragedy emerged slowly. The revenue cutter *Thomas Corwin* was able to visit Saint Lawrence Island only briefly the following year, 1880, but received a concurring report on the cause of the starvation. The Eskimos told the men from the *Corwin* that the weather had been "cold and stormy for a long time, with great quantities of ice and snow, so that they could not hunt walrus and seal" and that they had made it through the winter by eating their dogs and the walrus hide coverings on their boats and houses.

It was not until 1881, when the *Corwin* returned to Saint Lawrence Island, that they learned of the extent of the devastation—the deaths of possibly 1,000 persons out of a population of about 1,500 and the extinction or abandonment of six villages. Edward William Nelson, an ethnographer traveling aboard the *Corwin*, described the scene at one village on Saint Lawrence Island:

> In July I landed at a place on the northern shore where two houses were standing, in which, wrapped in their fur blankets on the sleeping platforms lay about 25 dead bodies of adults, and upon the ground and outside were a few others. Some miles to the eastward, along the coast, was another village, where there were 200 dead people. In a large house were found about 15 bodies placed one upon another like cordwood at one end of the room, while as many others lay dead in their blankets on the platform.
>
> In the houses all the wooden and clay food vessels were found turned bottom upward and put away in one corner—mute evidences of the famine. Scattered about the houses on the outside were various tools and implements, clay pots, wooden dishes, trays, guns, knives, axes, ammunition, and empty bottles; among these articles were the skulls of walrus and of many dogs. The bodies of the people were found everywhere in the village as well as scattered along in a line toward the graveyard for half a mile inland.
>
> The first to die had been taken farthest away, and usually placed at full length beside the sled that had carried the bodies. Scattered about such bodies lay the tools and implements belonging to the dead. In one instance a body lay outstretched upon a sled, while behind it, prone upon his face, with arms outstretched and almost touching the sled runners, lay the body of a man who had died while pushing the sled bearing the body of his friend or relative.
>
> Others were found lying in the underground passageways to the houses, and one body was found halfway out of the entrance. Most of the bodies lying about the villages had evidently been dragged there and left wherever it was most convenient by the living during the later period of the famine.

All of the observers of the tragedy agreed as to its terrible extent, but disagreed as to the cause. Those who were connected with the Revenue Cutter Service attributed the cause of the starvation to the effects of alcohol (despite the report received by the *Corwin* from the natives). They claimed that in the autumn of 1878, the trading vessels landed such a large quantity of spirits at the island that the hunters were incapacitated throughout the winter. These reporters were not entirely unbiased. At the time the Service needed funding for its expanding operations. One of its primary responsibilities in the north was patrolling the trade in contraband. Exaggerating the problem might have helped gain the much-needed appropriation. Nevertheless, even if an extraordinarily large amount of alcohol had been traded to the natives, it is difficult to believe that the quantity would have been sufficient for them to remain crippled from the effects of drink all winter.

One must also be skeptical of the traders' reports (which make no mention of alcohol) for it was well known that they sold alcohol and it was not in their interests to have the revenue cutters closely patrolling their activities. On the other hand, Ebenezer Nye and both of the traders gave the only reports of visits to the island close to the time of the disaster. They emphasized the lack of food, although they disagreed as to the cause.

A house on Saint Lawrence Island, 1880. From C. L. Hooper's Report of the Cruise of the U.S. Revenue-Steamer Corwin . . . 1880

Despite their disagreement, it is likely that these three observers were correct in their assessments of the starvation among the Saint Lawrence Islanders. The walrus population had been cut to the quick and a tragedy was looming. When the stormy weather and ice conditions made productive hunting extremely difficult, these island people had no land resources to fall back on. The presence of a large quantity of alcohol might have been a contributory factor, but probably not a central one.

Other native groups also suffered severe hardship at about the same time because of the loss of the great walrus herds. In 1880 William Healey Dall visited Bering Strait on a hydrographic survey and reported that groups of natives from East Cape had moved to Archangel Gabriel Bay, north of Cape Navarin, and to Cape Olyutorski because of the "growing scarsity of seal and walrus about the strait."

The lack of walruses in the winter of 1879-80 apparently drove a number of King Island natives from their home; the following winter they were living at Sledge Island, where Edward William Nelson encountered them. "They were a strong energetic set of men," he wrote, "and being bold and dishonest, did not hesitate to bully and otherwise terrify the more peaceable villagers into supplying them with food."

Arthur and Aurel Krause, German ethnographers who traveled on the Chukchi Peninsula in the summer of 1881, reported that the natives starved in Plover Bay in the winter of 1879-80, when the Saint Lawrence Islanders were again reduced to eating their dogs and boat covers. They also mentioned that the natives at Indian Point spent little time walrusing because there were so few walruses left to catch. Six years later Herbert Aldrich, a journalist traveling with the whaling fleet, reported that the Indian Point people had to go all the way to Saint Lawrence Island to find walruses.

In 1882 Captain Leander Owen, one of the most effective walrus hunters, brought the steam bark *North Star* into North Head, Saint Lawrence Bay (Cape Nunyamo): "Natives came on board and [we] gave them flukes & blackskin & lean of the whale which they seemed very glad to get and needed much for food, and they complained very much of the ships catching the walrus and causing them to suffer for food."

Almost a decade later, in the winter of 1890-91, disaster struck the King Island people who, as fully as the Saint Lawrence Islanders, depended on walruses as their mainstay. Perhaps adverse weather conditions again made it impossible to hunt what few walruses were available. In any case, when the revenue cutter *Bear* reached their tiny, vertical-sided rock in the summer of 1891, they found the number of people reduced by two-thirds. The Eskimos had caught only two walruses that year, and the survivors had been living on their dogs and seaweed.

The people north of Bering Strait suffered, too. Lieutenant Patrick Henry Ray, the leader of the U.S. Army's Signal Service expedition to Point Barrow in the early 1880s, reported that the Eskimos told their children "of the happy days before white men came to drive away the whales and walrus, and when food was always plenty." During the winter of 1881-82 "food became very scarce. . . . Walrus hide and pieces of old boat-covers were considered delicacies."

By 1890 Henry D. Woolfe, a former shore whaler and the enumerator for the eleventh census, considered the walrus to be "almost extinct." "Now the natives along the coast from Point Hope to Point Barrow consider it a very lucky catch to shoot 10 walrus during the season, where formerly 500-600 were obtained."

Although articles began to appear more frequently in the nation's press describing the needless slaughter of the walrus herds and the destitution of the natives, it was not until 1908 that the United States prohibited the commercial killing of walruses. By then, however, the whalemen had quit hunting walruses and after 1883 they had taken very few. The price of oil had been slowly sinking since the end of the Civil War. Eventually there were simply too few walruses left to justify the effort spent in hunting them.

A small commercial walrus hunt began again in the early years of this century, but it did not last long. However, the Soviet sealing fleet again brought the population to its knees with a severe harvest that lasted from 1930 to late 1962.

It is ironic that today, despite an excessive and wasteful harvest by the Eskimos who kill animals merely for their ivory, the walrus population has not only fully recovered from its suppression, but may now have reached a high level, greater than or comparable to that in the early 1800s before the whalers began their slaughter.

Strangely enough, this dramatic increase may be the direct result of the severe reduction that began with the whalemen. The walrus population may have been reduced to such a low level that large areas of the animal's former feeding range may have gone undisturbed for as long as half a century. These unexploited areas apparently allowed an abundance of food for the recovering population after the severe harvests ended. Like some terrestrial mammal populations that have been introduced to a new food supply, this walrus population, because of time lags in birth and death rates, may today have increased beyond the carrying capacity of the ecosystem.

8 / *The 1870s, Disastrous Decade*

The 1860s were benign in comparison to the havoc the 1870s brought to the Arctic whaling industry. Apart from war-related losses, the 1860s claimed just nine vessels. In the succeeding decade fifty-seven vessels were lost, despite the fact that there were far fewer whaling cruises to the western Arctic. This ten-fold increase in unit shipping losses came about in part because the weather and ice conditions were significantly worse in the seventies. But what made the whaleships especially vulnerable to these poor conditions was the lack of whales.

The pelagic whalers had reached one-third of their total number of bowhead kills in the first five years of the fishery and two-thirds by 1869. The whalers' very success made it more difficult to catch bowheads within this severely reduced population. To make matters worse, however, the price of whale oil began to decline from its high point in 1865 when cheaper petroleum oil began to make inroads on the oil market. To maintain profits the whalemen had to take more whales from a shrinking population. One response was to harvest the walrus population heavily. The other, more dangerous tactic was to push farther and farther north into the Arctic Ocean in search of the increasingly elusive bowheads.

During the 1860s the whaleships had cautiously pushed farther and farther north. Such explorations were effective, for during that decade ice conditions were exceptionally favorable, but the deceptive openness of the arctic seas proved to be a preamble for disaster.

When the whaling fleet returned to the Bering Strait region in the late 1850s they began to develop a pattern to their hitherto relatively random cruising. In those years before the Civil War, once the whaleships made their way through Bering Strait, they avoided close contact with the ice as much as possible, concentrating instead on hunting whales in the southern Chukchi Sea until August. Then they moved to the waters north of Cape Lisburne, waiting for bowheads to appear. The year 1857 was good for the few ships that returned to Bering Strait from the Okhotsk Sea, but the following three seasons were poor. The whalers' fortunes improved in 1861; good weather, open seas, and plentiful whales led the whalemen to conclude that the whale stocks of the western Arctic had not been fished out by the massive assault of the fifties, as they had suspected.

Conditions were even better in 1862. In that year the whaleship *John Howland*, cruising in advance of the fleet, sighted both Herald Island and "Plover Island," a mysterious headland that had been sighted indistinctly by a British naval vessel only a few years before.

The following season was even more ice free and at least three vessels sighted

"Plover Island." The remarkably open water permitted other ships to sail near the uncharted land in 1865 and 1866 as well: in 1865 the *Oriole*'s mate wrote to the *Pacific Commercial Advertiser* that in staying out of sight of the *Shenandoah* he had reached 72°20′N and "having run 50 miles off the chart, concluded to work south again."

However, 1867 proved to be the best year of all. On their way north the whalemen learned from the natives that there had been almost two months of southerly winds during the winter. Almost from the start the whalemen found very little ice to bar their way. The *President* sailed through Bering Strait without trouble on June 4, more than three weeks earlier than in a normal year, and others may have preceded her. By July 16, forty-five vessels were cruising near Herald Island. But as free as the seas were of ice, they were equally free of whales, causing one of the officers aboard the *Martha* to write on August 24, "Saw Herald and Plover Islands, 37 sails, and 1 whale." Another claimed that the fleet—more than eighty vessels—had taken only twenty-five whales in August.

With such a swarm of ships criss-crossing the same waters and with the seas so ice-free, captains started looking for better hunting. The bold ones went north, choosing to break away from what the rest perceived to be the safety of numbers. The *St. George* reached 73°10′N, and Captain Owen Williams's *Massachusetts* probably achieved the record for a sailing vessel in the Chukchi Sea when he attained 74°30′N at 173°W in the Hole on September 21. He reported no ice in sight to the north but some to the east.

Eventually Captain Thomas W. Long of the New London bark *Nile,* frustrated by empty seas, sailed west in company with several other ships and discovered Wrangel Island. Though in earlier years a few ships had reported seeing an island that they called "Plover Island," like Captain Henry Kellett of the Royal Navy who first sighted this headland in 1849, they were confused about its size and shape. Long correctly delineated its eastern and southern shores.

On the first of August the *Nile,* in company with another New London bark, the *Monticello,* left the fleet and steered westward out of the Chukchi Sea. Searching for whales, they followed the mainland shore of Siberia west, traveling through the strait that would eventually bear Long's name. On August 8 they reached the mouth of Chaun Bay, well into the East Siberian Sea. They were the first foreign vessels to enter the East Siberian Sea from the ocean.

The *Monticello* kept on west for another sixty miles, but Long turned north. He saw no whales and steered back toward the east on August 10. On the fourteenth at 9:30 A.M., a lookout at the masthead spotted the western promontory of the island. Long sailed east along Wrangel Island's shore for the next two days and at times approached as near as fifteen miles. Long, a well-read man, proposed that the land should be named in honor of Baron Ferdinand von Wrangell. In the 1820s Wrangell had explored the Siberian coast on foot and had learned from the natives that "one might, on a clear summer's day, descry snow covered mountains at a great distance to the north."

The *Reindeer, Nautilus, Corinthian,* and *Massachusetts* cruised near the island that summer. Like the *Nile* and *Monticello* they sent out reports of their discovery, and

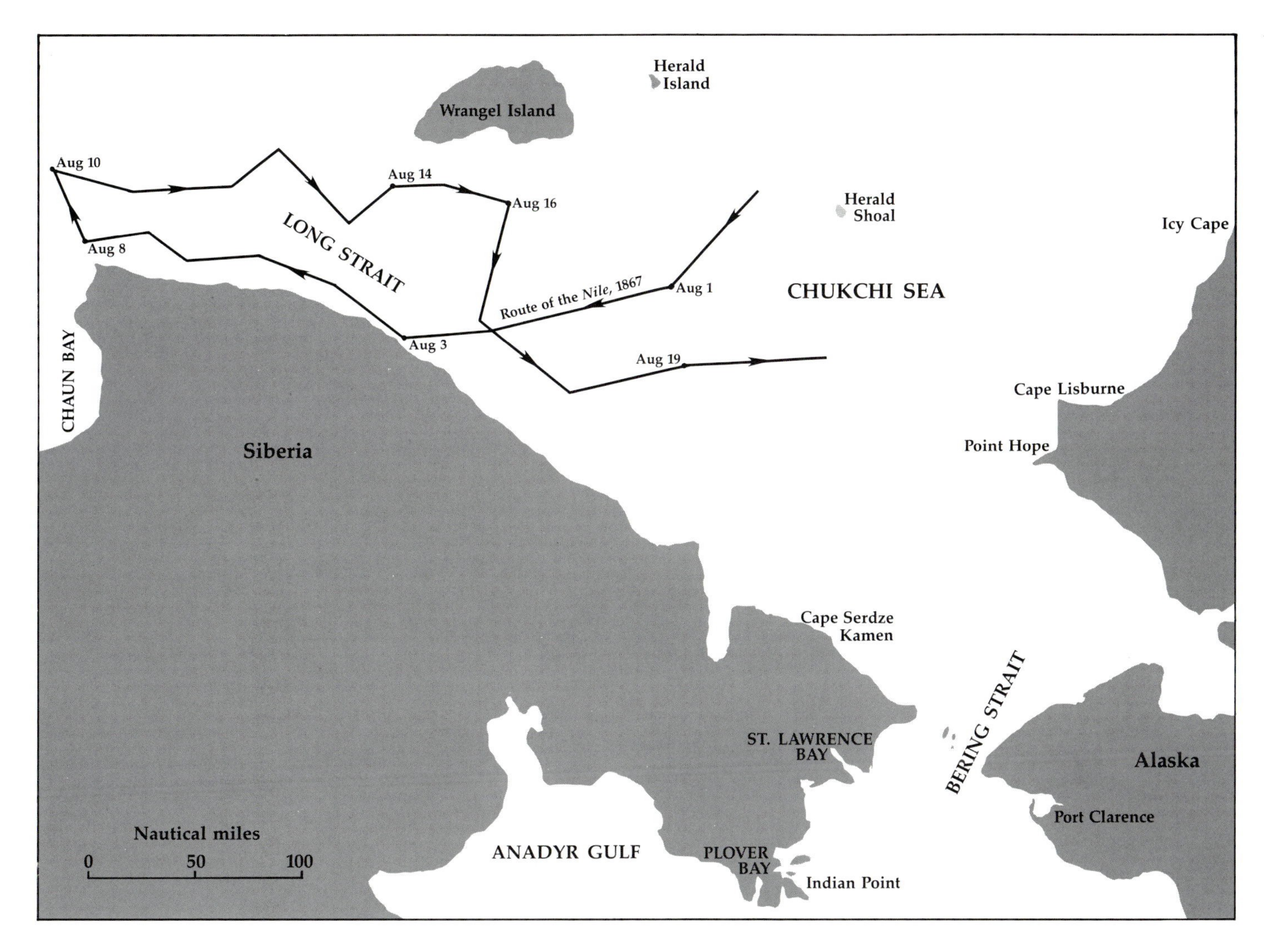

Thomas Long's cruise in the Nile, *1867*

like the *Nile* and *Monticello,* none sent a party ashore; they were searching for whales, not geographical precedences. Most important to them, however, they found no whales. Long wrote that "from 175° to 170° east there were no indications of animal life in the water. We saw no seals, walrus, whales, or animalculae in the water. It appeared almost as blue as it does in the middle of the Pacific Ocean, although there was but from fifteen to eighteen fathoms in any place within forty miles of the land."

The whalemen had learned by trial and error that crystal-clear waters contained no whales; murkier seas often indicated the presence of "animalculae," or plankton, upon which the oceans' food chains are built and upon which the whales feed. As soon as the whalemen were confident that the waters surrounding Wrangel Island

Thomas Long's view of Wrangel Island, 1867

contained few, if any whales, they deserted the area. Only rarely in the next half century would whaleships come within sight of Wrangel Island.

Until 1867 the fleet generally had confined itself to the southern and western Chukchi Sea, waters where a strong current clears the ice early and carries plankton and nutrients north-north-westward out of Bering Strait and into the Arctic Ocean east of Herald Island. But in 1868 the whalemen found no whales on the "west shore." Believing that waters farther to the west were worthless, they began to probe the northeastern Chukchi Sea, the only area of the sea where they had not hunted intensively.

The reasons for not working there before were sound: good whaling was to be had elsewhere, and from the time of Captain James Cook's voyage to Icy Cape in 1778, the northeast Chukchi Sea had a reputation among explorers as being treacherous—full of shoals, currents, and menacing ice. Captain Cook had named Icy Cape because he found heavy impassable ice there, and fifty years later, Captain Frederick William Beechey, seeking to pass that point, had grounded his ship, HMS *Blossom*, on the shoals off shore. In 1861 the *Cleone* had narrowly escaped being destroyed when her captain drove her beyond Pt. Franklin. The ship had been swept north by a strong current while simultaneously the ice pack began to close down on the land. The captain desperately drove the *Cleone* south, tacking his way out in the narrowing path of water. In the fog and rain and dark of midnight on August 25 the ship struck a few pieces of ice. The crew thought nothing of it as they raced south against the current and the closing pack. But the following noon, when the mate looked into the after hold, he saw water sloshing around the oil casks. The ship was stove so badly that it took the crew one thousand strokes of the pump each hour to keep up with the leak. Only with the assistance of the French ship *Caulaincourt* did she escape. In a fleet that shared information vital to one another's safety this incident was quickly broadcast, and until 1868 ships traveling that coast tread softly indeed.

The season of 1868 was at first disappointing. The fleet took only a few whales south of Bering Strait and found only heavy ice. In July the weather turned sour. Several ships spent their time poking about for whales in the small bays that rim the waters near Bering Strait, hoping no doubt to find whales there as they had in the Okhotsk Sea; some spent their time hunting walruses or gray whales near Cape

Serge; many simply left in late July to go right whaling in Bristol Bay. Then at the end of July several made their way to Cape Lisburne and began working northeast in the wide corridor of water between the ice pack and the land, entering waters that had not been visited by a whaling fleet since the exploratory cruise of 1854. In late August and early September most of the ships took ten or more bowheads along the coast between Icy Cape and the Sea Horse Islands. The whaling was so good that the master of the *Ocean* wrote to the ship's owners, "I don't suppose there was ever such whaling seen in the Arctic Ocean since the five first seasons. . . . Anybody who knew enough to lay a boat round could get one of those whales after the 1st of September."

But, as usual, the fleet paid a price for challenging the Arctic in the autumn.

In autumn, when the violent northwesterly gales begin to blow, the northwest coast of Alaska becomes a vast, shoal, lee shore. The first whaleship to be lost was the *Corinthian* of New Bedford. The *Corinthian* was cruising northeast of Icy Cape on August 29 in water where her crew had recently taken half a dozen whales. Late in the evening a wind began to rise out of the northwest. By midnight it was howling a strong gale with a thick snowstorm. Knowing that the ship was on a lee shore and that it had to pass Icy Cape to get a little more sea room, Captain Valentine Lewis began to run southwest as best he could. It was a gamble all the way, but he had not counted on the current that the gale had blown up; the ship was forced ever closer to shore and ever closer to the ugly spur of Blossom Shoals, shallows that reach out in a series of parallel spines as far as ten miles northwest from Icy Cape.

All night long the ship bowled along toward danger. The crew could see nothing in the thick snow. They were running blind at a terrific rate down an alley with the shore on one side and the ice on the other, and the spray and wet snow had frozen the running rigging so that the crew could barely work the lines.

At 6 A.M. the man on the lead line found seven fathoms—shoal water.

> The Captain came on deck and immediately gave orders to wear ship and make sail; on account of everything being frozen solid, had to carry hot water aloft to thaw out the brace blocks before we could wear round; wore ship in five fathoms, heading off shore; again sounded and found eight fathoms; stood off till we came to the ice; sounded and found eight fathoms; in fifteen minutes sounded again and found six fathoms; and in about two minutes after struck the shoals; struck three times, with the space between of about ten seconds. . . .

The *Corinthian* had run down in the trough between two of the parallel shoals and when she wore around, she was driven right onto the inshore shoal.

> The ship struck . . . and immediately fell over on her beam ends. We cleared away the larboard boats to save them. The Captain then gave orders to cut away the masts, which was done, and after starting some of the oil we had on deck, the ship righted up.

The captain then gave orders to abandon the wreck. A few men chose to remain on board, however, and after the captain with his wife and others had left, they found the ice packing in around the hulk, calming the seas and allowing her to ride easier. The next morning the men began to throw overboard the remainder of the deck oil and all of the blubber. They were assisted by the Eskimos who had come aboard to plunder the ship, and by mid-day the wind had swung into the southeast, moving the ice away from the hull. At four in the afternoon the men felt the hulk begin to move; soon she had drifted off her strand and they anchored her in seven fathoms. To their great joy, shortly thereafter six whaleships hove into view.

The men were taken on board the ships just as the pack ice, now released from the pressure of the southeasterly, began to come back down onto the shoals. Prudently all of the ships at once began to stand away—except for the *George Howland*: while salvaging the *Corinthian*'s whalebone, her crew had paused to take a whale themselves. "After getting the bone and crew out," wrote Captain James H. Knowles,

> I started to abandon her, but finding my officers and crew all willing to help save her oil on salvage, I thought I couldn't do better for my owners than make a trial. So I put my second mate on board to slip her cables; got a hawser to her and took her in tow. The next day I went on board, and with some spare yards rigged jury-masts, and that night by taking my storm-sails from my own ship, got a fore, main and mizzen staysail set on her, and found I could work her so well, I made up my mind to take her to St. Lawrence Bay.

With his mate and men aboard the *Corinthian*, Knowles set out towing the hulk. Not wasting an hour, he had the men on both ships boiling and stowing as they made their way south. They reached Saint Lawrence Bay on September 19, and Knowles estimated that with the tacking he had done to work his ships south he had towed the wreck 1,500 miles by the time he got there.

He immediately set the men to transferring the *Corinthian*'s oil to his ship because he realized that with an early autumn upon him he stood no chance of re-rigging the *Corinthian* sufficiently to sail her south that year. He put her on three anchors and made a deal with the natives, who promised to look after her during the winter in return for the ship's bread and molasses.

Knowles and his crew made a good profit on the 1,150 barrels of oil he salvaged, but they saw nothing else from their work on the wreck, for when the *Ada A. Frye*, a small trading schooner from San Francisco, hurried to Saint Lawrence Bay the next spring with the idea of wrecking the hulk, the crew found the *Corinthian* under three feet of water.

But the Arctic was not yet done with the fleet that season. On September 21, 1868, a few ships were still cruising inside the ice along the coast between Point Franklin and Point Barrow. In the morning a strong wind blew up from the northeast, sending in a thick fog. The ships began feeling their way south as best they could. On the twenty-second the *Active* and the *Reindeer* had anchored off Point Franklin near Sea Horse Shoal when the ice began to come down on them fast. Point Franklin and Sea Horse Shoal closely resemble Icy Cape and Blossom Shoals in shape and menace. Seeing that they had no chance of escaping by running west—the pack extended too far to the southwest—the ships could only run east, away from the ice,

The George Howland *towing the* Corinthian *in the Arctic, 1868. Watercolor by Benjamin Russell*

hoping to find safety in the bight behind the shoal. But first they had to reach around the shoal. The *Active* didn't make it.

The ships were moving east into ever shoaler water, trying to keep in four fathoms, when the lead man shouted that they had only two and three-quarters. A moment later the *Active* struck. As night fell the men found themselves aground in two and a half fathoms with a rising wind and the ice closing in on them. At daylight the wind had increased to a gale and the bark was lying broadside to the wind with the ice crowding in and forcing her over.

By noon the swell had increased and she was pounding on the bottom. The captain ordered the crew to throw the blubber overboard as well as their hoop iron, barrel heads, and any other heavy articles they could lay their hands on. By 4 P.M. she had lightened somewhat and swung head into the wind. The crew worked like madmen, cutting up the blubber and throwing it overboard. They threw out 400 barrels of blubber and another hundred barrels of oil as well as "three casks of shook heads and hoops, four casks of iron hoops, three casks of towlines, one hundred bbl of empty casks, one try pot, lot of blubber tubs, lot of empty barrels and various other small articles that was in the way." They put out kedges and began to work the *Active* "out of the bed [in the sandbar] she had worked herself into."

Amazingly enough, because of their back-breaking action, they got clear. Finding that the bark had not been sprung, they began to try out the remaining forty barrels of blubber—all that was left of the five whales they had on board when she grounded. On their way south from the shoal they passed the Honolulu bark *Hae Hawaii.* She had not been as lucky. Her anchors had dragged. She had gone ashore and was bilged. Captain John Heppingstone and his crew had to haul their boats over the ice to reach clear water. They spent the night on shore, under their boats at Point Franklin, and on the twenty-third they reached the safety of some ships that had been able to ride out the gale.

The seasons of 1869 and 1870 were similar. The ships that stayed near Point Barrow until late September made good catches at the very end of the season. A few even stretched their cruises farther by moving to Herald Island and Herald Shoal for a week or ten days before passing Bering Strait about the first of October. When the brig *Kohola* reached Honolulu in 1869, Captain Tripp reported that when he had left Point Barrow September 20, "the water was literally filled with whales and the eight or nine ships in sight were all boiling down."

The following year, 1870, the *Eugenia* went to Point Barrow with at least thirty-four other ships and did well there at the end of August; it was the first time a large fleet had lingered in those waters. The *Henry Taber,* which had not seen a bowhead between May 20 and August 16, took seventeen of them near Point Barrow from late August to mid-September and left there directly for Honolulu on September 21.

As in 1868, however, the successes of the two following years were marred by the autumn losses of several ships. In 1869 the bark *Eagle* of New Bedford left Point Barrow, bound south, on September 30. Working through the twilight Captain James McKenzie judged that he was well past Point Franklin. To his surprise, he struck bottom and grounded on what he assumed to be Blossom Shoals. A rising sea soon was pounding the bark hard on the bottom. She was quickly bilged and filled half full of water. When dawn broke McKenzie discovered that he had grounded on Sea Horse Shoal, "although we had been running 13 hours before a strong wind with all drawing sail set and supposed ourselves 50 or 60 miles from that place." The crew were taken off by the bark *John Carver.* Captain McKenzie had failed to take into account the northeasterly current that can run along that coast at speeds of up to four knots and which, no doubt, he had been bucking all the way.

The following year at the end of August the *Almira* and *Hibernia,* both cruising off Point Barrow, struck scattered ice and stove holes in their bows. Both were lost, although Captain Thomas W. Williams, who had lost the *Jireh Swift* to the *Shenandoah,* was able to run *Hibernia* aground and to sell her at a hastily convened auction for $150. All hands were saved by the large number of ships in the area.

In the clear light of hindsight it is easy to see that all the elements in the losses of 1868, 1869, and 1870 were present in the great disaster of 1871—when the men abandoned thirty-two vessels. After three good whaling seasons near Point Barrow, the whalemen had grown bold and ignored the dangers that were attendant in those waters. The ice pack could act like a hammer on the anvil of the land, there were strong currents, and often a lee shore. Nevertheless, profits were a powerful incentive to overcome prudence; many in the fleet would have sympathized with the *Ocean*'s mate, J. W. Thompson, when he wrote to her owners at the end of the 1868 season and complained that Captain Barber was afraid to go north of Icy Cape in September to hunt the abundant whales that the rest of the fleet was catching there. The cautious masters without doubt understood the dangers of the twin spurs of Blossom and Sea Horse shoals: the coast from Point Barrow to Icy Cape was, in fact, a 150-mile ships' graveyard.

The 1871 season opened inauspiciously. On the assumption that the pickings would be slim near Bering Strait from April through July, a number of ships spent the first half of the season in the Gulf of Alaska or in Bristol Bay hunting right whales. Those that chose to take the traditional cruise in the western Arctic, arrived north of Cape Olyutorski in late April and found the ice heavy and closely packed. It took more than a month for the first ships to worm their way through the pack to the comparatively open water on the "north shore" in Anadyr Gulf, and the first ship could not reach Indian Point until June 6.

When the bark *Oriole* reached Plover Bay, Captain Hayes found the little twenty-three-ton schooner *Hannah B. Bourne* there. She had wintered in the bay to trade whiskey and manufactured goods to the Eskimos in return for their furs and baleen. Hayes decided to charter the schooner, probably to assist in the walrus hunt and then to serve as a tender to tow his boats ahead of the ships early in the season up the narrow shore lead north of Icy Cape. He must have borrowed this technique from the Okhotsk Sea and Baja California whalers that regularly used small tenders to set up camps on shore far beyond the reach of the pelagic fleet.

Hayes's plan might have been successful, but in early July the *Oriole* ran into a piece of ice near Saint Lawrence Bay, badly stoving her bow below the waterline. The damage went unnoticed "until a boatsteerer took off the fore hatch and jumped down into the hold to get an iron when he landed in water up to his waist." Captain Hayes was lucky to have the schooner along then, for he used her to sail out among the fleet, getting help for his stricken ship. The crews of several other ships came aboard. Their help with pumping and with jury repairs enabled the *Oriole* to limp to Plover Bay for a survey of the damage. She was hove down on the beach there and found to be beyond repair. The captain of the *Emily Morgan* bought the ship for $1,350, salvaging the spars and all the gear to sell throughout the fleet.

As the walrus season drew to an end in July 1871 the fleet, expecting to find the open conditions of the previous years, began to gather near Cape Lisburne. To the surprise of the men, they encountered heavy ice south of Point Lay. By August 3 they were still three miles from Icy Cape and the next day they had to retreat because the ice was coming back down on them. Not until August 6 did the ice lift enough the allow the ships to sneak around Blossom Shoals and tack northeastward in the strip of open water between the land and the heavy, closely packed ice four or five miles away. By noon the next day they had reached the limit of navigation a few miles north of Wainwright Inlet. The wind died and stayed light. By August 10 there were twenty ships waiting there to work their way to Point Barrow.

In a normal or even in a moderately bad year for ice on the Alaska coast this would have been a reasonable procedure. Ordinarily it was possible for a sailing ship to reach Point Barrow in the early days of August. A warm sunny spring, combined with a couple of brief violent gales to churn up the ice, followed by off-shore winds, will complete the job of clearing the ice off the Alaska coast that is begun by the warm current running northeast along the shore out of Bering Strait. The whalemen could not have known that several times a century the atmospheric conditions directly contradict this pattern. In the spring a stationary high pressure system can develop over northeastern Siberia, resulting in cold onshore winds that retard the ice decay. When such a situation develops, the current is the main factor in opening the ice near shore while the main pack ice remains relatively solid. It was into such a cul-de-sac that the ships sailed in 1871.

On August 11 the wind swung into the northwest and began pushing the ice toward shore. It closed quickly and the ships were helpless to do anything but to scurry around for protection under the lee of heavy grounded floes. At the same time the ice compacted, closing all its small leads and catching many of the cruising whaleboats as many as six miles from their ships amid the tightening pack ice. The crews were forced to abandon their boats temporarily and to make their way on foot to their ships over the broken and piled floes. The ice closed in on the *Seneca*, farthest to the north, and her crew had to use their blubber hooks and lines to warp their way through the floes into the open water near shore, which in many places was only a mile or less wide.

During the next two weeks the wind dropped and the ice loosened somewhat, allowing another dozen or more ships to work up to the coast near Wainwright Inlet and Point Belcher. All of the ships lay at anchor within twenty miles of one another with the ice ahead, abeam, and astern.

During this fortnight none of the men doubted that the ice would soon be forced open by an off-shore wind as usual, enabling them to reach Point Barrow. Being the capable whalemen they were, they did not let the time go to waste; instead they adopted the tactics they had used in the Okhotsk Sea for working inside the ice toward Penjinsk Gulf and Shantar Bay. Many of the ships "mated" with one another and outfitted two boats apiece with several days' provisions. The quartet then sailed in the very shallow water near the beach as far as the Sea Horse Islands where several caught whales.

When they killed one, the boats towed the carcass to the ice and cut it in from the ice edge, rigging makeshift cutting tackle to roll the carcass over. The blubber being buoyant, they made rafts of the blanket pieces and then sent one of the boats back south, towing the blubber to the two ships, while the other three boats continued cruising. The men spent their nights on shore. They cooked over driftwood fires and slept under shelters made from their sails.

The situation changed on August 29. A southwest wind sprang up in the

Satellite photograph of the pack ice 25 miles off Wainwright Inlet, July 16, 1977. Icy Cape is near the center; Point Franklin, upper right

U.S. Government, ERTS Data Center

morning; it was light at first but by evening it was fresh enough to start the ice moving.

One may wonder why a southwest wind would create dangerous ice conditions on a coast that runs in the same direction. It would seem that the sea ice ought to be pushed along shore parallel to the coast and not encroach on it. But, in fact, sea ice that is driven by the wind does not move with the wind. In the northern hemisphere, because of the effect of the earth's spin (the coriolis effect), sea ice moves at an angle of about thirty degrees to the right of the wind direction. Consequently, the southwest wind, instead of moving the ice along shore, torqued it onto the land.

The New Bedford bark *Lagoda* was cruising near Point Belcher when the wind sprang up. Captain Stephen Swift immediately stood to the south under all sail, narrowly escaping from the "large body of ice to the westward" that he saw closing down on him. The edge of the pack was jagged and a lobe first struck shore to the north of Point Belcher. Scattered and loose at first, the ice quickly closed around the ships anchored there. The ships ten miles to the southwest saw the pack coming down on them. They moved closer to shore, some getting into the lee of large grounded floe-bergs. Throughout that day and the next the men aboard the ships watched the open water amid the floes shrink smaller and smaller as the pack ice tightened its grip. The five northernmost ships, the *Roman, Comet, Concordia, Gay Head,* and *George,* were completely surrounded. Slightly to the south the *John Wells, Massachusetts, Contest, J. D. Thompson, Henry Taber, Fanny, Monticello,* and *Elizabeth Swift* were not as tightly gripped, though their crews saw pieces of ice washing past their ships and filling in between them and the land.

As the ice caught them, several of the ships began to drag their anchors. Some hove up their ground tackle and got under way, working to the southwest through the broken ice toward more open water. The *Fanny* got clear and anchored very close to shore in only four fathoms. The *Monticello* and *Elizabeth Swift* also got clear but had not gone far before they ran aground on a shoal. The crews from many of the nearby ships went aboard to help shift their cargos and get them afloat.

By August 31 the ice had stopped moving—temporarily. Several of the ships still had their boats off cruising for whales, but some of the officers in the fleet had begun to understand the possible outcome of their situation. Captain Timothy Packard of the *Henry Taber* wrote, "Oh how many of this ship's company will live to see the last day of next August? God only knows. I will trust to his wise hand."

The next day, September 1, it began to blow out of the southwest. It was a light wind, but enough to set the ice in motion. The *Roman,* northernmost of the fleet, was anchored off Point Franklin, protected by a piece of grounded ice and cutting in a "stinker," a decomposing whale. The southwest wind began to raise the water level, as it does on that coast. (On the northwest coast of Arctic Alaska the tidal range is only about half a foot; winds have far greater effect on the height of the water, westerlies raise it and easterlies lower it.) Soon the *Roman* was naked, her protective ice floes having drifted north in the current. The men moved her behind another grounded floe. They moored here there in fifty-two feet of water, but the current soon broke that floe apart and at 11 A.M. the drift ice drove in with such force that it not only stove in her stern but it went right under the hull, raising the ship, driving her over and stoving in her whole side. The crew worked frantically. They lowered her three port boats onto the ice and dragged them out of the way as her spars fell. Forty minutes after the first contact, only a little of the ship was left above the ice.

The crew panicked. Surrounded by a jumble of grinding, moving ice, without any open water nearby, they lost all sense of coherent action. Captain Jared Jernegan knew they must get the boats to open water if they were to reach the safety of another ship. He put sixteen men on each boat to haul them over the steep pressure ridges, and, with a pistol in hand, he kept them at it. He led them over the broken ice, testing its thickness with a blubber spade as he went. Once they reached open water he distributed his thirty-eight men in the boats and sailed twenty miles to the nearest ships not yet enclosed in the ice.

The *Comet* of Honolulu, the ship nearest the *Roman,* was the next to go. At one the next morning "she was pinched until her timbers snapped and the stern was forced out, and hung suspended for three or four days." When Captain Packard of the *Henry Taber* saw the *Comet's* ensign at half mast he went to her in his boats to offer assistance. He found Captain da Silva and his crew abandoning ship. Captain Knowles of the *George Howland* bought the entire ship and cargo for thirteen dollars.

Nearby, things were not much better. "The main body of the ice pack kept steadily advancing toward the land, driving the grounded ice over the shoals," wrote William Earle, the first mate of the *Emily Morgan.* "We now have no more than a quarter of a mile of land water." By then it had become clear that of the thirty remaining ships there, twenty-two, the "northern fleet," were in greater immediate danger, strung out as they were over about eight miles in the narrow shore lead from Point Belcher southward. Half of them were already surrounded by ice. The remainder of the imprisoned ships, the eight vessels of the "southern fleet," were grouped less tightly, a little farther off shore, from Wainwright Inlet southward. In between the northern and southern fleets a lobe of ice was pressed close to the shore, making it impossible for the northern ships to work their way south. Even the *Lagoda* at Icy Cape, far to the south of the southern fleet, was "surrounded with ice." The men knew that they desperately needed a northeasterly to free them.

On Thursday, September 7, the current began to run northeast at the speed of a knot, carrying the ice along with it, and those ships that could still maneuver tried to work their way inshore into whatever water remained to avoid being crushed or carried away to the north. The next day a lobe of ice pressed down on the southern fleet, grounding in five and a half feet of water to the south of them. It stove the *Awashonks,* nipped the *Julian,* and snapped the rudder off the *Eugenia.* The *Awashonks* was sold at auction for sixteen dollars.

On Saturday, September 9, the captains in the northern fleet faced the likelihood that their ships would not escape and that they must act quickly to save themselves. As far as ten miles south of their position their scouts could find no clear water deep enough to move one of the larger ships. Holding a meeting, they decided to lighten the smallest vessel in the northern fleet, the 270-ton brig *Kohola* of Honolulu, to try to get her south inside the ice. The *Kohola* would then go in search of the remainder of the Arctic whaling fleet. The captains assumed other ships must be cruising in ice-free water outside Blossom Shoals.

The captains sent their men down to the *Kohola* to put her water, oil, stores, and other movables on board the *Carlotta*. When they had finished and got her going, she was able to cover only a few miles before she grounded in six feet of water on the sand bar that projects off the mouth of Wainwright Inlet and on which the ice had grounded in five and a half feet. There was nothing they could do.

The next day some of the captains sent their men four or five miles farther south to the San Francisco trading brig *Victoria*. As she was only 149 tons, they hoped she could be lightened enough to work around the lobe of ice that rested on shore south of the southern fleet. They failed. With their hopes dashed, the men knew that they would have to abandon their ships. Wintering would be impossible on this exposed coast, provisioned as they were for only a few more months. They would have to make their way south in their whaleboats at least as far as Icy Cape, more than sixty miles from the northernmost vessel.

On board all the ships the men set to work bagging provisions and clothing, making boat covers and mounting boards above the boats' gunwales to keep out the spray. They also mounted false keels on the hulls to keep them from being stove as the boats were dragged over the ice. The hulls were quickly sheathed with copper to keep the ice from wearing through the planks. This last precaution was most important because young skim ice acts like sandpaper and can wear through thin planking in a matter of hours. To make matters worse, that night was cold and at dawn the men found an inch of young ice in places on the water inshore.

That day, the eleventh of September, while the ice was "slowly narrowing the already narrow strip" of water, the captains began to send their boats south to relay loads of provisions to Icy Cape and to tell the ships in clear water that more than a thousand whalemen—as well as some women and children—needed their help.

The same day Captain D. R. Frazer of the ship *Florida* of San Francisco, commanding a group of three whaleboats, reached the *Lagoda* ten miles from Icy Cape. He learned that there were seven ships nearby, two in clear water southwest of the shoals and the rest in easy ice inside the shoals and likely to get out soon. The word was sent out among the seven. All agreed to stand by to take the crews aboard. The men on the seven ships understood full well their own precarious situation in this risky industry. If not now, then sometime in the future it was likely they would have to rely on others for similar help.

"Tell them all," said Captain James Dowden of the bark *Progress*, "I will wait for them as long as I have an anchor left or a spar to carry a sail."

On Tuesday, September 12, Captain Benjamin Dexter put off from the *Emily Morgan* to take his wife to Icy Cape; while William Earle, his mate wrote, "All hope of saving this ship or any of the others has entirely vanished. We can only expect to escape with our lives, if even that."

Captain Frazer returned with his news to the northern fleet at 10 A.M. As the wind rose in the southwest and the ice crept in closer, the captains met aboard the *Florida*.

The whaling fleet trapped by the ice, 1871. From Harper's Weekly

They drew up the following document.

Point Belcher, Arctic Ocean, Sept. 12, 1871

Know all men by these presents, that we, the undersigned, masters of whale-ships now lying at Point Belcher, after holding a meeting concerning our dreadful situation, have all come to the conclusion that our ships cannot be got out this year, and there being no harbor that we can get our vessels into, and not having provisions enough to feed our crews to exceed three months, and being in a barren country, where there is neither food nor fuel to be obtained, we feel ourselves under the painful necessity of abandoning our vessels, and trying to work our way south with our boats, and, if possible, get on board of ships that are south of the ice. We think it would not be prudent to leave a single soul to look after our vessels, as the first westerly gale will crowd the ice ashore, and either crush the ships or drive them high upon the beach. Three of the fleet have already been crushed, and two are now lying hove out, which have been crushed by the ice, and are leaking badly. We have now five wrecked crews distributed among us. We have barely room to swing at anchor between the pack of ice and the beach, and we are lying in three fathoms of water. Should we be cast on the beach it would be at least eleven months before we could look for assistance, and in all probability nine out of ten would die of starvation or scurvy before the opening of spring.

Therefore, we have arrived at these conclusions: After the return of our expedition under command of Capt. D. R. Frazer, of the Florida, he having with whale-boats worked to the southward as far as Blossom Shoals, and found that the ice pressed ashore the entire distance from our position to the shoals, leaving in several places only sufficient water for our boats to pass through, and this liable at any moment to be frozen over during the twenty-four hours, which would cut off our retreat, even by the boats, as Captain Frazer had to work through a considerable quantity of young ice during his expedition, which cut up his boats badly.

Masses of whaleboats were already heading south, relaying provisions to the ships in clear water, when on Wednesday, the thirteenth, the ice crushed the bark *Fanny* and dragged the bark *George* past the *Gay Head,* smashing her jib-boom on the way. There was no improvement on the fourteenth. At noon the remaining crew on board the *Emily Morgan* paid out all the chain on both her anchors, and at 1:30 P.M., wrote William Earle, "with sad heart [I] ordered all the men into the boats and with a last look over the decks abandoned the ship to the mercy of the elements."

The *Emily Morgan*'s men worked their way south in light airs with a hundred other whaleboats in sight. All the ships they passed were already abandoned or their crews were in the last stages of leaving them. The last ship they saw inshore was the brig *Victoria,* "hard aground and lying well over on her side."

As night fell the wind increased and the heavy, lowering clouds became blacker and blacker until the men in the boats could see only a few feet ahead and were in danger of stoving their fragile craft on the chunks of ice floating in the channel. Sighting fires where some other crews had hauled their boats up, they went ashore. After making coffee they shoved off again into the darkness, despite the fact that the wind had continued to rise and was driving in a cold rain. They worked on under double-reefed mainsails, sounding continually lest they go aground in the dark and break up their boats.

Although the wind began to moderate, at 1:30 A.M. one of the boats was stove, hitting a piece of ice and they had to go ashore for some quick repairs. By 7 A.M. in better light, they found themselves twenty miles from Icy Cape. They reached there at 10:30 to find twenty-five or thirty other boats waiting out the wind that had again come up strong.

Eager to go to the ships the *Emily Morgan*'s boats pushed off into the wind after a short rest. Alternately beating under double-reefed sails and rowing, they made their way inside the ice six or seven miles south of the Icy Cape. At 3 P.M. they stopped to eat dinner on a large grounded floe.

The wind was stronger than ever but they pushed off again, picking their way through broken ice, and ran into open water for the first time in more than a month. Inside the ice they had been protected from the rough water that the strong wind was driving up, but once outside they took its full blast and the boats were tossed violently about. Heavily loaded with passengers and provisions, they were low in the water and, despite the raised gunwales, the crews had to bail constantly to keep themselves afloat. Cold and frightened and covered with a glaze of frozen spray, they fought their way out to the waiting ships that were themselves holding on straining anchor lines. The *Arctic* and the *Lagoda* both parted their chains as they rolled and bucked in the steep, rugged seas.

By the afternoon of the seventeenth the seven ships had taken aboard 1,219 men, women and children. The *Europa* carried 280; the *Arctic,* 250; *Progress,* 221; *Lagoda,* 195; *Daniel Webster,* 113; *Midas,* 100; and *Chance,* 60. They were packed in shoulder to shoulder on the ships. "Pushed together as bad as any Irish emigrants," wrote one. Aboard the *Progress,* Captain Dowden turned his tiny cabin over to three captains and their families, and rough bunks were knocked together between decks for the sailors and boatsteerers. Captain Thomas W. Williams of the *Monticello* was aboard the *Progress* with his wife and two children. His young son, William F. Williams,

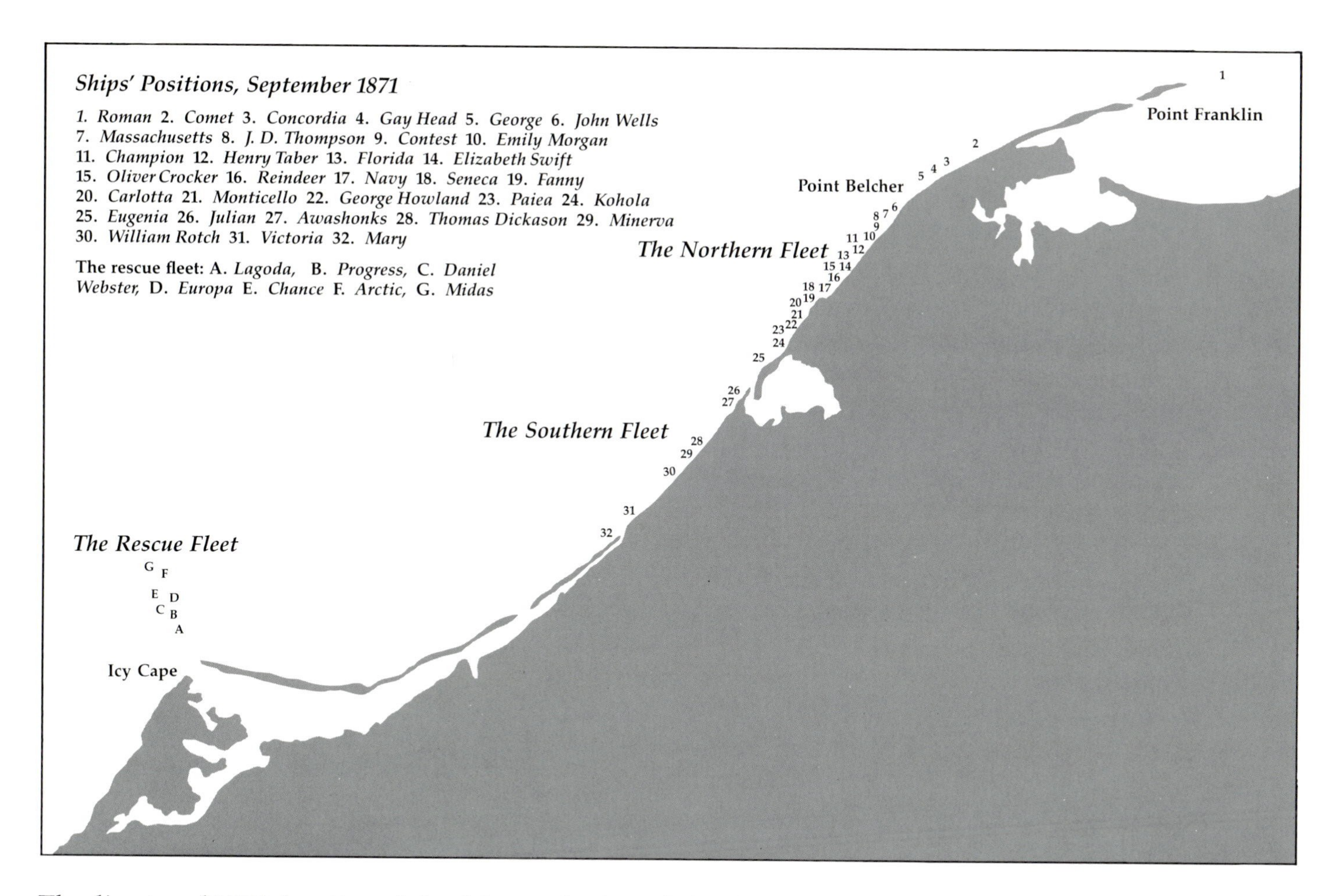

The disaster of 1871: location of the ships at the date of abandonment

recalled the tight conditions, noting where each group slept. On concluding his inventory, the only person he could not account for was Captain Dowden, "I never did know where he managed to sleep."

The seven ships headed south directly, most stopping to take on water in Plover Bay before bearing away for Honolulu, where they arrived before the end of October.

The news of the massive loss of ships and cargos and the evacuation of more than twelve hundred people without the loss of a life created a sensation. Out of forty vessels in the 1871 Arctic whaling fleet, only seven returned that autumn. More important were the far-reaching effects on the whaling industry. The 1871 season had accounted for the loss of thirty-three ships, valued with outfits and cargos at more than $1,600,000 (approximately $13,000,000 in 1982 dollars). Twenty-two of these were from New Bedford and were valued well in excess of a million dollars.

Although the blow must have been staggering for the New Bedford owners and their insurers, I am unaware of any bankruptcy or financial ruin resulting from the 1871 season. On the other hand, few of the lost ships were replaced. Since the late

Abandoning the George, Gayhead, *and* Concordia *off Point Belcher, September 14, 1871. From* Harper's Weekly

1840s, as the whaling grounds of the world slowly declined, many whaling merchants had been gradually withdrawing their ships and putting their capital to work in more lucrative and less risky ventures. A substantial part of the money that was paid by the insurance companies in 1871 was put to work in textile mills and other enterprises closer to home.

Ironically, the owners and crews of the seven vessels that stood by at Icy Cape did less well. The government of Hawaii paid them for the passages of Hawaiian subjects, but the owners had to petition the U.S. government for the loss of the business they suffered when they carried the passengers to Honolulu. Congress was slow to act, possibly because the five American vessels made optimistic estimates of their losses (the two others, the *Arctic* and the *Chance,* were from Honolulu and Sydney, Australia, respectively). The five claimed damages varying from $50,000 to $71,000 for potential catches lost; the minimum estimate was 900 barrels of oil and 16,000 pounds of baleen. Not until 1880 did Congress seriously consider the matter and not until 1891 were the American vessels paid—an award of $138.89 per U.S. citizen carried. These awards were paid to the owners to be dispersed among themselves, the officers, and the crew; it is doubtful that, twenty years after the fact, many of the latter could be located to receive their portion.

Once the news of the disaster was out, the skeptics began to question the wisdom of abandoning the fleet so hastily. The masters must have known the question would arise. Their unanimous declaration on September 12, 1871, was intended to counter this criticism. The critics were gratified to read the following year that when the fleet reached Point Belcher the men learned from the Eskimos that a gale had blown the ice off shore about two weeks after their departure. On the other hand, although a few ships would probably have been able to break free, had the crews remained aboard, it would have been an excessively risky gamble to have waited in such an abnormally cold year for that possibility, for the ships carried only a few months' provisions.

The editor of Honolulu's *Friend* responded quickly and effectively to the criticism:

> It is an easy matter in Honolulu, with the thermometer at 80°, to criticize the actions of men who have faced danger and starvation under the shadow of icebergs, and while the icy barrier was momentarily pressing a fleet of ships on the barren shores of Siberia. We have no doubt that the owners and agents of whaleships and Insurance Companies in New Bedford, seated before a good coalfire, will express their *deliberate* opinion that the fleet was abandoned too soon. We have been permitted to read the private journal of one of the shipmasters, whose ship was saved, and it tells a story of anxiety that ought to silence all foolish censure of those shipmasters who were compelled to leave behind them their hard-earned wealth. The idea that thirty-three shipmasters and their crews abandoned their ice-bound vessels, except from stern and dire necessity is not to be entertained for one moment.

As quickly as the *ex post facto* criticism came, so also came the salvors. Knowing well the amount of oil, bone, ivory, and equipment he had left behind, Captain Thomas W. Williams joined the former mayor of Oakland, Samuel Merritt, and three others to form a salvage company to bring back what they could. They outfitted the bark *Florence* for a whaling and salvage voyage, hoping to beat the rest of the fleet to Icy Cape.

Other groups had the same idea. The Alaska Commercial Company fitted out the handsome little schooner *Eustace* under Captain Elijah Everett Smith, who had been master of the *Carlotta.* The bark *Francis Palmer* was sent north by others, and a number of whalers and traders no doubt hoped also to gain some salvage. One

The crews from the abandoned ships reaching Icy Cape, 1871. Lithograph by Benjamin Russell

curious incident occurred when the small schooner *Cygnet* of Santa Barbara was on its way north for salvage. The captain decided to go via Hokkaido and the Kuriles, looking for fur seal rookeries and sea otters on the way. They never reached Bering Strait, for they came upon some undiscovered sea otter beds in the southern Kuriles and made a small fortune in a few weeks.

Most of the salvage fleet left San Francisco in late May, reaching Icy Cape well after the middle of July. There the ships were blocked by ice. The men aboard the ships had seen the Eskimos with inordinately large amounts of baleen to trade and they assumed it had come from the wrecks. When they reached Icy Cape the natives told them that three vessels were still afloat.

The salvors all knew that even a few hours' head start would pay big rewards. Captain Williams hastily fitted out two boats with salvage gear and a month's provisions and sent them up the coast under the command of his first mate, Edward Perry ("Ned") Herendeen. They worked their way inside the ice, reversing the route that they had traveled the previous September.

Whaleboats in heavy weather reaching the rescue ships, 1871. Lithograph by Benjamin Russell

They were close to Wainwright Inlet before they saw the wrecks. It must have been an eerie sight for the whalemen to visit this ships' graveyard: empty hulks, lying at unnatural angles, broken spars and slack rigging, the shore a mass of sea wrack. They first came to the *Mary* she had been pushed ashore and crushed. Then they found the *Minerva* over on her beam ends in shallow water near the inlet. Apart from some chafing from the floes and a hold full of solid ice, she was sound. They came upon the *Awashonks* beached in only two feet of water; her masts were gone. The *Thomas Dickason* was hard ashore, high and dry with 800 barrels of oil in her hold, and the *Kohola* was completely crushed. The *Reindeer* was bilged and over on her beam ends. Next they found, upside down, two hulks that they could not identify. Captain Williams found the bow and stern of his *Monticello* half a mile apart. The *Emily Morgan* had such a large hole that the tide was ebbing and flowing right through her. Twenty-five miles north of the *Mary,* the *Seneca* was frozen into a cake of ice. She was in relatively good condition with her masts standing, but with her "bowsprit gone, bulwarks stove and rudder carried away." The *Florida* had been

Remains of the 1871 disaster: Point Belcher in the 1920s Private collection

dragged up the coast to the Sea Horse Islands and had been burned there by the Eskimos. The shore was a jumble of litter: "spars, timbers, planks, staves, casks, hoops, boxes of soap, bread," and countless other things.

The Eskimos had also burned the *Gay Head* and *Concordia* and stripped many of the remaining ships. Many of the natives of Icy Cape stayed inland in the summer of 1872 for fear of retribution. Saddest of all, as they rummaged through the ships searching for supplies of alcohol, most of which the whalemen had destroyed before leaving, they broke into the ship's medicine chests and a number are reported to have died from drinking the contents of the bottles.

Ned Herendeen set to work at once on the best ship, the *Minerva.* They chopped the ice out of her hold, took out what remained of her cargo, and thus lightened, they got her afloat and at anchor. They refitted her with some new gear and sails, bought all the baleen they could from the natives, and collected the oil that had washed ashore in casks from the *Reindeer* and *Champion.* The other ships took some gear and cargo from the wrecks. No records remain of the *Eustace*'s and *Francis Palmer*'s salvage.

Captain Williams put Herendeen in charge of the *Minerva* and then moved on to the *Seneca,* just as the ice surrounding her was beginning to break up. He took the *Seneca* in tow and was heading south when a strong northwesterly came up. Williams had to cut the *Seneca* loose to save his own ship. The *Seneca* went ashore and was lost; nevertheless, the *Florence* and the *Minerva* arrived in San Francisco in October with a combined cargo of 1,300 barrels of whale oil and $10,000 worth of baleen—as well as walrus oil and ivory. Thus the *Minerva* had been twice lucky: she had escaped the *Shenandoah* and she had escaped the season of 1871, the only one of the thirty-two that were abandoned.

One man, a boatsteerer, had chosen to stay behind when the rest abandoned the fleet. He had planned to salvage the baleen from the other ships and to live aboard the *Massachusetts* during the winter. In 1872 he was found at Point Barrow, "about used up." He reported that the *Massachusetts* had drifted around the Point during the winter and that he eventually got to shore there. He claimed that the Eskimo men had taken away most of his baleen and had planned to kill him but that he was saved by their women and that finally he was taken care of by an "old chief."

> He said $150,000 would not tempt him to try another winter in the Arctic. He said that four days after we left the ships last year the water froze over and the natives walked off to the ships; and fourteen days after there came on a heavy northeast gale and drove all but the ground-ice away [that never moved]. Shortly after there blew another northeast gale, and he said that of all the butting and smashing he ever saw, the worst was among those ships driving into each other during those gales. Some were ground to atoms, and what the ice spared the natives soon destroyed, after pillaging them of everything they pleased.

The unnamed boatsteerer told a sorry tale, but it is the survivors of the wreck of the *Japan* who deserve sympathy. The *Japan* had gone ashore at East Cape in 1870 in a brutal autumn storm just as the last ships were leaving Bering Strait. The men lived with the Eskimos for the winter. Many of them made their way to the *John Wells* in June 1871 and then were distributed throughout the fleet only to have to abandon their new vessels three months later. Some of the wrecks of 1871 are still visible today.

Clearly the whalemen would never have run the risks resulting in the disaster but for two important factors: the world's whaling grounds were becoming exhausted, leaving the western Arctic as one of the few reasonably productive areas, and the western Arctic itself was being fished out, forcing the ships to take greater and greater risks to secure oil and baleen for profitable catches.

If whale oil had been the only commodity available for illuminants, lubricants, and other industrial uses (among them, soap making, rope fiber dressing, and leather currying), its price would have risen to compensate for its scarcity. But because the whaling industry, even from the beginning of the nineteenth century, could not meet the American and European demand for oil, a number of substitute products were being developed.

From 1830 onward several important substitutes began to challenge the American whaling industry on the oil market. The first of these was an illuminating oil, camphene. Camphene was distilled from turpentine, an oleoresin derived from pine trees. Although camphene smelled strong and was far more volatile than whale oil, it gave off a brilliant light and was cheaper; consequently it came within reach of the common man, especially after inexpensive lamps were developed specifically for its use.

The whale oil merchants, and the industry in general, understood this challenge and sought to counter it. The *Whalemen's Shipping List* conceded that camphene gave off a bright light but claimed that it was not, in fact, cheaper than whale oil because camphene was consumed faster. The *Shipping List* also reported, with a certain amount of smugness, fatal accidents and explosions connected with camphene's use.

> We have often thought that the State, which recently seems to take such care of what we shall eat, and what we shall drink, ought to interfere in this matter. We have statutes for the storage of gun-powder. It is put away, out of the reach of hap and hazard. Yet every day we hear of this house burned, of that child killed, by the explosion of what is called "Camphene." It seems to us that the State might find some ground for its legislation, even here.

To make matters worse, cottonseed oil and fish oils began to be used for a variety of industrial processes including the manufacture of paint and rope. Coal was being distilled to create naptha, lubricating oils, and illuminating gas. New Bedford itself had a gas works in 1853 and a kerosene distillery in 1858 and San Francisco's Stanford Brothers began both refining whale oil and distilling camphene in 1859.

But the American whaling industry's principal adversary was the American petroleum industry. In the mid-nineteenth century a group of speculators hired Benjamin Silliman, Jr., professor of chemistry at Yale University, to analyze the economic potential of refining the crude petroleum oil ("rock oil" or "earth oil") that was seeping from the ground in Titusville, Pennsylvania. Silliman estimated that at least 50 percent of the crude oil could be distilled into an illuminant that could be used in camphene lamps. Speculators combined this knowledge with the idea of drilling for oil, and in August 1859, Edwin L. Drake had the first oil well.

The discovery was coupled with new methods of refining and bulk transport, and by the turn of the century the American petroleum industry was producing two hundred products that duplicated every use then known for whale oil. At the same time its output was nearly sixty million barrels a year.

The whalemen understood at once the significance of Drake's well. In 1860 Captain Willis Howes aboard the *Nimrod* near Bering Strait wrote in his journal:

> Spoke bark Cynthia. Capt. Low . . . came on board. . . . Talking about pumping up oil at the rate of 90 bbls. per day. In fact coal and its offspring oil was the . . . topic of the day. . . . In a word this infernal pursuit of extracting oil from the bowels of the earth [has caused men to

"Grand ball given by the whales in honor of the discovery of the oil wells in Pennsylvania." From Vanity Fair, *1861*

sell] their wives to buy a pump and to be run in troops marching in search of that fatty el dorado.

The next year the *Shipping List* reported that the emerging petroleum industry had brought "a most ruinous effect" to the whaling industry. The publication added, however, that "the whales themselves will undoubtedly be grateful for the discovery of oil which is fast superceding [sic] that hitherto supplied by themselves."

Within a decade of its founding the petroleum industry had a substantial impact on the whaling industry. In its annual review of the whaling industry for 1870 the *Shipping List* reported that the price of whale products, "which at the opening were considered unremunerative, steadily declined throughout the year, closing at the lowest quotation of any year since 1861." With such a gloomy prospect for the industry the owners of the abandoned fleet did not rush to replace their capital in whaleships. In fact, in 1872 another twenty ships were withdrawn from the American fleet.

Those who chose to stay with the industry found that the remainder of the decade was an unlucky time to be engaged in the Arctic fishery. By that time the whalemen had learned that only the late season was a really productive part of the year. For the rest of the decade some vessels chose not to go north until the middle of summer. They spent their time instead in Bristol Bay or the Gulf of Alaska, estimating that the best place to find whales in concentration was at Point Barrow in August and September.

When the ships arrived north of Bering Strait in 1872 they were confronted with another icy year, though not as bad as the previous one. By the middle of August the fleet had gathered at Point Barrow and found the whales coming out of the Beaufort Sea. Perhaps they thought that at Point Barrow they were beyond the most dangerous part of the coast, having, as they did, the option of going east as well as south to avoid the ice. But the waters near Point Barrow are equally, if not more dangerous than those to the southwest, for the current is stronger at the Point and can move the pack with amazing speed and force.

On August 18 most of the fleet that was not salvaging the 1871 wrecks was anchored or tied to grounded floes southwest of the Point. The pack ice was surging by in the two to three knot current. Gradually the *Helen Snow,* at anchor, was surrounded by loose ice. The ice tightened its grip and then dragged her north into the pack. The men were desperately frightened. Against Captain George Macomber's orders, the crew abandoned ship, scrambling their way to the safety of the ships nearby in open water. The next day the moving pack took the *Roscoe,* also at anchor, and swept her down against the grounded ice, crushing in her starboard side. At the same time the *Sea Breeze* was caught by the pack and taken north. Her crew abandoned her and started over the ice, but the pack loosened. The crew returned to her, and succeeded in working her out and back to Point Barrow.

To the chagrin of the *Helen Snow's* men, in the same easing of the pack that allowed the *Sea Breeze* to escape, the *Jireh Perry* was cruising north of the Point and saw the *Helen Snow,* abandoned, deep in the pack. Captain Leander Owen, who with his wife and child had abandoned the *Contest* the year before, was able to get a salvage crew to the *Helen Snow.* Eventually they freed her, bringing her to San Francisco for sale. In the same nip the *Live Oak, Arnolda,* and *Joseph Maxwell* were also badly stove.

This brush with the ice did not frighten the fleet away from Point Barrow. The ships returned the following year, 1873, to find light ice and the pack receding quickly, but few whales. By the end of July, twenty whaleships had worked their way up to the Point and lay at anchor. Seeing no whales, the captains decided to move to the east, the direction from which the whales arrived in August, to intercept the bowheads in waters that were barely charted and where no whaleman had ever been. In a crowd they worked their way east 150 miles to Harrison Bay. Seeing no whales, they turned back, passing Point Barrow in early September and then proceeded on what was to become a regular route, cruising westward along the ice edge to Herald Shoal and up into the Hole near Herald Island. They finished the month of September in those waters before heading south to Honolulu and San Francisco.

KEROSENE OIL!

KEROSENE OIL!!

Clear Oil, Clear Light!

NO WEAK EYES.

FARRAR'S BOSTON REFINED

Clearer and Whiter than any in use.
Stands higher Chemical Test.
Emits no Odor, and Warranted to burn in any

KEROSENE and all

ARCAND BUTTON-TOPS,

CAMPHENE AND SYLVIC OIL

LAMPS!

Burns slower, and gives a Clearer Light as may be proved by actual test.

FOR SALE AT EXTREMELY LOW RATES, BY

A. S. CLEGHORN.

P. S.--LIBERAL DISCOUNTS TO THE TRADE.

Also to Arrive

100 Cases Downer's Oil!

From Boston.

432-1m

Advertisement in Pacific Commercial Advertiser, *Honolulu, early 1870s*

Despite the lessons of 1871 and 1872 the open conditions of the 1873 season emboldened the whalemen; many thought they could have pushed on to the Mackenzie River that season. They believed that a mother lode of whales would be found somewhere between Point Barrow and Banks Island. They were correct, but it would be two decades before they tested the assumption.

The 1874 season was as mild and open as the previous one, and the fleet pushed past Harrison Bay. The following year was even better. In 1875 some ships went 300 miles east of Point Barrow to Barter Island, becoming thereby the second group of foreign vessels to traverse those waters, and the first since HMS *Enterprise* had passed there twenty years before. Most of the ships made good catches, and a number of bowheads were taken east of Point Barrow. Judging from newspaper reports, the industry was relieved that the Arctic fishery had again proved itself lucrative. In retrospect, however, it can be seen that the state of things in 1875 was very similar to that in 1870: both seasons were the third of three open seasons, both had provided good catches in new waters and both preceded major disasters.

In 1876 only twenty vessels comprised the Arctic fleet, reflecting the misfortunes of the industry. Most ships went through the usual routine of whaling in the early season at the ice edge of the Bering Sea, then walrusing in the western Chukchi Sea during the middle season. In the latter part of July many moved toward Cape Lisburne to begin to work their way inside the ice to Point Barrow.

The season had already started badly. South of Bering Strait in May the *Marengo* collided with the *Illinois* and sent her to the bottom in fifteen minutes. The Hawaiian bark *Arctic,* which had survived the 1871 season, had taken advantage of offshore winds early in July and run boldly and foolishly up the shore lead as far as Point Franklin. The wind changed on July 7 and crushed the bark twenty miles from shore. After thirty hours of back-breaking work, the crew reached shore and eventually all were saved by the bark *Onward.*

Offshore winds sprang up again and, despite heavy ice, fourteen of the eighteen whaleships left in the fleet were able to work up to Point Barrow by the middle of August. There they found plenty of whales. On August 18, however, the wind changed to the northeast. For protection, the ships ran around to the southwest of the Point to get under its lee. This was the only maneuver open to the ships, but, in fact, they were entering an even more dangerous trap; a huge body of ice was still far to the southwest of the Point and, unknown to the men, was being carried north by the current to trap them from behind.

Gradually the open water narrowed around the ships. To avoid being crushed, all except the *Rainbow* and *Three Brothers,* which were cutting whales in a bight just under the Point, frantically tried to beat their way south to open water. Trying to run against the strong northeast current was a mistake. The *Marengo,* under all sail and leading the fleet, worked her way thirty miles down the coast to within a few ships' lengths of open water but the current and the ice stopped her there. Fifteen miles behind her the *Clara Bell* was nipped, lost her rudder, and Captain Thomas W. Williams was forced to run her into a deep bay in the shore-fast ice for safety. Gradually all ten other ships were surrounded by the broken and moving floes and carried back northward. The only one to escape was the *Florence,* commanded by Captain Lewis Williams, Thomas Williams's brother. Although the *Florence* had lost her rudder, Lewis Williams broke her through into a small lead, near Cape Smyth, about fifteen miles south of Point Barrow, and she was saved.

The pack swept its captives past the *Clara Bell* in her bay in the ice, then past the point where the *Rainbow* and *Three Brothers* lay at anchor, boiling. Aboard the *Three Brothers* Captain Leander Owen understood the danger and sent a whaleboat to the edge of the pack to pick up Mrs. Hickmott, wife of the captain of the *Acors Barns.* She scrambled three miles over the ice to reach the boat.

The captains of the trapped ships certainly realized their danger, but no doubt they were loath to re-enact the 1872 abandonment of the *Helen Snow,* which after her salvage had returned to the fleet as the Hawaiian bark *Desmond* and was, at that moment, caught with them.

At Point Barrow the current branches; one leg runs northwest into the Arctic Ocean and the other, turning east around the Point, follows the Beaufort Sea shore. The trapped crews were exceedingly lucky to be caught by this easterly branch, for there would have been no chance of return if they had gone northwest.

The ten ships were deep in the pack, twenty miles or more from shore off Smith Bay, when the captains decided to abandon ship. Taking spare food and clothing and, from each ship, two boats that had been reinforced for dragging over the ice, three hundred men set out. They left behind more than fifty who chose to stay, probably in hopes of salvaging the cargoes. The men traveled for two days on the ice, camping under the boats and cooking over fires fueled by the boats that were battered to pieces as they went. They reached shore exhausted and sailed on to Point Barrow where they found that the *Three Brothers* and the *Rainbow* had now been trapped inshore. They made their way around them and then south to the *Florence*—only to find that she too was trapped, the pack having closed in on shore well to the south of her.

The men learned from the Eskimos that the pack was hard ashore from Cape Smyth to Point Belcher and probably even farther south (they later learned that it was ashore all the way to Icy Cape). They did not deceive themselves about the possibility of a mass rescue as in 1871. Although Captain Lewis Williams offered to share the *Florence*'s provisions with the three hundred, he had supplies sufficient for only three months for his own crew. As the ice began to thicken around the ship, assuming the worst, the men divided themselves into parties to prepare for the winter. The best whalemen were sent out in nine boats to get whales for food, another group was sent along shore with dog teams lent by the Eskimos to collect driftwood for timber and fuel, while another began building huts for the winter.

To their surprise and joy, on September 14 the pack began to open. They piled on board the *Florence* and began sawing and warping through the grounded ice, working the ship into clear water. She had sailed as far as the Sea Horse Islands before they found the pack hanging on the shoal there. The next day, as the pack continued to open, they were joined by the *Three Brothers* and *Rainbow.* Once they reached clear water Bernard ("Barney") Cogan, captain of the *Rainbow,* wanted to

HARPER'S WEEKLY.
A JOURNAL OF CIVILIZATION

VOL. XX.—No. 1038.] NEW YORK, SATURDAY, NOVEMBER 18, 1876. [WITH A SUPPLEMENT. PRICE TEN CENTS.

Entered according to Act of Congress, in the Year 1876, by Harper & Brothers, in the Office of the Librarian of Congress, at Washington.

The trapped ships being swept north near Point Barrow, 1876. The Florence *(foreground), with her rudder gone, was able to break out of the drifting pack ice. From a sketch by Captain William H. Kelley of the* Marengo

continue his cruise, but Captain Owen had had enough. Owen took half the men from the *Florence* and sailed with her to Saint Lawrence Bay, where they took on water and rearranged their passengers: 190 wanted to go aboard the *Florence* to San Francisco; the rest to Honolulu on the *Three Brothers.*

Although the number of ships lost in 1876 was only a third the number abandoned in 1871, no lives had been lost then. The safety of the more than fifty men who chose to stay behind was a compelling concern throughout the winter following the 1876 disaster. The *Helen Mar* reached Point Borrow in 1877 and learned from the Eskimos that two of the abandoned ships had drifted close to shore during the winter and that five men had reached Point Barrow. Only three of them survived the winter. Nothing was heard of the rest. Of the ships, only the *Clara Bell* was found. She was near where she had been abandoned, and although she had been stripped by the natives, she was still sound. The schooner *Newton Booth* salvaged her and towed her to San Francisco.

The season following 1876 was not kind to the fleet. The ice sank three ships, two of them near Point Barrow; one was Leander Owen's *Three Brothers.* A gale drove Ebenezer Nye's *Cleone* ashore in Saint Lawrence Bay, where he abandoned her.

One may ask why, faced with these appalling losses, did the whaling merchants continue to send their ships north. Judging from the *Whalemen's Shipping List,* it appears that by the mid-1870s the fishery had become so marginally profitable that it was difficult for merchants to sell their ships. They chose to keep their equipment working in hopes of a modest profit and concentrated on cutting costs, primarily the costs of refitting and transporting the product to the market.

These factors, coupled with increasing costs in foreign ports, especially in the Hawaiian Islands, and the completion of the transcontinental railroad in the United States, led to the rise of San Francisco as the Pacific whaling industry's center in the final decades of the nineteenth century.

Whaleships had been visiting the Hawaiian Islands to take on water and fresh food, to make repairs, and to recruit new sailors since 1819. The number using the islands grew quickly: in 1824, 66 ships touched there; in 1834, 156. The islands soon became the principal foreign port of the Pacific whaling fleet, situated as they were centrally between all its major whaling grounds.

A few vessels chose to refit at Hong Kong, the Australian or New Zealand ports, Hakodate, Talcahuano, Tumbes, or San Francisco. After the 1849 gold rush, however, the ships gave San Francisco a wide berth; not only were prices inflated by the boom, but for the ordinary seamen the lure of gold was far stronger than the lure of oil—as whaling masters learned the hard way. Several masters, refitting their ships in San Francisco, awoke to find their ships staffed only by skeleton crews. Soon after the ship *Massachusetts* of Nantucket reached San Francisco in late 1849, Captain Seth Nickerson, Jr., found that he had only six persons aboard—including himself and his wife. Eventually he was able to hire three sick miners and the nine of them worked the ship to Lahaina where he was able to collect a crew.

The merchants and ship chandlers of San Francisco perceived that they were losing a potentially large market, and by 1854 one of the shipping papers there announced that when the New Bedford whaleship *Mogul* visited the port, "she did not lose one of her crew by desertion or otherwise." By 1856 San Franciscans were claiming that their prices for wood, water, and food were cheaper than Hawaii's.

The merchants in Hawaii, particularly Honolulu, knew well the value of the whaling industry and understood the nature of this threat to their market. As early as 1844 one writer claimed that without the whaling fleets the islands would return to a state of "primitive insignificance." It was estimated that each ship spent $4,000 in the islands annually, and in 1856, 250 whaleships visited their ports. The *Pacific Commercial Advertiser* responded to San Francisco's challenge, pointing out that while Hawaii had legal authority as a foreign port to prevent crews from jumping ship, San Francisco did not. "It will be a sorry day for any who chance to go to San Francisco,"

The retreating whalemen camped on grounded ice, 1876.
In Harper's Weekly, *from a sketch by Captain William H. Kelley*

wrote the *Polynesian,* "as, aside from the fact that no recruits can be procured there at any rate, the brilliant prospects offered in the gold region will induce every man to desert."

Despite the fact that the California legislature passed a law in 1855 exempting whaleships from pilot charges, which had been nearly ten times those of Honolulu, it was not until well into the 1860s that the whalemen seriously began to look for a port other than Honolulu for refitting—and this had nothing to do with San Francisco's inducements; rather it arose from difficulties in Hawaii. As the whaling industry declined, sugar cane became the principal industry of the islands. The cane planters, competing with the whalers for the island's limited labor supply, successfully sponsored a law requiring whalemen to post a bond of $100 for each Hawaiian they shipped for a whaling cruise. Once the bond was established, the price rose higher and higher until it stood at $300 in the mid-1860s. Combined with other shipping taxes, whaleships often left Hawaii having posted bonds totaling three to ten thousand dollars. A group of New London owners sought to get around the charges by registering their ships in Hawaii, thereby avoiding the costs and many bothersome details required of foreign ships. But added to these problems were the charges levied by the United States consul, considered "extortionate" by the whalemen; port charges of more than $130; difficulties in insuring cargoes for transshipment home; unfavorable rates of exchange; the long period of time it took to communicate with the owners in New England; and what the whalemen considered "harassment" by the local police.

The Florence *escaping from Point Barrow, 1876.*
In Harper's Weekly, *from a sketch by Captain William H. Kelley*

San Francisco had its problems, as well. Desertions remained a major drawback. The *Martha* of New Bedford lost fifteen men there in the autumn of 1868. San Francisco was also, according to the captains, home to a particularly lupine collection of lawyers. Using charges trumped up by the sailors, the lawyers freely preyed on the officers and captains, they claimed. Although the New Bedford whaling merchants, Swift and Allen, opened an office in San Fransico in 1866, the number of whaleships visiting that port remained very low—often only one or two ships outfitted there annually until the 1870s.

But in the late 1860s things began to change in San Francisco's favor. The transcontinental railroad and telegraph brought San Francisco much closer to the East. In Hawaii the continuing larcenous behavior of the U.S. consul and renewed effort on the part of the cane planters to retain the nearly five hundred Hawaiian whalemen on the islands caused the *Pacific Commercial Advertiser* to warn that unless conditions were improved, the whaling business would be lost to San Francisco by default.

In 1870 ten whaleships used San Francisco; in 1872, twenty-two. The number of whaleships visiting Hawaii and San Francisco then see-sawed back and forth. But gradually, as the size of the American Pacific whaling fleet shrank throughout the seventies, and San Francisco developed a modest domestic market for whale oil, and the lure of the gold fields subsided, more ships turned to that port. With the rise in the price for baleen in the late seventies, the Pacific whaling industry finally centered in San Francisco. Eastern merchants could ship baleen from the city quickly and cheaply by rail, enabling them to sell it at the most favorable moment. A substantial amount of new capital investment then began to come from San

Francisco, and Hawaii's long association with the Pacific whaling industry came to an end.

The gloomy decade closed as it had begun. Like 1870, 1878 was an open year in the Arctic, preceding a disaster. The fleet was able to push on as far as Camden Bay, 250 miles east of Point Barrow, and it is possible that Ebenezer Nye worked the *Mount Wollaston* even farther, approaching the Canadian border. The fleet pressed ahead toward the Mackenzie River to search for the whales they thought they would find there. They paid a price for their haste: the *Florence* and the Hawaiian trading brig *William H. Allen* were lost early in August at Point Barrow.

In 1878 the whales did not make their appearance until late in the season. The fleet did not catch more than seventy-five. Many of those were captured in September and October near Herald Island, and the ships that lingered longest there made the best catches. Because of the autumn season of 1878 it must have seemed that the way to make a good catch was to remain very late in the autumn in the northern Chukchi Sea, deep in the cul-de-sac called the Hole.

The owners understood that they were asking their captains to push their ships to the limits of their technology. Shortly after 1875 they began sending tenders north to meet their ships in July. The tenders provided the whaleships with fresh supplies and communications from the owners, but their essential task was to carry south the oil and baleen that the ships had caught to that date and thus to avoid loss should the whaleship herself be lost in the dangerous late season.

In 1879, when the ships found heavy and dangerous ice still near Icy Cape at the end of August, Captain Frederick A. Barker, who had lost the *Japan* at East Cape in October 1870, addressed himself to the whalemen's dilemma of making good catches at the expense of safety: "It is infectious," he wrote, "this running all danger that five years ago not a man in the fleet would dare to run. Now one starts blind and the rest follow, giving him courage to go ahead & the rest that follow, the same."

In that statement he foretold the tragedy of October 1879.

In the autumn, a great triangular tongue of ice usually projects down into the center of the Chukchi Sea from the Arctic Ocean. On either side of this central tongue there is a band of clear water, kept ice-free by the warm northerly currents flowing up from Bering Strait. One branch of this current flows north over Herald Shoal, then past Herald Island and on into the Arctic Ocean; the other flows northeast along the coast of Alaska. This tongue is larger or smaller depending on the warmth of the summer and it wags east or west depending on which way the wind forces it. The whalemen called the tongue's southern tip "Post Office Point" because it was there that whaleships cruising back and forth between Point Barrow and Herald Island often met and exchanged mail and information.

The fleet moved around Post Office Point in September 1879 hoping to make a good catch out of an otherwise unproductive cruise and to duplicate the catches made there the year before. In former years the ships had always left the Chukchi Sea by the end of September to avoid the advancing ice, violent gales, and increasing cold of late autumn. They knew that in September the tongue began to move southward and that the southeasterly current running along the north coast of the Chukchi Peninsula toward Bering Strait often brought with it large quantities of young ice. With strong north winds this ice could be forced right into the Strait itself, plugging it like a cork in a bottle.

Nevertheless, in 1879 many of the ships lingered near Herald Island well into October in deepening cold, searching for whales in waters which, in a period of calm and intense frost, could be covered by six inches of ice in twenty-four hours. One by one the ships bore away to the south until on October 16 only the *Mount Wollaston, Vigilant, Mercury,* and *Helen Mar* remained. For safety the *Mercury* and *Helen Mar* formed a pair, keeping an eye on one another, as did the other two. On the eighteenth Captain George Bauldry aboard the *Helen Mar* began preparations to go south. He was anxious to leave, for the nights were by then very long and very cold. He held up for a day, however, to let the *Mercury* finish cutting a whale.

The two stood south on the nineteenth but as they neared the Chukchi Peninsula, they found a vast field of young ice blocking their way. Assuming they could work around it, they steered northeast to prepare to come down to the Strait on the other side of the sea. Two days later they reached Herald Shoal and then steered southeast toward Point Hope. They first found a few scattered pieces of ice. Later these increased in number. Then young ice began to form between the pieces. First it was just a scum, but it gradually thickened and on the twenty-second it brought them to a halt. At the same time, the northeasterly current began to carry them away from Bering Strait. Hoping the ice would wash past them, they dropped their anchors, but they had been caught: the northern pack, the tongue, had met the western pack and they were caught in the Arctic's maw.

Captains Bauldry and Hickmott knew that two ships together had a better chance of survival than one on her own. When the *Mercury* began to drag her anchors, they made her fast astern of the *Helen Mar* using cutting falls to warp her in. Soon, however, it began to blow hard from the northeast. Overjoyed, they hove up their anchors to stand before the wind, hoping to break out of their prison. They found the *Mercury* could only move about two ship's lengths and the *Helen Mar* could not move at all.

One can imagine the terror the crews must have felt, isolated as they were so far from land. The night was calm and very cold, and the next morning the captains decided their only choice was to prepare to winter where they were. They had only a few months' provisions aboard the ships, but it was better than nothing; and they still had a great deal of blubber on the *Mercury*. They found that the new ice was so thick by then that it would support a sled with a ton or more on it. They worked like demons for two days, fearing that the *Mercury* might be sunk or swept away from them. On the twenty-fourth they had most of the supplies transferred and Captain Hickmott and his wife and young daughter moved aboard the *Helen Mar*.

The temperature dropped even further.

"I had never before imagined such cold as we now experienced," said Captain Bauldry. "We had a fire in the forecastle, two in the blubber room, and one in the cabin, yet every night frost an inch or more thick would form on the inside of the ship."

Captain George Bauldry of the Helen Mar

The next day, miraculously, the wind swung into the north and began to blow a gale, breaking up the ice. They cut themselves free of the *Mercury* and set all sail. With the spars and rigging straining, slowly the *Helen Mar* began to gather momentum. For two days they madly forced her ahead through the slush and the floes, moving at only a knot or two in the gale, pushing the ship to its limits and giving it a fearful battering. She was covered in ice, and it was the ice sheathing that saved her from smashing in her bows or simply wearing the planks away.

The Helen Mar *in the Arctic, ca. 1891* Presbyterian Historical Society, Philadelphia

Near Point Hope they finally broke out into clearer water, but almost at once a new gale, stronger than the first, struck them. It was so cold that the frost smoke rising from the water made it impossible to see from the deck. Bauldry went to the foretop to direct her from there. The temperature dropped to three degrees Fahrenheit and the ice coating on the ship grew thicker. "We could only steer with the sails, as the rudder was held solid by ice," Bauldry remembered. "The stem was a mass of solid ice even to the catheads, and around the ship at the water's edge was a regular platform of ice, so that you could walk around the vessel."

On the twenty-eighth they saw the Diomede Islands in the middle of Bering Strait twenty-five miles to the southeast, and, in apparent safety, they waited for a favorable wind to push through the Strait against the strong northerly current. Bauldry described their exhaustion.

No human being can imagine what we had gone through, or what anxiety of mind I suffered with all these lives dependent upon me. Neither the mate, Mr. Carter, or myself had had a bit of sleep for five nights or days, and I feared he would die, he was so nervous. To induce sleep, I gave him a large amount of laudanum, and he lay with both eyes wide open, yet in the deepest sleep. Had there been fifteen miles more of ice to go through, we certainly would have gone to the bottom. The ship could not have stood the strain. As it was, she was almost cut through on the bluff of both bows.

When the wind looked good for passing the Strait, Bauldry gave orders to begin beating south and, at the limit of his endurance, fell into his bunk. Soon the ship entered a thick snow storm. Unknown to the men, the strong current began to set them toward the shallows that run north from Cape Prince of Wales for thirty miles. at 6 A.M. the ship struck hard on the shoal, ripping off her false keel and forefoot. Bauldry woke at the first shock, knowing what had happened. He fought his way on deck as the crew raced around trying to lighten the ship, frantically throwing overboard everything they could get their hands on, while the *Helen Mar* pounded harder and harder. She began to list steeply, and Bauldry could see that at that angle she was drawing less water. He ordered the crew to back her sails, and with the wind screaming around them they backed her off the shoal.

They passed through Bering Strait on October 30, one of the latest passages a vessel ever made in the nineteenth century, and reached San Francisco November 26, a month after they had been given up for dead.

When they hauled out the worn and racked *Helen Mar* for inspection they found that her bow timbers had been badly started, her bottom timbers shaken, her false keel, forefoot, sheathing, and copper had been torn off, and at one place in her bows the ice had cut through all but an eighth of an inch of her planks.

The *Mount Wollaston* and *Vigilant* never arrived, and nothing was learned of their fate the following year. It was not until the summer of 1881 that the story emerged. A party from the U.S. revenue cutter *Thomas Corwin* reached Cape Wankarem on the north shore of the Chukchi Peninsula and learned from the natives that in the autumn of 1880 a dismasted ship had drifted close to shore. The natives boarded the wreck and noted that the masts seemed to have been cut away for firewood. Below they found four blackened, emaciated corpses. They collected a few articles and intended to return the next day but that night the wind changed, and in the morning the wreck had disappeared from view. From the description of the ship—a rack of reindeer antlers was fixed to the jib-boom—and from a few of the things the Chukchis had carried away, the Corwin's men identified the vessel as the *Vigilant*.

The *Vigilant* had been carrying enough supplies to winter over at Point Barrow, and it was assumed that she and the *Mount Wollaston* had been trapped farther north than the *Mercury* and *Helen Mar*. It seemed likely that the men had abandoned the *Mount Wollaston* to live aboard the *Vigilant*, for they identified Captain Ebenezer Nye's spectacles.

9 / *Trade*

When Thomas Roys reached Bering Strait in 1848, although he was after whales, he found that the Asiatic natives were "well supplied with valuable furs" and eager to trade them for manufactured goods. Roys assumed at first that the natives had been totally isolated, but they were in fact already experienced in foreign trade.

For at least two thousand years a well-integrated trade network had existed, linking peoples at great distances from one another in Asia and America via the conduit of Bering Strait. Iron and other metals found their way to Alaska and presumably were traded for furs, ivory, and other sea mammal products. However, when the European world began to come into contact with the Bering Strait region at the end of the eighteenth century the volume of the trade intensified. In 1789 the Russians established a trade fair at Anyui on the Kolyma River, 800 miles west of Bering Strait. At that time the fur production in Siberia was falling off, and Anyui's purpose was to obtain American furs—fox, marten, lynx, bear, wolf, and wolverine. Because of the great distances involved, groups of native middlemen grew up, passing the goods from one to another, until in extreme cases European and Asian products, principally beads, reached the Eskimos of the Mackenzie River delta.

The experiment at Anyui worked so well that in 1816, for instance, when Captain Otto Von Kotzebue of the Imperial Russian Navy entered Kotzebue Sound searching for a northwest passage, the Eskimos there asked for tobacco and appeared to be well acquainted with trading. In 1778 Captain Cook had found the American Eskimos ignorant of tobacco.

Kotzebue's voyage had other results. By that time a number of American vessels were criss-crossing the Pacific for the fur trade, and at least two of them capitalized on Kotzebue's information, as the Russians discovered in 1820 when another of their discovery ships reached the Sound. The Russians were surprised to find that the Eskimos asked them for gunpowder and lead in return for their furs. One of John Jacob Astor's ships had been there the year before, trading firearms. The Eskimos attacked the Russians, and after shooting a "cloud" of arrows, they shouldered muskets and began firing. Only the Russians' cannon kept the Eskimos from overwhelming the ship.

By that date, the natives living near Bering Strait had established themselves as the principal middlemen in the trade for Anyui. But their control of the flow of foreign goods in that area did not last long. In 1833 the Russian-American Company, which held a trade monopoly from the czar for all of Alaska, established a fortified trading post at Saint Michael on Norton Sound, only 250 miles from Bering Strait.

Situated as it was, near the mouth of the Yukon River, the post was well placed to intercept the furs that were carried down the river and then on to Bering Strait.

The Bering Strait natives immediately felt this incursion on their monopoly. In response, a large number of American Eskimos attacked the fort in 1836, unsuccessfully trying to kill the interlopers who had disrupted their trade.

From the date of the establishment of Saint Michael, it is probable that the Russian-American Company's annual supply ships continued on northward to trade with the Bering Strait natives. Thus when the first whaling fleet reached Bering Strait in 1849, the Yankees found natives who were far more sophisticated than they in dealing with foreigners.

Mary Brewster, the wife of the captain of the *Tiger* of Stonington, Connecticut, described one of the first encounters between Eskimos and whalemen near Plover Bay. She was just falling asleep at 11 P.M. on June 25, 1849, when she was jolted awake by one of the officers shouting: "The bloody indians are coming!"

We did not feel much pleased as we preferred to form our first opinion by day . . .

I asked husband what he should do. Said he should not allow them to come on board till he could see something of them—a few spades was got down—four old muskets was loaded (neither of them though would go off . . .) They came along side and when motioned to leave and go back, they went and we went to bed.

This morning several canoes came off and the first one which came said "good morning" with the word "English" [thinking, no doubt, that the *Tiger* was a vessel of the Royal Navy searching for Sir John Franklin's lost expedition; HMS *Plover* had spent the winter of 1848-49 there]. We said "America." They wanted "Towack" and not taking their meaning, a slight tap on their face with one of their number speaking more plain "Tobach," the tobacco was got and liberally distributed and the great joy exhibited by them showed they were well acquainted with the weed.

They had their women with them . . . and we all concluded they were harmless . . .

A day later eleven umiaks towed the *Tiger* to anchor in Plover Bay.

Plenty of native company. We have had company from the whole coast I believe: a chief by the name of Notochen [from Indian Point] with his wife and child who is the greatest man of the nation, I expect. We were caressed, touched noses together and as near as we can understand, they are to be our friends.

They were well dressed in seal skin trowsers and a coat made from deerskin made loose and belted round them. Women dressed the same fare. Their hair is long and braided. Two of the women were very pretty. Their skin, which had not been exposed, was as white as mine, black eyes and red cheeks. . . .

I suppose they had never seen a white female before. They brought presents of their garments and walrus teeth for which we paid them with tobacco. They are all smokers and chewers—even the children—and are extravagantly fond of it.

Their chief food is seal's flesh, walrus, and blubber with fish. The blubber is cut up into slits and I could not help laughing to see a child about 6 months old in the bottom of one of their canoes lying on its back with a strip of blubber as much as it could hold in both hands, sucking it and apparently very happy.

They stopped nearly all day. We gave them bread, combs, needles, thread and knives, and at 4 they left well pleased with their visit.

Clearly by 1849 the natives of the Chukchi Peninsula were comfortable with foreigners and showed no surprise at the presence of their vessels.

Their friendship was demonstrated that year when, on August 2, the whaleship *Richmond* of Cold Spring, Long Island, went aground and was lost twelve miles south of Saint Lawrence Bay. The natives helped the whalemen in their unsuccessful attempt to refloat her and then cared for them for four days until they were rescued. When in 1852 the *Citizen* of New Bedford was lost near East Cape in a terrible autumn gale, the crew was housed and kindly treated for the entire winter.

Despite this amity between whalers and natives, Mary Brewster foresaw problems. On August 3, 1849, near East Cape she recorded:

> Several of the natives came on board. I saw many a familiar face amongst them, having seen them down in [Plover Bay]. They brought several articles of no value and a few common furs, which we bought for tobacco. They are very sly and will cheat in trading. Will show all their poor stuff first and if they can not pass it off, they will soon show their best. They will not let a thing come up to the ship till they are paid for it. Their demand is usually for rum but as we have no liquor, they have to get it somewhere else.
>
> I think if the whaling continues good here they will not be the happy people they are now. They will naturally learn many ways which will not be for their good. Liquor will only cause them to quarrel. It's a great pity that they should . . . get accustomed to the use of it and I wish we had more temperance ships and more temperate commanders so the influence could be better.

The whalemen are usually accused of having introduced alcohol to the Bering Strait region, but Mary Brewster's remarks show that the Eskimos had a well-developed taste for rum before the first whaleships arrived there. How they developed this taste is more difficult to determine. Probably not a great deal of alcohol was carried 800 miles from the Anyui trade fair, and the Russian-American Company prohibited the sale of alcohol in Alaska. Although HMS *Plover* spent the winter of 1848-49 in Plover Bay, it is improbable that the ship was the source of a significant amount of spirits. Rather, it seems likely that in the 1840s a number of trading vessels from southern ports began visiting the Bering Strait region, seeking furs in return for the cheapest cargo they could carry—rum.

The first vessel to winter in the Bering Strait region was the trading brig *Swallow* of Hong Kong. In 1850-51 the *Swallow* wintered at Saint Lawrence Bay, suggesting that the traders had knowledge of the geography and native inhabitants before the whaling fleet first reached the area.

In any case, by the summer of 1851 at least six small trading vessels were operating in Bering Strait, the *Lady Franklin* and one other schooner from Australia, the *Eliza* and the *Rena* whose ports are not recorded, the brig *Juno* of Honolulu, and the *Koh-i-noor* of Hong Kong, which was "trading for walrus teeth and furs, with plenty of rum on board."

For most of the whalemen the early encounters with natives on the coast of the Chukchi Peninsula were pleasant. Again and again the logbooks record that the natives appeared friendly. The whalemen, who often did not care to trade, simply provided the natives with needles, pins, iron, and other small manufactured goods *gratis*. To the native inhabitants of the area, where a finely developed ritual of reciprocity existed, this behavior was at first puzzling; later they grew to consider it their right to receive gifts when they visited a ship. If they did not receive such things, then they asked for them—and thus grew up the foreign perception that most of the natives of the Bering Strait region were bothersome beggars.

Those who traded with the natives in those early years usually bought furs, ivory, native clothing, fish and birds for food, blubber, and whalebone. The natives' first requests for payment were often for rum, tobacco, and knives. By the summer of 1852 it seemed to many whalemen that the natives' main desire was for rum, but it is unlikely that the whalemen were able to supply much alcohol in this early period. At that time whaleships carried, at most, very small quantities of alcohol, mostly private stocks of the captains and officers.

Nevertheless, rum was the cause of the first ugly incident between whalemen and natives in the Bering Strait region. On July 10, 1851, the ship *Armata* of New London was running with all sails set along the edge of the ice on the east coast of the Chukchi Peninsula near Big (Arakamchechen) Island. At about 10 A.M. she struck hard on Brooke Bank. The crew was unable to free her and the ship began to go to pieces. Her main and mizzen masts fell the next day. Several ships soon gathered around her. On the thirteenth, while an Australian ship was helping to salvage her gear and cargo, the *Armata* was boarded by a large group of natives who, in the confusion, immediately made for the captain's supply of rum. The natives may well have been drunk on arrival, for the *Koh-i-Noor* had recently sold them "plenty of rum." In any case they were soon visibly drunken and began plundering the wreck.

One of the sailors forced one of the natives to give up something he was carrying off, and in return the native stabbed the whaleman with his dagger, mortally wounding the man. Immediately a fight broke out. The whalemen defended themselves with the only weapons at hand: harpoons, lances, hatchets, and cutting spades. When it was over, eight Eskimos and one sailor lay dead. The whalemen buried their shipmate on the shore of Seniavin Strait.

This story spread quickly and the whalemen were on their guard, but some whalemen or traders seem to have ignored the lesson. There were a number of descriptions in logbooks from the years following the *Armata* incident that report natives on the Asian coast coming aboard whaleships "drunk" and "saucy." The captain of the *Julian* wrote in 1859:

> Come near having some trouble with [the natives] while my boats were off, after a couple of chiefs went on board of the Good Return and Capt Fish gave them some rum and they got a little tight and wanted more rum and [he] wouldn't give them anymore, and they come on board of my ship and wanted rum. I saw they were half drunk and told them no and made motions that they were drunk. They said no and coaxed a long time but no use, I wouldn't let them have any and they told me if I didn't let them have some rum, they would go on shore and cut my raft rope and destroy my [water] casks. I went down below and loaded my revolver and went on deck and told them to go and cut it if they wanted to and showed them my revolver. I heard no more about rum or raft ropes after that day. If they had done it I should have shot everyone that had a hand in it.

In the early years of the fishery, despite unfortunate encounters like these, trade continued between the whalemen and the natives of the Chukchi Peninsula at a generally low level—profits were easier made at sea from the abundant bowhead population. The bulk of the foreign trade in the Bering Strait region at that time was carried on by about half a dozen brigs and schooners, trading vessels from Hawaii, San Francisco, and other ports.

Most of these vessels cleared port as whaleships, but in fact they were primarily fur traders. They touched at all the settlements in the Bering Strait region, trading rum, tobacco, and hardware. Unfortunately, there is little information on the extent of their activity. It may have been considerable, however. In 1854, for instance, the *Pfiel,* a little German-owned schooner from Honolulu, brought home 4,000 pounds of ivory, 3,000 pounds of whalebone, and "200 or 300 skins—marten, sable and others." In 1858 the brig *Agate* of Honolulu landed 22,000 pounds of ivory and 1,200 skins. If a small vessel took a whale as well, then the oil and baleen covered the costs of the voyage and the results of trading were pure profit. Rum was an essential item for this trade, as one whaleman remarked at East Cape in 1852: "one canoe of indians came along side to trade. They had skins, walrus tusks, etc. but want nothing but rum, once in a while some tobacco but will take nothing else if they can get rum and you can get anything they have got for a drink of rum. Two good drinks will get them drunk and then they are happy." Six years later, at Saint Lawrence Island, a whaleman from the *South America* found that "the bone from a bowhead was packed alongside of the [chief's] hut which he would not trade, unless he could obtain rum."

When the whaling fleet deserted the Bering Strait region from 1855 to 1857, a handful of brigs and schooners maintained contact with the region, but when the fleet returned in the late 1850s the complexion of the whalemen's trade had changed. The price of baleen had doubled and many more captains found that trading for it made sense. In 1859 one captain discovered that he could "buy a hundred dollars' worth of bone or fur for a gallon of rum which cost 40 cents at home."

One result of this discovery was that the whalers' trade, which had been more or less restricted to the Chukchi Peninsula, spread to the American side of Bering Strait as well. They had previously avoided trading on the East Shore. The whalemen believed that the American Eskimos were truculent, irascible, and treacherous. As many early explorers found, the American Eskimos' aggressive behavior increased in direct proportion to their numbers. When they outnumbered the white men, the American Eskimos often tried to take their goods by force.

At Point Hope, for instance, before 1880 the Eskimos were considered "quite wild and very rude and had to be watched." There were at least four instances when large numbers of Point Hope Eskimos surrounded boats' crews, pricked their throats with knives, cut the buttons off their clothes, stole tobacco from their pockets, and the like. At least once they tried to take a whaleship by overpowering the crew—as they did aboard the *Cornelius Howland* about 1870. A number of visitors to Point Hope echoed the remarks of Captain Calvin Leighton Hooper of the Revenue

Marine: "They are lazy, filthy, worthless, and dishonest and require constant watching."

The Russian-American Company had been under no illusions about the intentions of the Bering Strait Eskimos. In the mid-nineteenth century its vessels were armed with cannons, guns, and sabers. American whaling vessels, by contrast, carried few, if any, weapons. Although cautious about these bold natives, the whalemen pressed ahead with the trade and by 1859 six whaling vessels, at the very least, traded at Port Clarence for furs and ivory in return for "guns, rum, powder, knives." It was about this time that the whalemen began trading firearms, finding them second in demand only to rum and tobacco.

The whalemen then began expanding their trading activities northward, gingerly touching at Point Hope and beyond—places they had previously avoided. They found, however, that from Point Hope onward the quality of trading goods was generally low—perhaps the natives were still transferring their best goods at the native trade fair at Kotzebue Sound where Asian Eskimos and Chukchis met American Eskimos from all over northwestern Alaska. When the *George and Susan* touched at Icy Cape in 1859 and became one of the first whaleships to attempt to trade there, the men found that "the inhabitants had no furs & but little of anything except canoes and dogs." Similarly, at this date there are no records of the Alaskan natives north of Kotzebue Sound asking the whalemen for alcohol or firearms. The trade must have been restricted to the Bering Strait region.

Nevertheless, it did not take the American Eskimos long to develop a taste for manufactured goods. Port Clarence, formerly the site of a native trade fair, quickly became a major trading rendezvous, a place for the natives to meet whaling and trading vessels each summer. The price that the white men charged for their manufactured goods in the western Arctic soon stabilized because the competition for bone, ivory, and furs was intense and the competition worked in the natives' favor.

The captains also found that unless they arrived early in the spring the best of the trade would already be gone. To get the better pieces some of the captains began outfitting native traders with goods so that they could trade for valuable items all winter. Captain Frederick A. Barker of the schooner *Leo*, for instance, in 1878 staked "Cooley," a native of Plover Bay, with thirteen barrels of rum and more than seventy pounds of tobacco. A few white men simply began living ashore for the winter in settlements on the Chukchi Peninsula or at Port Clarence.

Initially, a more effective way of getting both whales and trade goods was to winter a ship in the Arctic. No doubt the whalemen had learned from the natives that from late October to early December the bowheads travel south from the northern Chukchi Sea and then swim along the coast and often into the bays of the Chukchi Peninsula. In the spring, from April to early June, on their return to the Arctic the whales pass back along the coast.

From the winterings of British naval vessels in their searches for Sir John Franklin's lost Arctic expedition (1848-54), it was well known that a ship could be safely frozen into the ice of an Arctic harbor and that the crew could live through the winter without extreme hardship. In any case, the new Bedford bark *Cleone* and the German-owned brig *Wailua* of Honolulu entered Plover Bay in September 1859 to begin the first wintering voyages by whaleships in the western Arctic. By then a number of whaling masters had considered the idea of wintering either on the Chukchi Peninsula or at Point Barrow, but most had rejected the plan, believing that the risks of scurvy and hostile natives outweighed the chance for modest catches.

Lacking the experience and resources of the Royal Navy, these pioneers suffered appalling hardships. Beginning in mid-October, the whalemen saw a number of whales but the young ice prevented the *Cleone*'s crew from getting more than two (one on November 18 and one on December 8). The *Wailua* took none. One of the *Wailua*'s boat crews was blown away in a late autumn gale, and when a search party found the boat, only three men remained alive.

During the winter at Plover Bay the temperature fell to -29° F. Even with four stoves going constantly on each ship, the ice formed everywhere on the inside of the hulls. Scurvy then appeared and progressively debilitated many of the men on both vessels. To make matters worse the whalemen ran out of firewood during the winter and had to burn their spare spars for warmth. When the whales finally began passing again early in April, the men found that the open water was many miles from the ships and that they were incapable of dragging their boats to the lead edge. The *Wailua* lost fifteen men from scurvy; the *Cleone,* two. The death-toll would have been much higher were it not for the advice and help of the natives. The providential arrival of several whaleships in mid-June allowed the men to pull through. The whaleships sent medicine and fresh vegetables to the wintering vessels across two miles of solid ice.

The ships' crews began to cut channels to the open water in early July. Even so, the ships did not get out until July 14. By then the whales had passed them. Together, the two ships took only 200 barrels of oil, some of which they had been forced to trade to the natives for food.

The whalemen digested this gloomy news over the following winters, but by early 1861, the bark *Coral* of New Bedford was ready again to try wintering in Plover Bay. Captain Sisson of the *Coral* was far better prepared and his strategy was different. Upon reaching the bay in the autumn he sent his summer's catch to Hawaii aboard the *Thomas Dickason* ensuring that should the *Coral* be lost, something would be salvaged from the voyage. Most important, he carried a full supply of trade goods which allowed him to buy food and fuel from the natives. By the end of the winter he had bought 9,500 pounds of whalebone—principally in return for rum and tobacco. The men saw plenty of whales that autumn, but there is no report of any having been taken. It did not matter; their "trade bone" amounted to the yield of four or five whales.

The cold of 1861-62 was not as severe as the winter of 1859-60 (it only reached -20° F.), but the weather was more brutal. Hurricane-force winds broke up the ice in the harbor several times and blew it away. No doubt this both destroyed the ship's thermal housing of snow and temporarily isolated the *Coral* from the shore, where one would assume her boats were stored. Nevertheless, the crew enjoyed good health all winter and the bark broke out of Plover Bay at the end of June.

Despite the *Wailua's* tragic winter, the idea of wintering voyages must have also appealed to the German shipowners who registered their vessels in Hawaii. In 1862 they repeated the experiment. In the autumn the bark *Zoe* of Honolulu (under the command of Captain John Simmons, who had commanded the *Cleone* on her awful winter) went into winter quarters at Plover Bay. At the same time the *Kohola* of Honolulu settled in to Saint Lawrence Bay. To keep their food requirements low both ships sent a number of men south to Honolulu aboard the trading brig *Victoria.*

For the *Kohola* the winter was marred by violence. A misunderstanding with the natives about the accidental death of a native leader resulted in the murder of Captain Brummerhoff. Barney Cogan, then first mate of the *Kohola,* described the tragic events:

> Soon after we anchored the sailors went ashore, stole some whisky from one of the native huts, got drunk, and came aboard and resolved to take the ship. We put them in irons and under guard. The native of whom they stole the liquor came aboard afterward and remained two or three days. Meantime a strong gale sprang up. Some of the sailors told him they had seen the wind carry off his hut and destroy everything. This set the fellow crazy. He insisted upon going ashore. We knew it would be impossible to land him, and tried in vain to dissuade him.
>
> While we were at dinner one day, he jumped overboard and started to swim ashore. I threw a line to him, but he brushed it away and started off. He had almost reached shore, when he encountered young ice, lost his strength and was drowned, his body never being seen again.
>
> After the gale subsided, his father and two brothers came aboard inquiring for him. We told them the truth, but the sailors, who sought revenge on the captain, told them that the captain had stabbed him and thrown the body overboard. They ignored our story and believed theirs. The Eskimos then told the captain they would kill him if he came ashore. On account of this affair, cheating in trade, giving poor rum mixed with pepper, etc. a strong hatred naturally sprang up among all the natives against him, and they refused to come aboard and trade.
>
> While I was on one of my expeditions, the captain traded for six deer. The natives took their trade—a keg of rum—in advance, and went ashore to get the deer, which were inland. I returned at this juncture, and the captain asked me to go and get the deer; but when I found what trade he had given them, I suggested that we wait until the liquor and its effects had disappeared. He then said he would go himself, and, inspite of warnings from the rest of us and the friendly natives, he started off.
>
> We followed him with the glass and soon saw there was trouble. We could see the team returning in great haste and a crowd following. Afterward we learned that the friends of the man that had been drowned and a few other natives, incensed at the captain's treatment of them, had followed his team and sought revenge. The captain fired at them with his revolver, then threw it away and fled toward the ship, but was soon overtaken, pierced by an arrow, and then stabbed to death. I endeavored to recover the body, but could not find a trace of it, the natives saying that it had been given to the dogs; but I recovered his clothing. The murderers then endeavored to induce all the natives along the coast to join in an attempt to seize the ship, but the plot never came to a head.

During the late 1860s and early 1870s a few very small schooners from San Francisco wintered on the coast of the Chukchi Peninsula, but the lessons were the same: the profits were not substantial enough to offset the disadvantages.

At the same time, however, the price of whalebone remained relatively high. The trade for whalebone, as well as for ivory and furs, became an important source of profits to the captains and officers of the whaling vessels, who usually were trading on their own account. The number of vessels working the Bering Strait that were primarily traders increased to a dozen or so. In addition to alcohol and tobacco, all types of firearms were traded (in the early years flintlocks, later percussion caps, then breech loaders) as well as a wide variety of other manufactured goods. The trade quickly spread throughout the region and by the early 1880s, for example, the Eskimos at Point Barrow had revolvers and the most up-to-date fifteen-shot Winchester rifles. By the same date the bow and arrow had become an historical artifact throughout the Bering Strait region.

The trade goods were carried far from Bering Strait via the well-integrated native trade networks. For example, the principal points of dispersal in Alaska were the native trade fairs at Port Clarence and Kotzebue Sound. There the ships met Eskimos from the river systems that drain into Kotzebue Sound, from the Seward Peninsula, from coastal points south and north of these regions, from the islands of Bering Strait, and even from coastal points on the Chukchi Peninsula.

The goods were then carried across the Brooks Range to the coast of the Beaufort Sea, where they were dispersed again at trade rendezvous at the mouth of the Colville River and at Barter Island. They were then passed on to Eskimos from farther east (and even to some Indians) who carried tobacco, pipes, blue and white beads, and large iron pots into western Arctic Canada. A small amount of these goods were possibly passed even farther, a thousand miles beyond the Mackenzie River delta to the Netsilik Eskimos of the Central Arctic.

From the Bering Strait region trade goods were also dispersed southward, into the Yukon drainage. The Siberian natives carried the goods westward to the Anadyr River and toward the Kolyma. They also carried on a subsidiary trade by bringing their colorful and highly mottled reindeer skins to the American shore, where they were highly prized (caribou, which is essentially an American reindeer, has only monochromal pigmentation). The Siberian traders often returned with muskrat, beaver, and fox furs, some of which had originated with the Koyukon Indians. These furs either were traded to the whalemen on the coast of the Chukchi Peninsula or they were carried westward to the Siberian trade fairs.

It appears that a few of the coastal natives on the Chukchi Peninsula maintained stocks of alcohol year round. There are reports of American Eskimos crossing Bering Strait in mid-winter, traveling over the dangerously moving ice floes to bring furs to East Cape to trade for alcohol. When the exploring ship *Vega* was wintering on the north coast of the Chukchi Peninsula in 1878-79, for example, A. E. Nordenskiold found the natives drinking American gin from a tin cup with the inscription "Capt. Ravens, brig Timandra, 1878." He was told that alcohol was available year round at Bering Strait.

In view of alcohol's well-known potential for causing problems, it is not surprising that serious trouble broke out between the traders and the natives. Throughout the

1860s and 1870s the ships' logbooks contain frequent reports of drunken and aggressive Eskimos coming aboard, but by all accounts the Eskimos from Cape Prince of Wales were an exceptionally aggressive group. John W. Kelly thought them to be "a band of hypocrites and shylocks, possessing a large share of brazen effrontery." Captain Calvin Leighton Hooper of the U.S. Revenue Marine considered them to be "the worst on the coast," "great bullies" who traveled "in large numbers compelling smaller bands to trade with them at their own terms." They were, according to Captain George Bailey of the same service, "a bad set, fond of rum." The ethnographer Edward William Nelson reported that "on several occasions the villagers of Cape Prince of Wales fairly took possession of vessels with small crews, and carried off whatever they wished."

This behavior, combined with their success at bullying foreigners, both native and white, may well have prepared the Cape Prince of Wales Eskimos for the tragedy of 1877. That year they had already tried to overpower two trading vessels. On one of them they succeeded in driving the crew below decks and then in robbing the ship's stores.

They met their match on July 5, 1877, in a tough group of Hawaiians. The trading brig *William H. Allen* of Honolulu had been working up the Siberian shore, and Captain George Gilley, "a half breed Kanaka" from the Bonin Islands, was taking her across Bering Strait when he was becalmed and surrounded by fog halfway between the Diomede Islands and Cape Prince of Wales. To avoid being set on Prince of Wales Shoal, Gilley dropped anchor. There were only twenty-four men aboard the *William H. Allen*, most of them Hawaiians. Gilley described what happened:

Soon a canoe load of Prince of Wales natives came along-side, and the chief waved a skin on a pole, indicating a desire to trade. When he got on board he wanted ammunition. I got some, and, after he had shot at cakes of ice for awhile, he asked me to give him five cartridges for his repeating rifle. This I refused to do, though I offered to trade. It was quite noticeable that he and some of his followers were under the influence of liquor.

The chief was about six feet five inches tall, by far the most powerful native we had ever seen. I knew that he was a murderous villain, and that his followers would do just what he told them to. This, in addition to the indications of liquor, put me on my guard. Meantime, two other canoe-loads came aboard, and with them was the chief that stood next in authority. All began to ask for rum, but I told them that I had none. They said that they knew I had, for all ships with two masts had it.

One fellow, apparently accidentally, fell overboard, and though a canoe was towing astern, the chief wanted me to lower a boat and rescue him. I said no. He then asked me to go into the cabin. Again I said no. At this, he grasped me by the throat, but when I drew a revolver, he let go, and stepping off, smiled as though it was a joke. Things indicated that there was to be trouble, so I stationed two men near me, each armed with a hand-spike. There was not a breath of air stirring, yet I ordered the anchor hove up. When the crew attempted to execute the order, the natives stopped them. Then the chief sent his wife with the other women and old men into the canoe. This meant a fuss. He then seized me again by the throat, and I told one of the men to tap him on the head with a hand-spike. The tap killed him.

The other natives were on the main deck and suddenly each drew a single-barreled muzzle-loading pistol and began to shoot and chase the crew about. Prompt action was necessary. I

George Gilley in 1887

knocked over the other chief with my revolver and called for my rifle. After the natives had emptied their pistols they resorted to big knives and stabbed one sailor in the back. I stood ready, and whenever I saw a native raise his knife I shot him. They had not expected this, and, balked in their attempt to take the vessel, they endeavored to escape. But their canoes had got adrift. A light breeze sprang up, and heaving anchor, we got under way. Their one desire seemed to be to conceal themselves, and all crawled under the t'gallant forecastle [sic]. I intended to take the survivors prisoners, carry them to the Siberian shore and land them, for they had been punished enough, but the instant I laid down my rifle, they tried to use their knives on me. Seeing no other alternative, I posted men above them, and when a native showed his head, he was clubbed and thrown overboard. Toward the last we hauled them out with gaff hooks. The three canoes had contained about twenty warriors, but not one of them escaped. Afterward I learned that these same men had looted Captain Jacobson's schooner a week before, and tried to take Captain Raven's brig.

When the fight was over thirteen Eskimos were dead. One Hawaiian had been killed and two wounded. In 1887 Gilley reported, "no attempt to take a vessel has been made since." Calvin Leighton Hooper thought that "the lesson taught them at that time seems to have had a beneficial effect." Nevertheless, for many years vessels gave Cape Prince of Wales a wide berth, and if they stopped there at all, they did not anchor and, instead, merely backed their sails while they allowed only a few Eskimos aboard at a time.

Nor did the Eskimos forget. In 1888 Charlie Brower, a shore whaler from Point Barrow, was nearly murdered in retribution. In 1893 the incident was likely a factor in the murder of Harrison R. Thornton, the first missionary to Cape Prince of Wales. And in 1897 Frank Boyd, a prospector working out of Point Hope was killed by an Eskimo whose father had died aboard the *William H. Allen.*

This sorry state of affairs spawned an attempt to control the traffic in liquor. In fact, by 1877 an attempt had been made—at least on paper.

Shortly after purchase of Russian America, Congress created a customs district for Alaska and "extended United States laws regarding navigation and commerce" to the new territory. An act of 1873 prohibited the sale of alcohol and breech-loading firearms to the natives. The latter restriction developed as a result of the Indian uprisings in the West. Nevertheless, there was no sustained effort made to enforce these laws until the U.S. revenue cutters began their annual patrols in 1879.

The Revenue Service men found plenty to keep them busy. About a dozen trading vessels departed annually for the Bering Strait region. The ships leaving from American ports were liable to be inspected by revenue officers when clearing port. To avoid legal difficulties, the captains either submitted a false manifest or claimed any alcohol and firearms were for trade on the Siberian shore (hence beyond federal jurisdiction). Many vessels simply called at Hawaii, then a sovereign nation, and took on supplies of contraband there. Their profits were assured because they could buy rum in Hawaii at 75 cents per gallon, then dilute it 50 percent with water, and trade it for a pound of whalebone, which in 1877, for example, was worth $2.50.

In the first years of their patrols the revenue officers were impressed by the pervasive strength of the maritime trade. At Saint Lawrence Island the natives seemed "perfect slaves to rum [who would] barter anything they possess[ed] to procure it and remain drunk until it [was] gone." William Healy Dall described the Plover Bay natives meeting a U.S. Coast Survey vessel:

> Two canoes . . . came out to meet us, it being the practice of the traders to give everybody a drink on their arrival "to facilitate trade." They asked for "lum" (rum) and were informed we had none. "No got lum! American schooner no got lum! (with rising indignation), Lie! Lum got!" was the exclamation. . . . However, much to their disgust, they were finally convinced we were not traders and departed, leaving us in peace.

At Kotzebue Sound in 1881 the officers were amazed at the sight of a small schooner "surrounded by umiaks three or four deep and the deck was crowded by a dense mass of the Eskimo. Tobacco, drilling, knives, ammunition and other small articles were used to buy from them the skins of reindeer, wolves, black bear, arctic hare, red, white, and cross foxes, etc." Although they found no alcohol on that particular schooner, they heard of one vessel charging one fox skin per drink.

When they seized the trading schooner *Leo* in Bering Strait they found in the hold twenty-six cases of alcohol, most of which was labelled "Bay-rum," "Jamaica ginger," "Pain-killer," and "Florida water." The schooner *Loleta* was seized at the end of her cruise in 1880 with twenty-nine Winchester rifles and 24,000 rounds of ammunition.

The annual Revenue Marine patrols were quickly effective. By 1884 they had succeeded in suppressing most of the flagrant whiskey trading on the American shore. This, however, merely gave the natives of East Cape, the Diomede Islands, and Cape Prince of Wales a better position as middlemen, for they frequently transported alcohol across Bering Strait to the American shore and traded it at Kotzebue Sound or Port Clarence.

At about the same time the Russian government became alarmed about the situation on the Siberian coast. Although the traders were being brought under control in Alaska, most of them redoubled their efforts to sell their alcohol on the coast of the Chukchi Peninsula so that they could arrive on the American shore with no contraband. This traffic was also in violation of Russian laws, but the trading vessels had little to fear from the Russians. The Russian patrols were irregular and largely ineffective.

Whether or not a particular ship made a practice of trading alcohol, by the early 1880s the natives on both shores of Bering Strait depended on whaleships and trading vessels to provide them with the manufactured goods upon which they had become dependent. Items such as firearms, ammunition, wood-working tools, needles, clothes, knives, and a hundred other things were by that date considered essential and a routine had developed in the trading procedures. Whaling and trading vessels annually passed northward along the coasts, beginning at Cape Olyutorsky on the West Shore or Nunivak Island on the East. If the natives wished to trade they set a flag or waved a shirt. The captain then either dropped anchor or hove to while the natives paddled out in their umiaks. When they reached the ship it was customary for the captain to provide each boat with a bucket of bread and some jam or molasses. Each wealthy native had a favorite captain with whom he preferred to deal, saving the best pieces for him and getting better prices in return.

The first request was usually for "lum" or "tanuk." "Tanuk" was pidgin, a generic word to encompass any alcohol. It probably originated with the wintering of HMS *Plover* in 1848-49 at Plover Bay. It is said that Captain T. E. L. Moore, when on the trail with his native dog-driver, would stop for a drink every day at noon, saying, "Come, Joe, let's take our tonic."

Many of the whaling captains refused to trade alcohol, but those who did usually provided the natives with a drink to make it easier to obtain their trade at the best price. The competition among ships was intense and the natives had developed such

The trading schooner San Jose *in the ice* Private collection

trading skills that they were able to drive very good bargains for themselves.

Next, the natives brought out their whalebone, ivory, furs, skin clothing and carvings—usually displaying inferior items first so that a price could be established before they presented their prime pieces. The variety of items traded on both sides was large. In 1887, for example, the whaleship *Francis Palmer* conducted the following transaction with the natives at Plover Bay. Sold were 5 boxes of tobacco from Honolulu, 2 packages of leaf tobacco, 1 Winchester rifle, 3 boxes of reloading tools, 2 dozen knives, 700 WCF rifle cartridges, 300 fifty-grain loaded cartridges, 2 pieces of white drilling, 27 bags of flour, 300 pounds of bread, 12 packages of matches, 4 hatchets, 2 axes, 1 saw, 1 mechanical toy, 1 bomb lance, 1 dozen thimbles, 1 dozen spools of thread, 1 dozen papers of needles, and 5 pounds of beads, in return for 237 white fox skins, 10 red fox, 1 polar bear skin, 1 brown bear skin, 92 pounds of ivory, 100 pounds of whalebone, 10 pairs of native boots, and 2 fur parkas. Considering that this is the report of just one whaleship out of the forty that were

Natives going out to a steam whaler at Port Clarence, Alaska, ca. 1886

north that year, one can imagine the volume of goods that were taken into and out of the Bering Strait region annually.

In this meeting of cultures a trade jargon grew up, allowing the foreigners and natives to communicate via a pidgin that incorporated words from English, Hawaiian, the Eskimo dialects, and many other languages. Here are some examples, drawn at random:

pickaninny—baby
kow-kow—food
pow—no, none
hana-hana—work
puni-puni—coitus
wahini—woman
myr-can—American
sabey—know

Sexual favors were also a profitable trade item, being one of the easiest ways for the natives to obtain alcohol. With the exception of the Diomede Islanders, who forbade their women to have intercourse with the foreigners, many native men on both sides of Bering Strait encouraged the practice to the extent that some observers reported that the native men had become merely pimps for their wives and daughters. Venereal disease became widespread throughout the area. By 1890 observers estimated that half the population of Point Barrow was "tainted" by secondary or tertiary syphillis.

Captain Barney Cogan and a Siberian friend aboard the Hunter, *1887*

Throughout the 1860s, 1870s, and 1880s, the native population of the Bering Strait region and northern Alaska was in general decline for a number of reasons, including the introduction of foreign diseases, the reductions of the whales and walruses, and natural declines in the caribou herds. With substantial reductions in numbers at most of the large settlements it became possible for a sole individual to accumulate enough power to dominate a settlement; whereas in earlier times, with a larger population, several men would have been in competition with one another, thus preventing the supremacy of only one.

In this period of time "strong men" are known to have consolidated power in a number of native groups, and some coastal settlements developed what might be called "trading chiefs," men who became the principal middlemen for the villagers' trade with all of the ships.

The most successful of these trading chiefs was Goharren of Indian Point, Siberia. At first he may have been staked by a whaler or trader, but soon he became a full trader in his own right. He amassed a great fortune in trade goods through both skill and deceit. At his zenith he owned three wooden store houses that were full of whiskey, firearms, ammunition, a wide variety of manufactured goods, as well as

Aboard the Alexander *at Port Clarence, ca. 1895*

whalebone, furs, and ivory. With these large stocks of foreign goods he was able to trade throughout the winter, obtaining trade at favorable prices, and often offering credit so that many Indian Point natives were in his debt more or less permanently. During the winter natives throughout the Chukchi Peninsula frequently made long journeys to trade with him, particularly for alcohol.

In 1891, for example, he had in his warehouses 200 sacks of flour, 80 boxes of tobacco, and ivory and whalebone worth between $5,000 and $8,000. He also sold second-hand whaleboats, outfitted four whaling crews, and owned 100 reindeer. At one time he had as much as $75,000 worth of whalebone stored away.

Chicanery was part of trading in Bering Strait, although it was practiced by comparatively few. Some captains carefully poured alcohol into containers that were partly full of water so that the alcohol rested on the surface. Others gave out samples of full-strength whiskey and then sold a heavily diluted mixture. In return, it was relatively common to find fox tails sewn on rabbit skins, or damaged fox skins cleverly patched with rabbit fur; broken walrus tusks riveted together with lead and the joints concealed by smeared reindeer fat; and stones set in the root canals of walrus tusks to increase their weight.

At Port Clarence, Alaska, ca. 1892

One trick among several that Goharren used was to put flat iron bars amid bundles of whalebone that he sold to the ships. On arrival in San Francisco a captain might find that his "trade bone" contained a wealth of iron sled runners.

Fate played a cruel trick on Goharren, however, for he lost much of his capital in an attempt to start his own pelagic whaling and trading operation. In 1886 Captain Benjamin Dexter reached Indian Point in his sixty-foot schooner, the *Henrietta*. Goharren at the time had two heads of whalebone (4,000 pounds of baleen worth almost $11,000—$120,000 in 1982). Goharren offered to buy the schooner from Dexter for the whalebone plus 800 to 1,000 pounds of ivory, 500 fox skins, and 3 polar bear

Goharren and his son, ca. 1892

skins. Dexter accepted the deal. Goharren had just set out on his first cruise, walrus hunting with six or seven natives aboard, when on May 17, 1886, a Russian gunboat, the *Kreisser*, appeared. She had been sent north to control the foreign trade. A party went aboard the *Henrietta* and found that it was a foreign vessel and that Goharren had not obtained permission from the governor in Vladivostok to be operating in those waters. The Russians confiscated the vessel. It was taken to the Sea of Okhotsk for use as a patrol vessel.

At Point Hope a man of power arose at the same time. Attungoruk was a man of exceptional size, strength, cunning, and gall. Like Goharren, Attungoruk became the middleman in much of the trade that went on with the ships. But unlike Goharren, who was ascendant for twenty-five years or more, Attungoruk's was a short reign, for he was far less mentally stable than Goharren. As Ernest Burch has said, his power corrupted him; he became a tyrant, murdering four or more people and taking other men's wives until he had five. He aspired to be as powerful as Goharren, about whom he had heard from the whalemen.

By the summer of 1881 it was already apparent to two observers that sooner or later someone would murder Attungoruk. Eight years later, on Valentine's Day, 1889, Attungoruk was assassinated in his house as he lay in a drunken stupor between two of his wives. His murder was part of a blood feud that claimed more than a dozen lives.

Oddly enough Attungoruk's death coincided with the beginning of a period of growing stability in the western Arctic. The nearly constant presence of the U.S. revenue cutters, as well as the beginning of missionary settlements in the villages made it more difficult for whiskey to be traded with impunity. At the same time many of the whalemen had concluded that it was simply not in their best interest to trade alcohol to the natives. By 1901 only two whaleships out of the ten that visited Indian Point regularly traded alcohol.

The decline in the whaleships' trading activity stemmed from several factors: the decline in the amount of whalebone that was being taken by the natives, a sincere spirit of temperance on the part of many captains, and hard and fast rules against the trade by some whaling companies, particularly the Pacific Steam Whaling Company. The natives had also developed new and more varied tastes, and they were able to satisfy their desire for alcohol by making it in stills.

Strangely enough, toward the end of the nineteenth century a substantial part of the alcohol that was traded on the Chukchi Peninsula began to be taken west and sold to the Russians and interior natives, reversing a trend that started more than a century before. Even more surprising was the fact that the natives began to sell moonshine to the whalemen. Walter Burns remembered one encounter when a Siberian Eskimo, "drew from beneath his deer-skin coat a skin bottle filled with liquor and sold it to us for fifteen hardtack. Wherefore there was, for a time, joy in the forecastle—in limited quantity, for the bottle was small. This product of the ice-bound North was the hottest stuff I ever tasted." He continued:

> The boat's crew found on the beach a little distillery in comparison with which the pot stills of the Kentucky and Tennessee mountains, made of old kitchen kettles, would seem elaborate and up-to-date plants. The still itself was an old tin oil can; the worm, a twisted gun barrel; the flake-stand, a small powder keg. The mash used in making the liquor, we learned was a fermented mixture of flour and molasses obtained in trade from whale ships. It was boiled in the still, a twist of moss blazing in a pan of blubber oil doing duty as a furnace. The vapor from the boiling mash passed through the worm in the flake-stand and was condensed by ice-

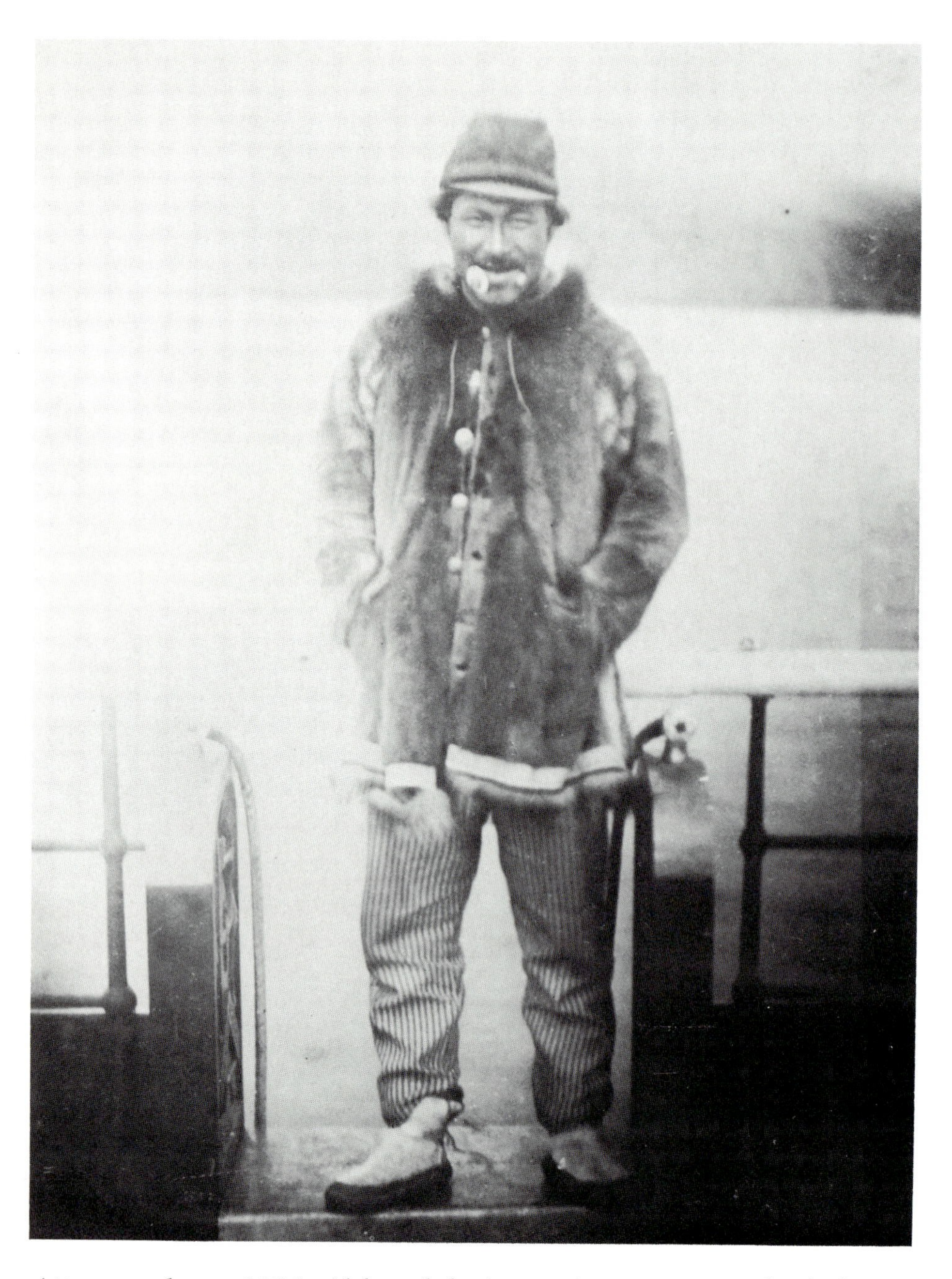

Attungoruk, ca. 1886. Although he is wearing western-style clothes, including mattress-ticking trousers, through his cheeks Attungoruk is wearing labrets, traditional adornment for Eskimo males

cold water with which the powder keg was kept constantly filled by hand. The liquor dripped from the worm into a battered old tomato can. It was called "kootch" and was potently intoxicating. An Eskimo drunk on "kootch" was said to be brave enough to tackle a polar bear, single-handed. The little still was operated in full view of the villagers. There was no need of secrecy. Siberia boasted no revenue raiders.

At the same time the natives of the region also acquired trade goods directly by working aboard whaleships. In the earliest years of the fishery some natives on the Siberian shore achieved considerable fluency in English as a result of their trading activities, but the first native to work aboard a whaleship was possibly Contine of Plover Bay. In the 1860s he signed aboard the brig *Victoria* as interpreter and "trading master" for the passage on the west shore. Later in a fit of remorse over having killed his father, he signed on for the entire summer's cruise.

During the heyday of walrus hunting in the 1870s a number of natives were signed aboard, often with their skin boats, to help in butchering the animals. In the 1880s many of the ships began taking several natives for the season to serve as ordinary seamen and occasionally as boatsteerers. During the 1890s, with many ships wintering at Herschel Island, whole families were signed aboard for two years or more, the women to serve as seamstresses, the men as hunters and dog drivers in the winter and as seamen in the summer.

If a native man signed on without taking his family along, the captain usually sent food ashore to his wife to tide her over for the summer. This advance would be deducted from the native's pay at the end of the cruise. Once the new native crew were aboard, the Captain broke out the slop chest and issued them with a sailor's outfit. They usually berthed in the "Masinker House," a special house the carpenter built for them on deck.

Then came the problem of nomenclature. The officers' ears were used to Anglo-Saxon or Iberian names, and if they could pronounce the natives' names at all, they usually mangled them in the process. To make things simpler, like the Kanakas, the northern natives were usually given more convenient English names: Sam Brown, Big Jim, Big Mike, Little Jim, Billy, Louie, Cockney, Riley, Samson, Harry, and the like. Others were given the name of their home: Prince of Wales, Cape York. Others carried the name of the vessel they had first served aboard: Charlie Newport, Jake Rainbow, Thrasher, Pacific, Bowhead, Herman, William Baylies. The origins of other names are obscure: Mudhook, Hoodlum, Black Jack, Santa Anna, Jags, Jawbone, Sam Lazy, Smart Boy, Matches, Chicken Powder, Shoo-fly, Heinie, Policeman, and Tommy Tough.

At the end of the cruise these "ship's natives" were usually paid off in trade goods, whaling gear, and most often, in used whaleboats, which were highly prized. Eskimos also traded directly for boats. In 1886 the price was 230 pounds of whalebone, more than $600. In the 1920s it reached $1,000. By 1900 Captain John A. Cook estimated that there were thirty whaleboats at Indian Point and Plover Bay.

When a whaleship landed its native crew at their village in the autumn the captain also gave them the flukes of the last whale they had caught. But in the heavy weather and growing ice of October, it was sometimes impossible to land the natives. To avoid being caught on a lee shore in an ugly season the captain occasionally had to keep on southward. In such cases the natives had no choice but to go to San Francisco with the ship. They would be returned to their homes six months later on the ship's next cruise northward. Some chose to go south in any event, relishing the chance to visit foreign shores.

These natives were usually returned to their homes in the spring of the following year. Having learned passable English, they were much in demand as interpreters. One native, Frank from Uelen, spent three years in San Francisco as a dishwasher in a restaurant. Another, Cornelius of Plover Bay, traveled to Washington, D.C., and New Bedford, in addition to wintering in San Francisco.

But the pace of life in the south was not to the taste of all. One native told the ethnographer Edward William Nelson that although he liked the United States, he thought "Merican too damn much work." "So," Nelson reported, "he returned to his squalid home on the shore of Plover Bay."

By the turn of the century those whaleships that remained in the fishery became more and more reliant on the profits of their trading activities as the bowheads became increasingly scarce. When the price of baleen collapsed in 1908, the remaining whaling firms converted almost exclusively to fur trading. They faced competition not only from the traders from the South but also from a mosquito fleet from Nome, small schooners that were primarily whiskey traders. In addition, a number of trading posts had sprung up along the Alaskan shore and John Rosene's Northeastern Siberian Trading Company established posts on the Chukchi Peninsula. Natives throughout the Western Arctic also bought their own schooners for whaling and trading.

In the eastern Arctic the withdrawal of the whaling fleet had left only the Hudson's Bay Company or Revillon Frères in each settlement. Being the only available purchasers for the furs, these sole buyers could usually dictate market prices. In the western Arctic, however, the situation was radically different. Many markets existed for the natives' trade goods. The result was that the competition for furs worked in the natives' favor, giving them good prices and a wide variety of merchandise to purchase.

The trading vessels became floating general stores. Shortly after the turn of the century, for example, Captain Hartson Bodfish, one of a number of traders, visited the small settlement at John Howland Bay on the Chukchi Peninsula. He sold the following goods:

> 49 rifles, 5 shotguns, 39,000 cartridges, reloading tools, powder, lead, shot, thousands of yards of ticking, drill, denim, calico, flannelette, foot sewing machines, hand sewing machines, needles for same, thread, thimbles, chewing gum, combs, canvas, twine, tobacco, matches, flour, bread, molasses, sugar, tea, baking powder, dried apples, prunes, rice, 3 phonographs, 110 records, phonograph needles, clocks, oak boards, boat anchor, brass kettles, primus stoves, dish pans, milk pans, enamelled pails, table spoons, serge cloth, coffee pots, canned milk, shovels, tacks, mirrors, scissors, darting-irons, cutting-spades, knives, harmonicas, files, drills, bits, breast drills, planes, hammers, hatchets, saws, axes, awls, coal oil, spy glasses, opera glasses, darting-bombs, shoulder-bombs, boat compasses, boat boards, screw drivers, cigars, beads, caps, suspenders, boys' clothes, leather belts, lady's coat, one 16-foot oar, old sails (these from the ship), whaleboat and gear, paint, paint brushes, playing cards, pepper, brooms, and one house, 30 x 20 feet, cut and fitted. The entire cost of this list of trade goods is set down as $6,030.86, and it was only a small part of the entire stock.
>
> In return we received 3553 lbs. of whalebone, 38 white fox skins, 6 deer, 1 seal coat, 1 pair of seal pants; but some of the articles had been ordered and possibly paid for in advance.

Eskimo store at Uelen, near East Cape, Siberia. At the left an American flag is flying Bancroft Library, University of California, Berkeley

By this time the entire western Arctic had become one cohesively integrated economic region organized around the maritime fur trade. Ultimately this trade was run out of San Francisco and Seattle. Its influence was particularly noticeable on the Chukchi Peninsula, where visitors expected to find broad evidence of the czar's rule. Instead they found none, for the Russians had developed no economic control whatsoever, and very little political control either.

The natives there were "almost wholly beyond the reach of Russian influence," according to the ethnographer Waldemar Bogoras. And one traveler there thought that their "higher cultural level" could "undoubtedly be attributed to their contact with American traders."

At about that time American trade goods were only half as expensive as those brought from Vladivostok or the interior of Siberia. By 1882 almost all the natives at Plover Bay and Indian Point could speak some form of English. They knew very little about the Russians and had a high opinion of the Americans on whom they were dependent for virtually all of their manufactured goods. In 1887 one Siberian Eskimo told Herbert Aldrich, "I b'lieve no whaleship, Masinker man all die."

In 1912 L. M. Starokadomskiy, a Russian hydrographer, wrote:

> Uelen left us with the impression that this settlement was located on the American rather than the Russian shore. Nothing here was Russian, not even the language. The local inhabitants, Eskimos, knew more English words than Russian words. In the only store, kept by the Eskimo headman, all the goods—guns, harpoons, knives, axes, gramophones, dress coats, brick tea, sugar, flour, and assorted statuettes—were of American origin.
>
> The Vladivostok merchant Churin, had attempted to open a store here, but had been unable to withstand the American competition.
>
> One should mention that the Americans had equipped the local inhabitants with good whaling boats with harpoon guns, and had supplied good hunting rifles, and a mass of essential domestic items. In one yaranga I saw a sewing machine, on which they sewed overalls, which they wore over their normal fur clothing, from multicoloured close-weave cotton cloth. One Eskimo had acquired a typewriter, on which he enthusiastically tapped out symbols that to him were quite unintelligible.

On the whole, the trade in alcohol seems to have declined somewhat on the Asian shore in the first years of this century—with the exception of the traders from Nome who were notorious for their sleazy dealings and whose activities were entirely unpatrolled during the Russo-Japanese War. When the Bolsheviks finally took firm control of the Chukchi Peninsula in the early 1920s, the coast was sealed. Only a few foreign traders who were licensed by the Soviet state were allowed to operate on the Siberian shore. The natives became demonstrably poorer in terms of material goods, although they were, of course, no longer being "exploited" by capitalists—their furs now had to be sold to the state monopoly.

In the same period the rule of law was extended to most Alaskan communities via the revenue cutter patrols, via resident magistrates and missionaries, and via ethical resident traders. The Alaskan maritime trade principally became the province of two men, Captain John Backland with his vessel, the *C. S. Holmes*, and Captain C. T. Pedersen, with first the *Nanuk* and then the *Patterson*. They served as freight haulers as much as traders, carrying cargo for rival traders as well as outfitting their own trading posts. Captain Pedersen retired in 1936 and Backland's service ceased in 1942, when his vessel was appropriated for the war effort. Since then the Bureau of Indian Affairs has outfitted the native cooperative stores along the Alaskan coast via its several vessels named *North Star*. Today all northern communities have regular year-round air freight and passenger service. For those settlements on the coast or rivers, summer-time tug-and-barge service is widely available. Personal needs, if they cannot be fulfilled in a native's settlement, can be quickly supplied by Sears or many other mail-order houses.

10 / *Steam Whaling*

By 1875 the outlook was dismal for the Arctic fishery and for the industry as a whole. The Arctic whaling grounds seemed on the verge of exhaustion. The fleet's catches were for the most part poor. More than forty ships had been lost in the preceding decade—the result of taking dangerous risks to increase catches—a fact that caused the insurance companies to place a 3 percent surcharge on the Arctic whalers' rates. Whale oil was selling for only 65¼¢ per gallon, whereas ten years earlier it had sold for $1.45. Baleen in the same period fell from $1.71 per pound to $1.12¾. And the North Pacific fleet, the size of which had approached three hundred ships less than twenty-five years earlier, stood at seventeen vessels. Although the annual "imports" (catches) of oil between 1865 and 1875 had fallen from about 76,000 to 35,000 gallons and of whalebone from 619,000 to 372,000 pounds, prices had fallen, too, because of the decrease in demand. In the contracting market the value of an average-sized bowhead had been halved, decreasing from about $7,100 to about $3,700 (from $44,730 to $32,523 in 1982 dollars).

Thus, by the mid-1870s the American whaling industry had reached a crossroad. The nation was in a period of economic recession. Faced with declining profits, many whaling merchants withdrew their ships from the fishery and invested their capital in textile mills and other industrial enterprises. The only recourse for those who remained was to increase their catches and reduce their expenses or to develop new markets.

Cost cutting was difficult in an industry that already figured its margins very finely. The major innovation of the 1870s had been to choose a refitting port in the Pacific with the lowest aggregate costs. The fleet see-sawed between Honolulu and San Francisco as each port offered inducements to draw the whalers. As the gold rush receded, San Francisco attracted more and more of the ships, and by 1877 it was servicing most of the fleet—not only because it was relatively close to the Arctic, but also because the captains could ship their cargoes to the East Coast quickly via the transcontinental railroad. Transcontinental rail shipment, faster by months than shipping cargo around Cape Horn or across the railroad at the Isthmus of Panama, reduced the time between investment and return and whaling merchants, by receiving their product more quickly, could choose the most advantageous moment to sell.

The task of increasing catches presented a formidable problem. The Arctic grounds had already been fully explored in all areas with the exception of the Canadian waters, which at that time were of unknown value. The owners turned

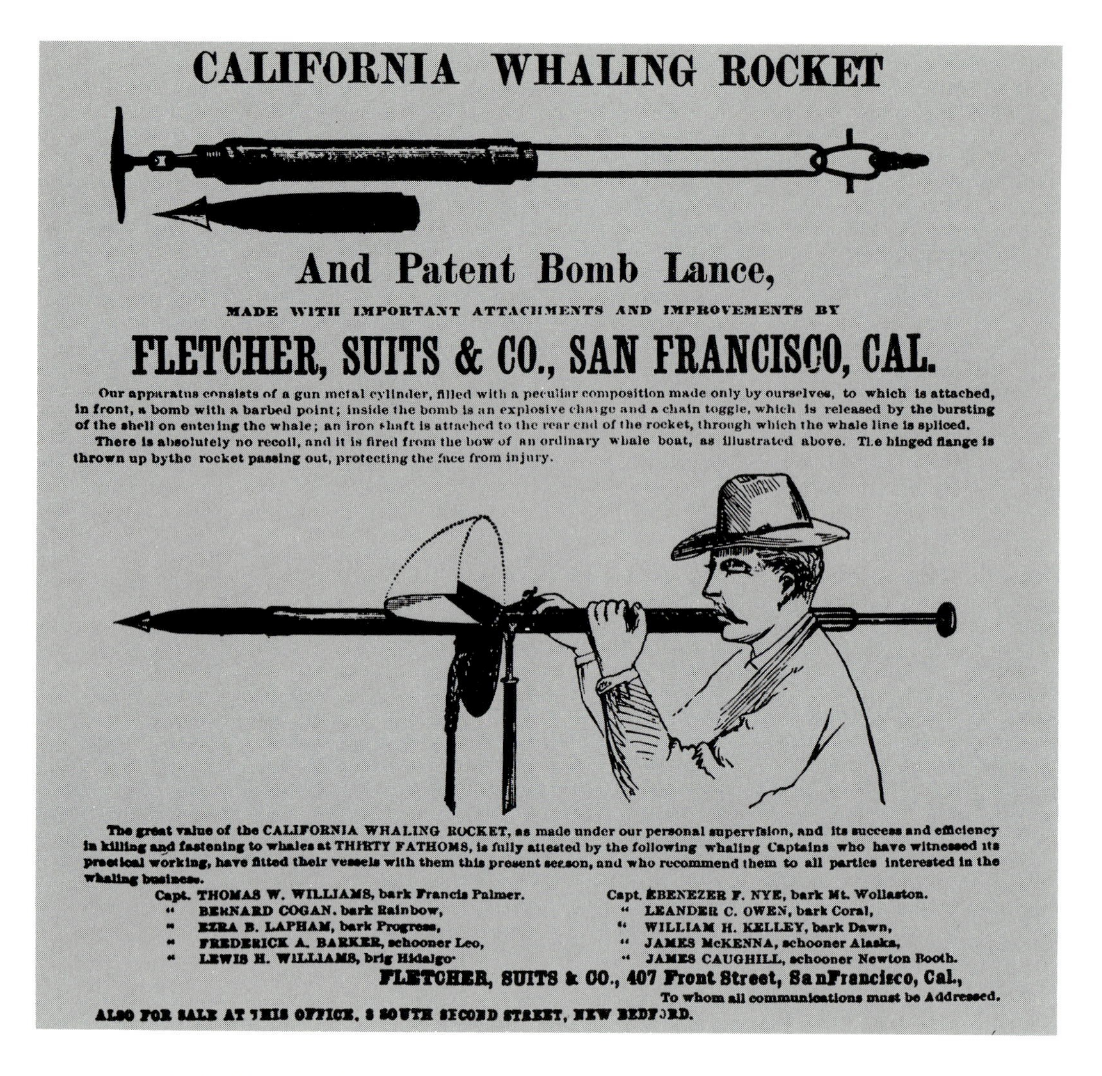

California Whaling Rocket. Advertisment in the Whalemen's Shipping List, *New Bedford, 1878*

instead to technological improvements as a means to increase their catches.

Until the late 1870s these technological innovations were largely restricted to perfecting whaling irons, darting guns, and shoulder guns—a trend that culminated in the Arctic with the experimental use of the "California Whaling Rocket," a primitive bazooka. Captain Thomas Roys developed this idea into a workable model in association with Gustavus A. Lilliendahl. The invention did Roys little good. He lost a hand in an explosion during his experimentation and eventually died destitute and "wandering in mind" in Mexico in 1877. The design was then adopted by John N. Fletcher and Robert Suits, who increased its weight and charge. But although the rocket harpoons were used experimentally in the Arctic by a number of whaling masters in 1878 and 1879, the harpoons were never widely accepted, apparently because the device was difficult to aim in the rough waters of the north.

Captain Leander Owen

Others believed the way to increase catches was to equip the ships with auxiliary power. As early as 1873 Captain Leander Owen, who in that year had taken the bark *Jireh Perry* more than one hundred miles east of Point Barrow to Harrison Bay, called for the use of a steamship in the Arctic fishery. "It seems an open season . . . ," he wrote, "and I have no doubt but a steamer could have gone to Mackenzie River, and I don't know but clear around the northwest passage." After the disaster of 1876 the advantages of steam auxiliary power must have occurred to several whaling merchants. In 1876, and in 1871, vessels had been wrecked when the ice closed in on them. A steamer, running through the broken ice, against wind and current, would have been able to escape the encroaching pack ice. By the 1870s steam power had already proved itself an effective method for moving large vessels in icy seas; the British whaling and sealing fleets had been using steamers since the 1850s.

"We keep the best and most popular makes of corsets."
From New Bedford City Register, 1892

Nevertheless, it was whalebone, not danger, that eventually built the American steam whalers. In the mid-1870s the fashion industry had begun a thirty-year trend, accentuating women's waists. Creating a wasp-waist silhouette required corsets to become "longer, tighter, and more generally constricting." Demand for baleen for corset bones and busks rose swiftly. By 1878 the wholesale price of whalebone was "without precedent," closing at $3.25 per pound in December, thus making the baleen from one average-sized bowhead worth close to $5,000 ($50,000 in 1982 dollars). It must have been clear to some that the price was likely to go higher, as Zephaniah Pease and George Hough put it in 1889: "while substitutes were found for other products of the whale-fishery, the inventive genius in vain has strived to supply an article which will fill the place of whalebone. Russian horn, celluloid—artificial and natural substances alike have been tried, but none will answer the purpose."

Pease and Hough identified the elemental change in the industry that ushered in the age of steam whaling. "It is within the remembrance of many old whalemen," they wrote, "when this bone, now so precious, was dumped over the ship's side as waste or only saved by the sailors for scrimshaw work."

USS Rodgers, *formerly* Mary and Helen, *in San Francisco, 1881. Shortly after this photograph was taken the* Rodgers *departed on a rescue mission to search for the lost Arctic exploration vessel* Jeannette. *While wintering in Saint Lawrence Bay, Siberia, she burned to the waterline* U.S. Navy

William Lewis initiated the experiment. Lewis was a former New Bedford whaling master who became a real estate speculator before returning to the whaling industry as an agent for Ivory H. Bartlett and Sons. Lewis dispatched Captain Cyrus Manter, an officer aboard one of the first whaling vessels to operate in the Bering Strait, to Newfoundland to examine the British whaling and sealing steamers from Dundee. After Manter's return, Lewis ordered the construction of the first of the class of American steam whalers, the *Mary and Helen*. Despite what was undoubtedly a favorable report from Manter, the venture was, for Lewis, very much of a speculation. To spread the cost of the vessel and the risk of the venture, he enlisted investment capital not only from his usual shipowning associates but also the men and firms directly concerned with the construction of the *Mary and Helen* and her power plant. Goss, Sawyer and Packard, the Bath shipbuilders, George H. Reynolds, who designed her steam auxiliary, and C. H. Delamater, who manufactured her power plant, all held shares of ownership.

The *Mary and Helen* was strongly built and carried steam donkey engines to assist in pumping, weighing anchor, and working the cutting tackle. With her fine lines and bark rig she was intended to conserve coal by making use of the wind on her passages to and from the Arctic grounds; her engine was for use during calms or for moving among the ice floes. George Reynolds, who had installed the engine on USS *Monitor,* designed the engine for maximum efficiency; it consumed only about four to five tons of coal per day if in constant use.

Considering Lewis's foresight in planning the *Mary and Helen,* it is puzzling that he designed her oil-carrying capacity conservatively. She carried her oil in casks, unlike the ships of the Dundee fleet, which carried blubber in iron tanks. Only a few years later San Francisco steam whalers also adopted the use of iron tanks. Hot oil could be put into the tanks directly from the tryworks without the intermediate stage of cooling, which had been necessary to prevent the wooden casks from shrinking.

Nevertheless, the $65,000 ($673,400 in 1982 dollars) invested in building the *Mary and Helen* signaled a fundamental change for the industry: baleen made the bowheads worth pursuing and only advanced technology would allow the whalemen to reach the whales. Suddenly the capital requirements for a first class whaleship had nearly doubled—but the chance for profits had increased even more.

The *Mary and Helen* was launched in Bath, Maine, on July 30, 1879. William Lewis and his partners immediately brought her to New Bedford for outfitting. On September 9, her tryworks built and supplies and whalecraft aboard, she sailed for the Pacific under the command of Captain Martin Van Buren Millard. After a stormy passage through the Straits of Magellan she reached Honolulu, where Captain Leander Owen joined her and took command for her Arctic cruise.

The ship, the partners, and the crew were lucky. Not only did the *Mary and Helen*'s power plant allow her to reach the whales easily, but the 1880 Arctic season proved to be one of such exceptionally compact ice that the whales were forced into open water. The entire fleet—twenty ships—made good catches, but the *Mary and Helen*'s news was astonishing.

She arrived in San Francisco full, with 2,350 barrels of oil and 45,000 pounds of baleen from 27 bowheads—one of the largest catches in the history of the Arctic fishery. With her cargo valued at more than $100,000, her investors had been repaid after only one season's voyage. Captain Owen reported that the vessel's maneuverability in calms and adverse currents had allowed him to stay in virtually constant touch with the elusive bowheads, consequently far outstripping the other Arctic vessels.

"We were, as whalemen say, blubber logged twice," reported Judson Cobb, one of her boatsteerers, "the first time, with twelve whales when we went and anchored off Point Franklin, boiled them out, stripped the whalebone and stowed all below. Then we went off shore and found the whales, again filled up our blubber room and decks, and on September first we started out of the Arctic leaving plenty of whales in sight behind us, but could take no more. We came through the Behring Strait in a blinding snow storm and put into Plover Bay to finish boiling and stowing down preparatory to our start for San Francisco. Every cask we

William Lewis

had was filled with oil; as there wasn't room enough below decks, we lashed 160 barrels on deck. As our coal was getting low, we put it all in one bunker and filled the other one with bone, and all the room between the top of the casks in the lower hold and between the decks was filled with bone too."

Even before the *Mary and Helen*'s return William Lewis organized another group of investors to build his second steam whaler, the *Belvedere*. When the *Mary and Helen* reached San Francisco he began construction of his third, the *North Star*, and arranged for the bark *Lucretia* to be converted to steam power. The *North Star* was launched in August 1881 and the *Lucretia* was ready to sail in late September of the same year, both financed by the profitable sale of the *Mary and Helen* to the U.S. government in the spring of 1881 for $100,000. She was urgently needed to search in the Chukchi Sea for the missing *Jeannette* expedition. To replace the star of his fleet, Lewis immediately ordered the construction of the *Mary and Helen II.*

The *Belvedere*'s maiden voyage, also under Leander Owen, was highly successful: she took 600 walrus, 14 bowheads, and a few sperm whales, and arrived in San Francisco in November 1881 with 1,800 barrels of oil, 34,000 pounds of whalebone, and 2500 pounds of walrus ivory. The *Belvedere* also brought down news of the loss of the USS *Rodgers,* formerly the *Mary and Helen.* On November 30, 1881, while frozen in Saint Lawrence Bay on the Chukchi Peninsula, she had burned to the waterline. The crew found shelter ashore with the natives.

But then Lewis's fortunes changed. With her heavy power plant the *Lucretia* proved to be a slow sailer and an unlucky vessel as well. She left New Bedford on October 1, 1881, and immediately ran into a hurricane near Bermuda. She barely survived the ordeal and lost her topmasts, bowsprit, many spars and all but one of her boats. Rather than endure a slow refit in a foreign port, Captain Thomas Mellen took her back to New Bedford, where her crew left the ship, claiming, legally, that the voyage was over when the vessel returned to her home port. Lewis repaired the ship, signed a new crew aboard, and the *Lucretia* put to sea again on December 14. Her bad luck held, however, and heavy gales prevented her from rounding Cape Horn, forcing her to reach the Pacific via the Indian Ocean. She arrived in San Francisco in November 1882, having missed the Arctic season. Her only cargo after more than a year was thirty barrels of sperm oil.

To make matters worse for Lewis, the *North Star* was lost on her maiden voyage. Leander Owen took her north. She reached the Sea Horse Islands, eighty miles southwest of Point Barrow, on June 25, 1882, the earliest date any vessel is know to have arrived there in the nineteenth century. As the pack ice descended on him, Owen drove the *North Star* on, hoping to find protection behind the great grounded pressure ridge of ice that marches parallel to shore near Point Barrow. On July 8, however, the pack closed on the shore-fast ice, and she was crushed in the vise near the Point, two-and-a-half miles from the beach. The ice drove in with such force that men on shore could hear her timbers cracking. The crew escaped over the ice to the U.S. Army's meteorological station.

For Lewis the following year was no better. The season of 1883 was a poor one for the entire fleet. The ice was so heavy all spring that some of the ships heading for Bering Strait had to work their way up the coast of Anadyr Gulf and right around the shore of Holy Cross Gulf. They moved ahead so slowly that they missed the bowhead migration at Bering Strait. When they finally got into the Arctic, they found few walruses and even fewer whales. Worse still, the pack ice did not allow them to go very far east of Point Barrow, and the autumn weather at Herald Island was too rough to allow much whaling. Some captains gave up in disgust and left early, trying for right whales in Bristol Bay before returning to San Francisco.

The loss of the *North Star,* combined with the fact that the *Lucretia* produced no income in 1882, and followed by the poor season of 1883, forced Lewis to sell the

Mary and Helen II to his competitors when she returned to San Francisco from her maiden voyage. Although William Lewis and his son would remain important figures in the industry for another three decades, the year 1883 marked the end of his brief but highly significant pre-eminence; as he declined, reciprocally, the merchants of San Francisco rose.

For the industry as a whole the year 1883 was pivotal. It marked the beginning of San Francisco's domination. Entrepreneurs in San Francisco had followed the fortunes of the first *Mary and Helen* with interest and were well aware of the advantages of both steam power and San Francisco's better location in relation to the Arctic grounds. In early 1881 Millen Griffith, the owner of a fleet of tugboats, George C. Perkins, the former governor of California, and Charles Goodall, ordered the construction of the *Bowhead,* the first of the class of steam whalers to be built on the West Coast. She cost more than $100,000 ($1,000,000 in 1982 dollars). At her launching on April 18, 1881, the *Alta California* reported the high expectations of her owners and correctly predicted the movement of the whaling industry's center from New Bedford to San Francisco.

> She is without doubt the strongest and most complete vessel ever built for the whaling business, and although our Eastern neighbors think that San Francisco will have to be satisfied with the skim milk, there is no reason why our people cannot secure the cream as well as New Bedford. Our enterprising men are determined to have it.

Like William Lewis before them, Griffith, Goodall, and Perkins knew the value of their new technology, and as soon as the *Bowhead* had been launched, they laid the keel of their second steam whaler, the *Orca.* Then in quick succession the group, with the notable addition of Josiah N. Knowles, launched the *Narwhal* and the *Balaena,* purchased the *Mary and Helen II* from William Lewis, and ordered the construction of the *Thrasher* at Bath, Maine—no doubt because the Dickie Brothers yard in San Francisco, the builder of their other steamers, could not cope with the volume of work. The hectic pace of activity at the yard was a marvel to observers. One newspaper's account of the departure of the *Balaena* on April 27, 1883, relates with pride the schedule of the building:

> The steam whaler *Balaena* sailed at 10 A.M. yesterday for the Arctic, completing one of the most remarkable jobs in ship building, engine construction and fitting out that has ever been performed in any port in the world. The keel of the vessel was laid on January 29th, and the builders, the Dickie Brothers, used their best endeavors to have her built in their contract time, three months, and as may be seen, they succeeded. As samples of the great and rapid jobs performed, we may mention that the engines were on the ground on Saturday, the 14th instant, yet on the following Saturday they were in position on board and the vessel was launched the same night and brought down to Vallejo Street wharf, the New Bedford of California. The rigger, Haversaith, rigged her completely and had everything in good order aloft and sails bent in one week. On the vessel reaching the wharf mentioned, although it was midnight, the work commenced, and early next morning preparations were all perfected for coaling and fitting out for sea. The stores etc. were put on board in a few hours, and coaling commenced at 5 P.M., and at 6 next morning she had her full allowance on board, 308 tons.

Balaena *in the stocks*

The San Francisco ships, like William Lewis's, were well built with fine, clean lines for fast passages to and from the Arctic grounds. But unlike Lewis's vessels, the *Bowhead* and the *Orca*, and probably the *Narwhal* and the *Balaena*, were fitted with a large, two-bladed propeller, from which the shaft could be withdrawn, allowing it to be hoisted in a sliding frame up into the vessel. A centerboard was then placed in the void to improve speed under sail. They carried extra whaleboats, too—the *Orca* carried seven—for more effectively saturating an area with boats during the chase. And the *Orca* and the *Thrasher*, and probably the *Narwhal* and the *Balaena*, were fitted with steam tryworks and iron oil tanks. The tryworks ("steam digesters") not only speeded the process of rendering oil from blubber, but they also left the oil with a lighter color and less odor than with the traditionally blubber-fired tryworks.

Josiah Knowles National Maritime Museum, San Francisco

The oil could be directly transferred to the iron tanks without first cooling it. The tanks also allowed the ship's load to be trimmed while at sea by pumping oil from one tank to another. With these new methods of rendering and storing oil, the vessels of the Pacific Steam Whaling Company anticipated the whaling factory ship of the twentieth century.

Thus, with only the successful maiden voyage of the *Bowhead* behind them, the partners launched the *Orca*, the *Balaena*, the *Narwhal*, and the *Mary and Helen II*—vessels valued at more than $500,000 ($5,180,000 in 1982 dollars).

But for the American whale fishery, certainly the most important event of 1883 was the incorporation of the Pacific Steam Whaling Company, an event that brought about the industry's greatest surge of innovations. On October 30, 1883, articles of incorporation were filed at the county clerk's office, forming the company

> for the building, buying, selling, owning, and operating of whaling and other vessels or other property, transportation of freight and passengers, trading, fishing, towing and salving; to purchase and own stocks in other corporations; to borrow and loan money; to issue bonds and conduct the business of a general whaling and trading company; to purchase operate and own coal mines . . .

Unlike the traditional method of individual shareholding in vessels, investors purchased stock in the company, which was capitalized at $2,000,000.

Except for Josiah N. Knowles, the partners of the Pacific Steam Whaling Company had no close contact with the business of managing whaling vessels prior to 1880. Unhindered by the traditions of the New England whale fishery, the company successfully introduced new practices and procedures. Knowles himself was more entrepreneur than whaleman, having been a successful and enterprising merchant skipper before becoming, first, a shipping agent, and later, a whaling agent for William Lewis. It was a measure of the partners' foresight and lack of traditional thinking that they designed a flexible organization; during slumps in the whaling industry the company equipment was diverted to other operations such as trading, salmon packing, or transportation to the Alaska gold fields.

Simultaneously with the incorporation of the Pacific Steam Whaling Company, the same partners established the Arctic Oil Works, capitalized at $1,000,000, and appointed Josiah Knowles as manager of both enterprises. Thus, with the creation of the first refinery for oil and baleen on the West Coast, the company had efficiently organized all the basic processes from catching to refining in one operation. Oil from the company's fleet could be refined and sold more cheaply and profitably on the Pacific Coast than if it were shipped to the East Coast for processing. The firm immediately retained as superintendent F. A. Booth, a New Bedford man experienced in the procedures of the East Coast refineries, and established their plant at the corner of Illinois and Center (now Sixteenth) Streets in the Central Basin area of San Francisco.

The Oil Works had the advantage of a 300-foot wharf extending into the deep water of the Bay. The whaleships could berth at the pier and pump their oil through a pipe directly from their iron tanks into the six 2,000-gallon receiving tanks at the Works. The raw oil could then be stored in the tanks—while the sediment settled out—until the owners judged the market to be favorable for refining and selling it.

The ability to pipe the oil from the ship and to store it in receiving tanks allowed a considerable saving over the methods that were still in use in New England, where the oil was taken in its casks from the hold of the ship and stored on the wharves

The Arctic Oil Works. Lithograph by Bosqui Engraving and Printing Company, about 1885. With its 300-foot wharf, the Arctic Oil Works had the advantage of allowing the Pacific Steam Whaling Company's ships (upper left) to unload directly at the refinery. Oil could be pumped into the 2,000-gallon tanks or be stored in casks in the flat-roofed warehouse, which had a clay floor kept damp to prevent the casks from shrinking. Refining was done in the three-story structure at the right. The one-story building in the center distance was the main baleen storage area and was fitted with double iron doors and shutters, presumably to keep out rats, and a system for flooding in case of fire Bancroft Library, University of California, Berkeley

under a heavy covering of seaweed, which had to be wet down frequently to keep the barrels moist to prevent them from shrinking and leaking oil. It was important to protect the casks on the wharf and at sea because the "ullage," the loss through leakage, could amount to a loss of 10 percent or more of the total volume of oil from the time of its trying out to the time of its sale.

When the time came to refine the oil at the Arctic Oil Works, the men pumped it directly from the receiving tanks, across the yard to the refinery building—unlike the procedure in New England, where it was sold to a refinery, ending the whaling merchant's control over it. The oil, which had been "recked off"—pumped away from its sediment in the large storage tanks—could then be sold in that state for use as cutting oil for machine tools or for lubricating the hemp and manilla fibers used in rope making.

Most of the oil, however, was fed into 100-gallon kettles, where it was heated to allow more of the sediments to precipitate out. It was then run into pits and frozen with ice. The congealed lumps of oil were next put into cloth bags and placed in a heavy hydraulic press. The pressure strained the oil through the cloth and thus purified it further.

At this stage the oil was classified as "winter strained whale oil," a term derived from an earlier method wherein the oil could only be frozen during the winter. The remainder in the cloth bags (the "foots") now resembled lard and could be reheated, refrigerated again, and strained to yield a less fine grade of oil, "spring whale oil." The remaining foots were then usually sold to tanners for softening leather. The oil, however, was often bleached again by heating it and then adding potash to leave it cleaner.

The oil was then usually pumped into shallow 500-gallon vats on the top floor of the refinery building where it was exposed to sunlight through glass windows for a few hours or days to continue the bleaching. The residue in the vats was usually made into "oil soap" and sold to citrus growers, who used it to wash their trees to protect them from insects.

The bleached oil was then pumped back down into the warehouse and stored in casks on a clay floor and under a louvered roof that could be opened to allow the rain in—to keep the casks from shrinking. After the loss of the illumination market to petroleum products, most of the winter and spring oil was sold to tanners, and some, mixed with graphite and paraffin oil, was used as a lubricant for axles on railroad cars.

The Oil Works had the capacity to process 300 barrels of oil per day, from which 150 barrels of refined oil could be produced. There was a rail spur into the Works yard, and the oil could be pumped directly into tank cars or loaded aboard in casks or in cases of tin containers.

The whalebone, which had been scraped, washed, dried, and bundled while at sea, was unloaded and taken directly to a building in the Oil Works yard. The "Bone House" was made entirely of concrete—roof, walls, and floor—with double iron doors and shutters to keep out rats. It also had a flooding system in case of fire. These features allowed the owners to avoid having to carry insurance on the often enormous value of whalebone in their yard.

The whalebone was cleaned again in the Oil Works yard to keep it from developing a foul odor. The bone was then separated into grades: "size bone"—slabs six feet or longer; and the less-desirable "undersize bone"—shorter than six feet. If any right whale baleen ("northwest bone") had been taken, it was kept separate, being coarser and less pliable than "arctic bone."

The bone was bundled once again and put back into the bone house until its sale.

Whalebone in the Arctic Oil Works yard

When a deal had been struck, the bone was loaded into boxcars for the trip across the continent to bone cutters in New York, Boston, London, Paris, or Berlin.

The cutters first trimmed the fringe hair off the slabs with scissors. The fringe was saved for upholstery stuffing and for the bristles in light brushes. The slabs were then put into large vats of water and soaked until they were soft and pliable, after which they were scraped and washed until smooth. The smooth slabs were then put into a steam box and warmed, making them very soft. A workman trimmed off the short bits at the base.

A bone cutter took the plate while it was still warm and drew it through a "splitter," a tool in which parallel blades could be set at any desired spacing, thus cutting the baleen into many parallel strips for whatever use it might be intended: very narrow strips for umbrella tines, wider for corset stays, and so on. The strips

Splitting whalebone

were then put through a "sider," which planed the edges smooth and parallel. The pieces were then polished with a kind of blacking and bundled ready for sale to various manufacturing companies.

Alexander Starbuck listed some uses for whalebone: "whips, parasols, umbrellas, dresses, corsets, supporters of various kinds, caps, hats, suspenders, neck stocks, canes, rosettes, cushions to billiard tables, fishing-rods, divining-rods, bows, busks, fore-arm bows, probangs, tongue-scrapers, pen-holders, paper folders and cutters, graining-combs for painters, boot-shanks, shoe-horns, brushes, mattresses, &c."

With so much capital at stake, the Pacific Steam Whaling Company's investors must have anxiously watched their fleet put to sea after the dismal season of 1883. The ships set forth armed with new tactics and procedures made necessary by the fact that the whalemen had already killed 15,000 bowheads in the western Arctic, 80

Cut and bundled whalebone, ready for use in corsets

percent of the number they would kill throughout the history of the fishery. With such a dramatic reduction, there were few bowheads to be found outside the ice edge in the spring. This development had already become so apparent in the 1870s that some vessels simply forfeited the spring season and spent their time instead right whaling on the Japan grounds or in the Gulf of Alaska and then tried to reach Point Barrow by the beginning of August to intercept the whales coming out of the Beaufort Sea. Some sailing ships continued this pattern in the 1880s as well.

The steamers, however, used different tactics. Like most of the fleet during the 1880s, they left San Francisco in November or December after a month in port. They often immediately headed south and went after humpback whales off the coast of Baja California for a few weeks before swinging west to Hawaii or the Japan Grounds for some sperm whaling prior to refitting at Hawaii, or, more rarely,

Hakodate. These three or four somewhat leisurely months allowed the officers plenty of time to break in the green hands and to carry out any maintenance on the ship that would have been more expensive if done by contract in port. Some ships also picked up their captain in Hawaii, for the winter months were considered unimportant enough to allow the first mate to command the ship.

Once the ships stood north, however, the relative nonchalance of the winter months vanished as the men prepared themselves and the ship for the Arctic cruise. In the 1880s most of the ships, moving north from Hawaii, entered the Bering Sea via "Seventy-Two Pass," reaching the pack ice at the beginning of April at about the sixtieth parallel. If there were no whales to be found there, the ships immediately went into the ice and began working ahead as quickly as possible toward the open water on the south shore of the Chukchi Peninsula. After a month or more of running through leads in the ice, of having the leads close on them, of being temporarily imprisoned, then released, again and again, the ships reached the open water near Cape Bering. They then turned east and worked along the south shore of the Peninsula toward Indian Point to intercept the last migrating whales "head on" as they were coming out of the pack ice of Anadyr Gulf on their way to Bering Strait.

The first ships to reach these waters stood the best chance for a good "spring catch"—and this is what the steamers were designed to do, but strong as they were, they lacked the power to bull their way right through the ice like a modern icebreaker. Instead they had to travel a route with the least resistance. Usually this was found on the "western route," from Cape Navarin to Cape Bering. But occasionally easterly winds prevailed throughout the spring, clogging the western route. In that case the ships turned east along the ice edge and often found open leads between Saint Matthew and Nunivak islands. There they turned north until they reached Saint Lawrence Island where they swung west to intercept the whales in the strait between the Island and Indian Point. Only rarely, and in times of exceptionally heavy ice conditions, did they pass east of Saint Lawrence Island.

The ships spent the month of June criss-crossing the waters immediately south of Bering Strait, searching for any straggling whales and devoting an equal amount of time to trading at the various coastal villages for Arctic clothing and boots to issue to the crew, as well as for whalebone, furs, and ivory. At the end of June, when all the whales had passed and the ice still blocked the ships' entry into the Arctic, the fleet paused to refit before resuming their northern cruise.

During the 1880s the Arctic mid-season refit took on greater importance. Since the beginning of the fishery whaleships had visited the three principal harbors of Bering Strait to ride out gales, to refill their fresh water casks, and to collect firewood from the masses of drift logs that wash down the Yukon River and are carried north by the ocean currents and then strand along the treeless seacoasts. In the 1850s a large number of ships used Saint Lawrence Bay, but in the 1860s and 1870s it seems that the majority visited Plover Bay instead.

In the first decades of the fishery whaleships seldom used Port Clarence even though it is the best harbor in the region. A few whaleships visited Port Clarence in 1851 and, in 1852, the bark *Harvest* put in, hoping to find a British naval ship willing to take several insubordinate sailors. By the mid-1860s whaleships visited Port Clarence more frequently because it was the location of a rendezvous for a number of native groups. They met there to barter among themselves and with the ships of both the Russian-American Company and independent operators. Trade had already become a sideline of a number of whaling captains and they were drawn there, too.

It was not until about 1884 that Port Clarence became the central summer refitting port for the Arctic fleet—and its rise had as much to do with the fleet's increased dependence on tenders as it did with its being one of the best harbors in the Arctic. Tenders, as we have seen, came into use in the late 1870s as the result of the rise in the price of baleen and the recognition by the owners that to secure profitable catches from a reduced whale population, their ships had to put themselves at considerable risk. A tender could not only resupply the ships in midsummer, but also take their cargoes south, thus preventing the loss of the spring catch should the ships be lost in the autumn. The New Bedford firm of J. and W. R. Wing in fact hedged their bets one step further by often keeping several of their ships cruising on the Japan and Okhotsk grounds, which were frequently less profitable but which provided the firms some measure of security should the ice take the rest of their fleet.

The first regular use of tenders in the Arctic may well have been made by the New Bedford firms of Ivory H. Bartlett and Son and William Phillips and Son. They sent north the barks *Legal Tender* and *Jenny Pitts,* in 1877. The following year the ship *Syren* replaced the *Jenny Pitts,* and the *Legal Tender* followed the fleet as far as Point Barrow. The experiment was successful. The *Legal Tender*'s responsibilities increased each year until in 1881 she was carrying cargo south for the entire whaling fleet.

The following year, 1882, the New Bedford firms sent the bark *Thomas Pope,* a converted whaleship, on the first of nine consecutive annual voyages. In July the *Thomas Pope* rendezvoused with the New Bedford ships near Cape Lisburne, a point where the ships could get water and wood and then run a short way up the coast to Corwin Coal Mine, the exposed low-grade coal seams near Cape Beaufort where the crews could blast out the coal with gunpowder. The steamers usually took aboard 100 to 150 tons for their boilers, and the sailing vessels took about 10 to 15 tons—for the fires in the galley and tryworks.

In 1883, however, Josiah N. Knowles and the Goodall, Perkins Company (soon to form the Pacific Steam Whaling Company) set the steamer *Bonita* to service their four steam whalers and the next year sent the *Syren* north carrying coal and the steam schooner *Beda* carrying supplies. They planned to meet at Cape Lisburne in 1884, but heavy weather forced them back to Port Clarence where the company established a coal pile for its ships. Thereafter the company annually sent a tender, and often a collier, north for its fleet. The sheltered waters of Port Clarence proved a better site for the rendezvous than the exposed waters near Point Hope and Cape Lisburne, a fact which was proved by the loss of the *Thomas Pope,* stove by ice at Point Hope in 1890. After the loss of the *Thomas Pope* the New Bedford ships also began using Port Clarence for their resupply, and for the remainder of the century virtually all the vessels of the fleet called there.

The whaleships developed an almost metronomic routine in their use of Port Clarence. Concluding their spring cruise, they entered that spacious and secure bay in the very last days of June to begin a short leisurely interlude—the only real pause in their six month Arctic cruise. Any maintenance that had been deferred during the spring season was taken care of while the captain sent parties ashore to collect and saw firewood. With so many other vessels nearby it was also a comparatively simple matter to "smoke ship" to kill rats. The men filled a deck pot one-third full of charcoal and set it between decks, lit the fire, pasted over any crack from which fumes could escape, then remained on deck or aboard another ship for twelve hours or more.

The ship's water supply also needed attention. Empty water casks were rafted together and towed to a stream at Cape Riley. Several of the steamers, however, carried very few casks because of their iron holding tanks. Their crews would usually take all of the whalecraft from two or three whaleboats, clean and wash the boats, plug up the centerboard holes, and then tow the boats to the river mouth where they would either fill them by bucket, or simply by capsizing them in the river. The filled boats were towed to the ship where the steam pumps drew the water directly into the ship's water tanks. Later the steamer captains discovered that it was deep enough at Big Diomede Island to allow them to warp the stern of their ships close to the island's precipitous shore. Then it was a simple matter to use the steam pumps to take on water directly from a stream on the cliffside.

The steamers usually carried a cofferdam and extra propeller blades because every year the ice took its toll of one or two blades in the fleet. When the men felt the dull snap of the prop hitting a piece of drift ice and then felt the heavy shudder of the shaft vibrating violently in its collar, they had no choice but to continue on under sail alone. In such a case the captain usually chose to finish his spring cruise before heading to Port Clarence for repairs. Once a ship with a damaged screw reached Port Clarence, the captain trimmed the ship to raise the stern as far as possible out of the water, then turned the shaft until the damaged blade was pointing directly upward.

Fitting the cofferdam was always a problem. The edges of the cofferdam had to be packed with oakum to keep them as watertight as possible. The men then put the dam over the stern, tightened it down with winches and pumped it out with the steam pump. A man then unbolted the broken blade from the hub and replaced it with a new one. If both blades had been broken, the crew had to remove the cofferdam, revolve the shaft one-half turn (so that the second blade was up), and then repeat the procedure.

When the tender and collier arrived at Port Clarence, the captains jumped into their boats to be the first aboard these ships. The arrival usually touched off a frantic race. The order in which the captains reached the tender determined the order in which they would warp their ships together to transfer cargo and supplies. It was the responsibility of the last ship to ballast the tender and collier, a chore that required several days of hard work hauling gravel from the beach.

The mid-season interlude was also an opportunity for the U.S. Revenue Cutter

Whaling captains at Port Clarence, 1887. Left to right, standing: William I. Shockley, Leander Owen, John B. Tobey, Joseph Whiteside, David B. Adams, Martin V. B. Millard, (?) Joseph Fisher; seated, *William B. Ellis, John Keenan, Joshua G. Baker*

Service to inspect the ships for violations of federal or territorial laws. Port Clarence was one of the first places the ships touched on the American side of Bering Strait. The fleet's concentration of ships there provided a convenient point for the Revenue Service officers to inspect the fleet for any contraband liquor and firearms that might be traded to the natives. The captains and officers of the whaling fleet at first resented the visit of the revenue cutter but soon grew to accept it and even welcome the presence of federal authority. It was an inescapable fact that the quality of foremast hands on whaleships had deteriorated markedly throughout the latter half of the nineteenth century, and with increasingly unruly men to handle, the officers welcomed the revenue officers who reinforced their authority over the crews.

The deterioration in the crews had set in after the American whaling industry's first flush of success following the War of 1812. In those years, when the whalemen had been able constantly to discover new whaling grounds (and voyages had been more regularly profitable), the foremast hands had been largely drawn from farm boys and others in the hinterland of the northeastern states who were often capable and well-motivated. As the 1840s wore on, however, the yield of most of the whaling grounds tapered off. The voyages became longer and less remunerative, their lure declined and New England boys often found more attractive opportunities in America's West or in the country's growing industries. As the result, the ships at first took on increasing numbers of Azoreans, Cape Verdeans, and "Kanakas," a term used by whalemen to indicate, in the narrow sense, Hawaiian natives, but which often included all Pacific peoples. The officers found that these islanders usually made good sailors and whalemen, but as the center of Pacific whaling shifted from Hawaii to San Francisco, they became less available, and the best sailors in San Francisco usually shipped aboard merchant vessels, where their pay was assured.

When a whaleship was preparing for sea in San Francisco, the commission merchants, "landsharks," went to work rounding up a crew. In January 1887 the *San Francisco Chronicle* described their methods:

> In hiring sailors and officers for whaling voyages, the services of men designated as shipping masters are called into requisition. Various systems are resorted to to obtain men. Plying with liquor of the vilest description, doling out sufficient money to enable them to keep within the clutches of the harpies who float around the Barbary coast and water-front region, and in some cases conveying desirable men into interior towns until the ship is ready to sail, are the methods in vogue. The classes of men composing the crews are of a most heterogeneous description; men who have never seen the sea, and to whom a ship is as unfamiliar as a rhinoceros, are to be found on board of a whaler.
>
> Besides these classes a whaling vessel has for a crew some of the greatest drunkards to be found in a large city, jail-birds, and thieves. In this mass of humanity, gathered within the confines of a forecastle, good leaven is small. The majority of those who can pull a boat at the outset of a whaling voyage are Kanakas, natives of the Caroline Islands, or men from the Azores. This class of men regard whaling as a profession.
>
> The lower grades of officers, such as boat-steerers and boat-headers, are nearly all colored men or Portuguese from Cape de Verde Islands. As is usual with ignorant persons placed in authority, their treatment of green hands before the mast is anything but kind. The mates and masters on the vessels are, with few exceptions, Americans, hailing from New Bedford or other Eastern whaling ports. Many instances are made public of the cruelty with which the sailors are treated by these officers but while there are, no doubt, occasions when brutality is displayed, in most cases the sailors' treatment is aggravated by their own conduct. When it is remembered that the crews of whaling ships are composed partly of a useless set of men and partly of a lot of vagabonds who speedily demoralize the others, it must be conceded that a strong hand is required to keep order and preserve discipline.

Once they had been mustered on deck out at sea, the foremast hands had to be immediately outfitted from the slop chest. The slop chest charges would be deducted by the managing agent from their earnings for the voyage, but because

Towing water casks ashore National Maritime Museum, San Francisco

many of these men, being green hands had been shipped on very "long lays," their proportion of the profits would often be less than their advances and slop chest charges. Furthermore, in San Francisco it was a standard procedure that before they sailed, the officers and men would agree to the unit price of any oil and bone they might catch. These prices were usually about half what the oil and bone would be worth—to account for culls and spoilage, the managing agent said.

For the foremast hand it was, consequently, a frequent occurrence to conclude a voyage in debt to the ship and to be signed off with one dollar—"in full settlement and compromise of all claim or claims I have or may have against agents, owners, and master of said bark or vessel for services rendered by me to date," the document read. The *San Francisco Examiner* reported one sailor's estimation of the situation:

"It's tough (said an old salt in explanation), but it's the usual run. I've been at it 8 years now, and I've grown to expect it."

"Why don't you drop the business then?"

"Blamed if I know. You see us fellows usually sail the same old familiar course, no matter where it leads to. We growl, of course, and do some swearing, but we're usually back in the ice again the next season. I suppose it's because we don't know anything else, and because it's so hard to strike a new course."

"You signed clear for $1.50, but how much did you owe the ship?"

"Fifty-seven dollars they made it. I've been 8 years at the business and I know enough to pull all I can out of the slop-chest while we are afloat. It never makes any difference; you only get a dollar or two when you are paid off anyhow, even if you never touch a rag. If the whaling bosses don't best you on the slop-chest they best you on estimating the catch. Go for the slop-chest, say I, and get a dud or two anyway. . . .

"Here we are after a 10-months' cruise turned ashore without a copper. We've got to live. . . . We are then in the boarding-house keeper's clutches, where the bunk bills run up. He turns us over to the whaling master and so squares our account with advanced money, and so we go the round."

The men were occasionally lucky enough to be aboard a highly successful ship and then did handsomely, but more often, with such meager prospects of pay, some contemplated deserting. During the nineteenth century whalemen jumped ship in the most improbable places throughout the world. In the 1850s and 1860s the half dozen desertions that are known to have occurred on the Chukchi Peninsula certainly qualify as lunacy. "It was a true saying," wrote Captain Jared Jernegan, "that if a ship was bound to heaven and should stop in Hades, some of the crew would desert."

In the 1880s, with the prospects of a paying voyage declining, the number of desertions increased, particularly during the layover in Port Clarence in the brief but surprisingly beautiful and balmy Arctic summer. The men almost always jumped ship during an expedition ashore for wood, water, or coal. It was a practice among the captains to take no deserters from other vessels aboard their ships—even if they themselves were short-handed. This practice developed in the interest of maintaining discipline among their own crews. It forced the deserters, if they wished to return south, to rejoin their own ships. The Revenue Cutter Service also supported the captains in this matter: if the men came aboard the cutter, they were put in irons until they could be returned to their ships. Only those who were in a condition of extreme destitution because of starvation or disease would be carried south on the cutter. Those who made good their escape no doubt realized the consequences of their action by the beginning of September, when the first snow began to fly. These unfortunates often threw themselves on the mercy of the natives or the shore whalers for the winter. After the discovery of gold on the Seward Peninsula in 1899, the desertions increased and the runaways, often termed "tramps" or "beachcombers" by the whalemen, became a greater problem for the ships and a greater burden on the year-round residents of the north.

The discipline aboard whaleships was always strict, but contrary to popular belief, usually no more severe than on merchant vessels of the time. Strict discipline was frequently defended on the grounds that with ships in difficult or dangerous situations (which was, of course, often the case in the Arctic), orders had to be obeyed instantly—and no doubt there was justification for this argument. Seamen had been protected against the cruelty of officers under various federal laws since 1790, and in 1850 flogging was outlawed. In 1872 a federal act codified all previous legislation, but as Elmo Hohman put it, for whalemen the act was "emasculated" two years later by a law which exempted from its coverage "any case where the seamen are by custom or agreement entitled to participate in the profits or result of a cruise or voyage." Despite the laws—and even before their "emasculation"—discipline at sea aboard whaleships tended to be a more or less rough and ready matter, doled out according to the captain's and officers' judgment as to the severity of the infraction. For example, when three deserters returned to the *Eliza F. Mason* at Saint Lawrence Bay in 1854, they were seized to the rigging and given eighteen lashes each. In the early years of the fishery floggings were also meted out for "saucy" behavior, or for assaulting an officer with a knife. But it was often the physical strength of the mates that enforced discipline. The following entry appears in the log of the *Emily Morgan* in August 1871:

> While catting the anchor, Frazier, a boatsteerer, having too much to say was told by the first officer to keep silence, which he would not obey. The order being repeated several times without compliance, the mate was compelled to use force to preserve the discipline of the ship and in so doing accidentally broke Frazier's jaw.

Generally, however, insubordination and all misdemeanors were punished by keeping the man on deck all day, or by sending him to the masthead for a number of hours, or by handcuffing him and keeping him in the blubber room on bread and water until he agreed to return to duty.

As the fishery wore on it appears that although the number of incidents of insubordination rose, paradoxically the discipline became somewhat more relaxed. It was up to the captain to determine the punishment for an offense, "so long as he keeps within the bounds of the law," as James T. Brown put it. It may well have been the presence of the revenue cutter throughout much of the cruise that brought this about, for while the revenue officers reinforced the authority of the captains, they also represented a powerful authority in the prevention of cruelty to sailors.

By the Fourth of July most of the ships had usually completed their refit, and they paused to celebrate the national holiday, dressing out the ship in flags, firing a salute and holding whaleboat races among the fleet. Shortly after the Fourth they weighed anchor and began working their way north through the now scattered ice. They reached Point Hope and Cape Lisburne by the middle of the month and a week or so later were at Icy Cape. But from that point the timing of the annual progression northward was increasingly variable, for the ships had reached the edge of the pack ice, the extent and movement of which were dictated by the warmth of the spring and the direction of the summer winds. In 1888, for instance, the steamer *Orca* reached Point Barrow on July 4, but in 1891 no vessels were able to pass Point Franklin all summer.

The ships usually reached Point Barrow in the first or second week of August. In the 1880s most of the sailing ships made it a practice to stay in the waters near Point Barrow, waiting to intercept the bowheads on their return migration out of the Beaufort Sea. A few of the more intrepid sailing captains began probing eastward, but their cruises were largely limited to the most ice-free years and usually only lasted a fortnight or so. The steamers, on the other hand, were built to challenge marginal conditions, and their passage east of Point Barrow became an integral part of their summer cruise. The steamers regularly reached "Return Reef," the low sand barrier islands 150 miles east of Point Barrow, and in favorable years they occasionally went much farther.

August at Point Barrow, however, is properly considered to be the beginning of autumn. By the middle of the month the pack ice, which usually has receded from the Point, begins its slow growth south. Sailing vessels usually left the area, bound to the west, by about the twenty-fifth of the month. The steamer captains usually planned to pass Point Barrow by September 10, although this could vary greatly depending on the ice conditions in a particular year. In 1888, for instance, an exceptionally open year in the western Beaufort Sea, the *Orca* passed the Point on September 20.

From there the ships made their way southwest along the margin of the pack ice until they reached "Post Office Point," the southern tip of the tongue of ice that is formed by the bifurcating currents that run north out of Bering Strait. From Post Office Point the ships turned northwest, moving over Herald Shoal and into "the Hole," the deep cul-de-sac east of Herald Island that is forced into the pack ice by the current. The bowheads feed there during September and October before turning south and returning to the Bering Sea. In the darkness and increasing cold of late autumn it was not safe for a sailing vessel to be north of Herald Shoal in late September. The steamers, however, often stayed until October 10. It was in these last few weeks of the season, when most sailing ships had already gone south, that steamers usually made their best catches in the 1880s.

The steamers consistently proved their value as an investment: throughout the 1880s the average catch of the steamers was slightly more than twice that of the sailing vessels, and the outfitting costs of a steamer were less than double the outfitting costs for a sailing vessel.

At the end of the season the ships cruised briefly on Herald Shoal, then steered southeast for Cape Lisburne, then south for Bering Strait, taking advantage of the more ice-free waters on the east side of the sea. Once through Bering Strait, the ships often put into Saint Lawrence Bay. There, in the temporary safety of the harbor that would not freeze for several more weeks, they took on water and tryed out any blubber that they might have taken in the previous week or two. They then steered west of Saint Lawrence, Saint Matthew, and the Pribilof islands on their way to Unimak Pass, the easternmost passage through the Aleutians and the shortest route to San Francisco, where they usually arrived in the first or second week of November.

11 / *Shore Whaling*

Once the steamers had been introduced into the whaling fleet there remained few innovations that could give the whalemen better access to the dwindling number of bowheads. The strength and maneuverability of the steamers allowed two or three weeks' more contact with the whales at both ends of the season, but, even so, the steamers, like the sailing whaleships, were unable to penetrate the dense pack ice of the Chukchi Sea in April, May, and June while the whales migrated toward their summer feeding grounds in the Beaufort Sea.

It had been recognized almost since the beginning of the fishery that once the whales entered the ice near Bering Strait they were safe from the whalemen for the spring months. Attempts were made to reach them during wintering voyages to the Chukchi Peninsula in the 1860s, but these were unsuccessful and no one tried again for more than a decade. In 1879 the *Vigilant* was outfitted with supplies to winter at Point Barrow. The experiment was never carried out, for she was lost in the autumn. In 1883, however, Edward Perry Herendeen reached San Francisco from the Arctic and commercial shore whaling in northern Alaska was born.

Ned Herendeen was a New Bedford whaleman who, as a young man, had been aboard one of the first whaleships to reach Point Barrow in the late 1860s. After that he made several Arctic whaling voyages, and in 1872 was first mate on the *Florence* when she went north to salvage the wrecks of the 1871 disaster. Most important, in 1881 he went north to spend two winters at Point Barrow as the interpreter and storekeeper for the U.S. Army's Signal Service Expedition, studying Arctic meteorology as part of the first International Polar Year.

In the spring of 1882 Herendeen saw Eskimos waiting at the lead in the ice for the whales to migrate north past them. There were two villages in the area: one right at the tip of the Point Barrow sandspit; the other, ten miles to the south on the first high land, a low bluff. When the lead began to open in April, eight or ten boats went out from each village. The Eskimos waited at the edge of the shore-fast ice for six weeks or more; when they sighted a whale, they launched their light, flexible umiaks (eighteen-foot boats shaped somewhat like a New England dory, made of driftwood frames and covered with sealskins) and paddled swiftly and silently toward it.

Although many of the boats were equipped with shoulder guns, darting guns and toggle irons, some Eskimos still struck the whale with traditional harpoons, eight feet long and tipped with a large bone toggle head that held a sharp slate blade. Whether they struck the whale with Eskimo or Yankee gear, once they were fast to the whale, instead of enduring what the Yankees called a Nantucket sleigh ride, they simply threw overboard a thirty-five-fathom walrus-hide line. Attached to the line were two or three inflated seal skins, each with a buoyancy of about three hundred pounds. The floats dragged on the line and tired the whale while simultaneously signaling its location to the Eskimos. As the whale surfaced to blow, other boats fastened their floats to it as well. Finally, when the whale was exhausted from trailing as many as twenty floats, the Eskimos could approach close enough to hamstring it by cutting at its "small" with a razor-sharp flint spade. Unable to submerge, the whale could be killed with a lance.

The Eskimos then towed the carcass to the nearest strong shore-fast ice and began to cut it in. They chopped stout bridges into the ice a few yards back from the water, then ran a thick walrus-hide line alternately through these and through slits cut deep into the whale's skin. Thus they formed an effective pulley with which they could haul the whale's head up high enough to cut it off. If the whale was small, they could pull it right out onto the ice. Large whales they cut in directly from the ice edge, using it as a sort of cutting stage and rolling the carcass over as needed. The Eskimos also patrolled the ice edge on foot armed with shoulder guns, wastefully shooting bomb lances at the passing whales, hoping against odds to get an instantly killing shot.

Despite the adoption of some Yankee implements, Eskimo whaling was an extremely conservative venture, shot through with "ceremony and superstition," as John Murdoch, the Signal Service expedition's ethnographer, put it. The captain and his boat were adorned with amulets and for success the captain possessed special songs. During the whaling season women were not allowed to sew, lest they drive off the whales; nor were women allowed to urinate on the ice, and menstruating women were kept ashore. If a death had occurred in a man's family within the year, he was not allowed to participate in the hunt.

The Eskimos of northwest Alaska had only used Yankee whaling gear since about 1870. In Point Hope the whale hunt supplied up to 50 percent of the winter food intake. At Point Barrow the importance of the hunt was probably similar. As a result, the Eskimos were highly ambivalent about Yankee whaling gear, a foreign element in the tradition-bound hunt that not only provided a major source of food but was a central annual cultural ritual.

In 1882 when they caught only one whale, their foreign weapons were thought to be the cause of the poor catch. In 1883 they reverted to their customary equipment but took just two whales. One can only imagine the confusion that the paradoxical situation must have caused the Eskimos.

Herendeen may well have considered the Eskimos too hidebound by custom to be effective commercial shore whalers, but he saw no impediment to Yankee whalemen wintering at Point Barrow and hunting the whales that the ships could not reach. Furthermore he knew that the Signal Service's building would be abandoned and that it probably could be used as the station for the shore whaling operation.

In late 1883 Herendeen approached Josiah Knowles, the manager of the Pacific Steam Whaling Company, with a proposal for a shore-whaling station at Point Barrow. Knowles must have seen the advantages of such an experiment. It would cost very little to rent the station and post a boat's crew there for a year or two; it could be run like one of the shore-whaling stations that had been operating on the coast of California since the 1850s, hunting gray and humpback whales. The California stations required very little capital: one or two whaleboats cruised offshore after the migrating whales, then towed the carcasses to a shore station for processing. Knowles knew that the price of arctic whalebone was on the rise, and if the men should succeed in taking one or two bowheads, then their net return would proportionally exceed that of a steamer. Equally important, the station could also be used as a trading post for fur and baleen.

In 1884 Knowles sent twelve men north on the company's tender *Beda*. Four were landed at Corwin Coal Mine, where they were to spend the winter, staking a claim on the coal seams and exploring the mine's commercial possibilities. The other eight, including Herendeen, were landed at Point Barrow.

The adventure at Point Barrow got off to a shaky start. The station's supplies were lost when the company's steam-whaling bark, the *Bowhead*, was stove and sunk by ice at Blossom Shoals. Herendeen managed to get a winter's outfit together from the other company ships. Despite Herendeen's efforts, during the winter his men began to quarrel and eventually refused to work together. In the spring whaling season of 1885 they had no success, and, as if to rub it in, the native crews caught twenty-eight whales. Five of the men left the station in 1885; for the next winter only Herendeen and two others remained. With this skeleton crew Herendeen was unable to "whale it" in the spring of 1886 and instead ran the station solely as a trading post. By that time Josiah Knowles had probably concluded that Herendeen's skills as a visionary exceeded those as a manager. In the summer of 1886 he fired Herendeen.

Knowles sent ten men to Point Barrow to replace Herendeen's crew. The leader was George Leavitt, who had managed the company's experimental venture at Corwin Coal Mine. Leavitt was assisted by two boatsteerers, Charles Brower and William Mogg. Charlie Brower had spent the winter of 1884-85 with Leavitt at the coal mine. Billy Mogg had served aboard several steam whalers in the Arctic and Sea of Okhotsk.

Leavitt, Brower, and Mogg apparently ran the station efficiently, but caught no whales in the spring of 1887, while the Eskimos caught twenty-two. They did, however, acquire 7,000 pounds of whalebone in trade. That summer Knowles sent up a little sloop, the *Spy*, to the station. She carried two whaleboats, which were almost as long as she was. She was to be used in the summer for cruising in the lagoons east of Point Barrow where whales had been seen feeding on their return from the eastern Beaufort Sea.

By the following spring, 1888, Leavitt, Brower, and Mogg had discovered that some of the Eskimo whaling techniques were more effective than theirs. They hired several Eskimos for a crew, some of whom had been banned from participating in that season's whale hunt because of recent deaths in their families. Charlie Brower had been a member of an Eskimo whaling crew at Point Hope in 1885, so he took on the task of organizing the crew. The Eskimos considered it extremely dangerous to tempt Nature by whaling without observing all their traditional ritual procedures:

The Pacific Steam Whaling Company's first station at Point Barrow, formerly the U.S. Signal Service expedition house

Brower proceeded to the lead edge with his umiak and men, armed not only with Yankee equipment, but also with powerful songs and amulets.

The season of 1888 at Point Barrow was marred by contrary, on-shore winds that kept the shore lead closed and forced the whales to pass far out to sea. Of a total of only four whales taken there that spring, Brower's crew caught one. It yielded 2,200 pounds of whalebone, worth more than $6,000 ($64,440 in 1982 dollars).

Leavitt and Brower had discovered the simplicity of using native crews, equipment, and procedures, and they thought they were poised on the threshold of a lucrative shore-whaling operation. However, when the ships arrived in the summer of 1888, to their surprise they learned that everyone at the station had been fired except Leavitt and Brower and that the company had sent up a group of Portuguese as replacements.

Brower resigned in protest and decided to stay at Point Barrow to form a shore-whaling partnership with several of the dismissed men: Patsy Grey, Ned Arey, and

Fred Hopson (left) and Charlie Brower at Point Barrow, ca. 1917
California Academy of Sciences

Conrad Seim. Later they were joined by Tom Gordon and Fred Hopson, who left, respectively, the *Orca* and the *Grampus* to join them.

At once Brower's group was faced with the pressing need for shelter and an outfit. Almost simultaneously Brower learned that the bark *Fleetwing* had been severely damaged in a violent gale on August 2, 1888, and had been abandoned by her crew. Brower wasted no time. He took an umiak and rounded up an Eskimo crew and eventually reached the broken hulk twelve miles offshore. They jury-rigged steering tackle and spliced lines to trim the sails. With the *Fleetwing*'s hold full of water, they managed to sail the sinking derelict to shore. They stripped the wreck of what gear they could, then managed to buy some supplies from USS *Thetis.* Almost providentially a few days later the little schooner *Ino* was driven ashore in another gale and abandoned. They salvaged her almost intact and found they could live aboard her where she lay.

As spring approached in 1889 Brower put together another native crew. He signed on two families who had arrived from Point Hope, as well as two local men who were available only because a family member had died. The 1889 season was a good one at Point Barrow: the natives caught twenty-two whales; Brower's group, three; and the Pacific Steam Whaling Company, three.

With this cushion of baleen, Brower hoped to increase his profits. He chartered the tiny sloop *Spy* from the company by agreeing to pay one quarter of any whalebone he took. He set off to the east, hoping to cruise as far as Herschel Island. Ice and wind worked against him and he returned to Barrow, clean, having gotten only about half way to the island. On his return Brower found his friend George Leavitt preparing to leave the Pacific Steam Whaling Company's station. Leavitt was due to be furloughed and he was being replaced by Henry D. Woolfe, a troublesome individual who had been a member of the company's experimental party at Corwin Coal Mine. Woolfe had brought a few men with him to replace the Portuguese who had previously replaced Brower's friends. "The squirt of the grapefruit is more predictable," Brower ruminated many years later, "than were the ways of the old Pacific Steam Whaling Company."

Because of the success of the 1889 season and because whalebone was fetching $3.50 per pound, the commercial shore-whaling effort greatly expanded in anticipation of the season of 1890. Antonio Bettencourt ("Antone Bett"), an Azorean, one of the Portuguese who had been fired by the company, chose to remain there to run his own shore station. Also the little schooner *Nicoline,* outfitted by Louis and Ned Herendeen to winter in the Mackenzie delta, had run into heavy ice on the way and had returned to winter in Elson Lagoon behind Point Barrow. The Herendeens planned to go shore whaling in the spring. All in all, in the winter of 1889-90 there were twenty whites living at Point Barrow where none had been a decade before.

In the spring there were 400 people actively engaged in shore whaling at Point Barrow. Fifty boat crews were sent to the lead edge, ten of them by whites; the rest by the natives, who were apparently using all available local personnel including very young men and any Eskimos who visited from other settlements. It was, however, another poor shore-whaling season. Only five whales were caught, four of them small. Brower and his partners caught two, one of them very large.

The Pacific Steam Whaling Company took no whales that year. Woolfe had manned his boats with inland Eskimos who were immigrants to Point Barrow. Caribou hunters who had moved to the coast from the mountains to escape famine caused by a natural cyclical decline in the caribou herd, they were ignorant of whaling and mostly incompetent. For Josiah Knowles, however, this was a minor disappointment: his profits from fur trading at the station were substantial and he had spent the winter devising a new strategy.

Knowles had originally planned to send George Leavitt back to Point Barrow, but

John W. Kelly at Point Barrow Presbyterian Historical Society, Philadelphia

Leavitt had arrived in San Francisco in 1889 afire with the idea of wintering a ship near the Mackenzie delta to hunt the whales that had just been reported there. Knowles consequently sent him north as second mate aboard the *Mary D. Hume* in the experimental wintering at Herschel Island.

For the Point Barrow station he chose John W. Kelly, "a bluff, level-headed prospector," according to Charlie Brower. Kelly had spent much of the preceding five years prospecting at Corwin Coal Mine and shore whaling near Point Hope. Knowles chose well, for Kelly was adaptable, capable, and energetic. He knew the Eskimos well and they apparently liked and respected him.

Knowles had realized that with the reduced whale stocks in the Arctic, merely having a few boats near the whales was not sufficient to insure their capture. To counter the increasingly sparse hunting, he added boats to his ships: the *Mary D. Hume* and the *Newport,* for instance, each squeezed another set of davits on their sides. Knowles, no doubt, told Kelly to carry out his plan for saturating an area with whaleboats.

Kelly followed his orders. On arrival at Point Barrow in 1890 he built a new and more spacious station, abandoning the dilapidated Signal Service Station. He then outfitted eight boats and hired a hundred Eskimos, men and women, using people from as far south as Point Hope. In a dramatic change from Eskimo custom he put up tents on the ice to allow his men to cook, rest, and dry their clothes. He used women to drive dog teams to relay supplies out to his crews, thus allowing the hunters to stay constantly on the alert.

The local Eskimos were extremely apprehensive about these innovations: they had never cooked or camped on the ice, and they feared if the women did not obey many customary practices, the whales would surely stay away.

Kelly changed shore whaling forever, and according to Brower, the change happened in 1891 when Kelly brought in a phenomenal catch. Eighteen whales were taken at Point Barrow that year. The Eskimos took one; Brower and his partner Tom Gordon, five; and Kelly, twelve. At the end of the season Kelly had $100,000 worth of whalebone in his warehouse ($1,074,000 in 1982 dollars), a profit approaching 1,000 percent. To the Eskimos, whalebone had become a currency for buying manufactured goods. They could not ignore the success of Kelly's methods. Although he had violated most of their customary observances, his success had been stunning.

The summer of 1891 was exceptionally icy and the ships had difficulty getting past Point Belcher. The *Orca* had to unload two years' supplies for the station there. Rather than haul tons of supplies seventy-five miles to Point Barrow by sled and umiak, Kelly simply moved to Point Belcher for the winter and built a temporary station. He hired eighty Eskimo men, many of them starving emigrants from Point Hope, to man about ten boats. Point Belcher was not as good a whaling site as Point Barrow, and Kelly's crews came home clean in 1892.

The whalemen at Point Barrow, following Kelly's lead, had also added crews. In addition to the now forty-or-so native boats, Brower and Gordon put out four and Antone Bett put out four. Knowles had now determined that Brower's and Gordon's station was hurting his business. Kelly returned to Barrow in the summer of 1892 with orders to buy the pair out, and Knowles sent word north that none of his company ships should trade with the two. Brower and Gordon accepted the inevitable and left Point Barrow on the USS *Bear* with $40,000 worth of whalebone and $3,500 for their station—a substantial profit on a very small initial investment.

Knowles must not have considered Antone Bett to be a threat, for Kelly made no offer to him. It is unlikely that Knowles's stifling the competition would have added to the Pacific Steam Whaling Company's catch—Brower and Gordon must have been cutting into the company's trade for whalebone and furs.

Kelly moved back to his station at Point Barrow in 1892 and embarked on even greater expansion. To add crews he hired one hundred Eskimo men (with their families this involved about five hundred people). It was, however, a poor season in 1893 and the local Eskimos took the only whale with commercial grade whalebone.

To Kelly's surprise, that summer the schooner *Jenny Wand* arrived at Point Barrow with his old friends, Brower and Gordon. They had formed the Cape Smythe Whaling Company in partnership with the H. Liebes Company, furriers of San Francisco. Brower and Gordon and Herman Liebes had decided to compete with the Pacific Steam Whaling Company by trading for fur and whalebone and to outfit whaling crews in the spring as well.

The following season, 1894, was a good one. Kelly put out twenty-two crews and took eleven whales; Brower and Gordon put out six crews and got ten whales. In all, 43 whales were caught at Point Barrow that year. At the end of the season the stations together had 21,000 pounds of whalebone. The Cape Smythe Whaling Company was off to a good start. It remained active in commercial shore whaling until 1908.

A similar chain of events took place at Point Hope. Peter Bayne, who had been one of the malcontents in Ned Herendeen's experimental wintering at Point Barrow in 1884-85, apparently convinced the S. H. Frank Company, a small operator of whaling and trading vessels, that the shore whaling plan was feasible at Point Hope, Alaska's other great whaling site. Point Hope, like Point Barrow, is a low sandspit that juts more than ten miles into the ocean. As at Point Barrow, leads in the ice are found very near shore, forcing the whales to swim nearby on their way into the Beaufort Sea.

Bayne arrived at Point Hope in the summer of 1887 aboard the little schooner *Ino*. He had eight men with him and began making trouble almost at once. He soon ran afoul of Attungoruk, the whaling captain that Charlie Brower had worked for in 1885 during a leave from Corwin Coal Mine. Attungoruk was tough and powerful. By gall and strength of character he had established himself as the middleman for most of the furs and baleen that were traded to the whalemen from Point Hope. The presence of a whaling and trading station at Point Hope of course destroyed Attungoruk's monopoly. At first he and Bayne merely quarreled. But deep suspicion grew between them. One day Attungoruk, holding his hands behind his back, approached Bayne to talk with him. Bayne immediately assumed that he was concealing a weapon, although in fact Attungoruk was only holding a pipe. Bayne drew his revolver and pulled the trigger. It misfired. Bayne then took it by the barrel and hit Attungoruk over the head so hard that he broke the butt. Attungoruk's strength was legendary. He stood nearly six feet tall and weighed 200 pounds: the blow only opened a gash on his head. He threw Bayne to the ground and sat on him "for quite a while," as Charlie Brower recalled.

The humiliation deeply angered Bayne. When Brower reached Point Hope by dog team early in 1888 to visit his friend Attungoruk, Bayne tried to persuade him to murder the Eskimo. Brower later learned from John W. Kelly, who that winter was prospecting at Corwin Coal Mine, that Bayne had tried to convince him to poison

Attungoruk with strychnine. Brower and Kelly both declined.

But Bayne won in the end. His troublemaking manifested itself in a far more insidious way. That winter he taught Attungoruk how to distill alcohol. A mixture of molasses, sugar, and flour was fermented, then water was added and the mess boiled in an old coal oil can. A gun barrel was inserted high on the side of the can and the edges of the opening packed with dough or clay to make them airtight. The gun barrel slanted downward through a block of ice or snow, and as the hot, escaping vapor passed down the barrel, the distilled hooch condensed on the side of the barrel. This rotgut dripped out of the barrel and was saved very carefully.

It was impossible to suppress these stills once the Eskimos learned how to make them. Before this the Eskimos had only been able to obtain alcohol from the ships during the summer months—months that were comparatively less important for hunting. But by March 1888 Attungoruk was making his own hooch and by the following winter Eskimos were operating stills three hundred miles away at Point Barrow. By the winter of 1894-95 it was reported that at Point Hope only three or four houses in the village did not have a still. Bayne characteristically tried to blame Brower for the introduction of stills to northern Alaska.

By the summer of 1888 Bayne had taken two whales and had traded for 1,500 pounds of whalebone as well as for a number of mink, bear, and fox skins. Despite his success some of the men at the station resigned that summer, among them was Henry Koenig, who returned south and protested to the government about Bayne's foul behavior. The next year Bayne took seven whales. In the summer he was furloughed, or possibly fired, and the S. H. Frank Company station was left under the care of only two men.

In the summer of 1889 another station was set up at Point Hope. Henry Koenig, who had quit Bayne's operation in 1888, had joined with his brother-in-law, John Hackman, and with George Marlin, who like Koenig, had been a cooper aboard whaleships. Each had put up $500 as had two other partners who remained in the south. They were moderately successful and Koenig ran a whaling station and trading post there for nearly two decades.

In that 1890 season only one small whale was taken at Point Hope and that, by the Eskimos. With such a small catch, food was very short, and about 150 people moved up the coast, to Point Lay, Point Barrow, and other places. Although a total of eight whales were caught in 1891 at Point Hope, none were caught in 1892, and most of the starving Point Hope Eskimos moved to Point Barrow to whale for John Kelly, Charlie Brower, and others.

The price of whalebone in 1892 stood above five dollars per pound and it stimulated another innovation in shore whaling. That summer Captain Benjamin F. Tilton warped the small schooner *Nicoline* through the tortuous entrance to Marryatt Inlet, a lagoon enclosed by the two arms of Point Hope sandspit. Tilton wintered there, using the schooner as a base for trading and shore whaling. The following season, he took her whaling.

In the autumn of 1893 the *Nicoline* was joined by two other little vessels: Peter Bayne returned north as master of the *Emily Schroeder* and the *Silver Wave*, owned by

James McKenna. Just as the schooners were preparing for the winter they were hit by one of the most violent storms to strike northern Alaska in a century.

On October 13, 1893, a brutal northwesterly gale hit Point Hope, sending in mountainous seas and raising the water level ten feet or more. Cold seas poured into the Eskimos' semi-subterranean houses. They fled for their lives, at times wading through waist deep water as they struggled toward the high ground eight miles away on the south shore of the peninsula.

In the lagoon all three schooners dragged their anchors and were driven ashore. None was so badly damaged that it could not be lived in that winter, but the *Emily Schroeder* would never float again. Bayne eventually sold her to Henry Koenig for a few hundred dollars-worth of whalebone. The other two schooners were eventually refloated.

The high price of baleen caused the number of stations at Point Hope to grow from two in 1892 to thirteen in 1898. Together the thirteen stations put out seventy crews. Few, if any, of the crews were run by local natives, most of whom had gone to Point Barrow to work for Kelly. As with the catches at Point Barrow, Point Hope's returns fluctuated according to whether or not the wind in a particular spring kept the lead open and allowed the whales to swim close to shore. The catches at Point Hope varied from thirty-three whales in 1896 and thirty-two in 1897 to none in 1892 and 1895.

With only a few exceptions by the mid-1890s most of the shore stations were small, poorly capitalized affairs staffed by only one or two men. Nevertheless the potential for significant profits was sufficient for stations to be contemplated at several other sites and a small shore-whaling operation was established at the northwest point of Saint Lawrence Island, a less favorable whaling site than the northwest coast of Alaska.

With so many crews at Point Barrow, Point Hope, and a few other places, one may well ask how the operators secured enough men to staff them. The fact is that for a few years most Alaskan Eskimos from Bering Strait to the Canadian border were involved with the whaling industry, and many from Saint Lawrence Island and the Siberian coast were involved as well.

In a sense, traditional Eskimo whalers became commercial whalers the moment they discovered that their baleen was a valuable trade item. Before, whalebone had been a by-product of the hunt for meat and blubber. When the Eskimos realized they could trade what had been surplus for manufactured goods, the whale hunt for Eskimos became a commercial enterprise.

In the 1890s, with the industry's pressing need for manpower, most Eskimos chose to work directly for the stations or the ships. From 1890 to 1898 whaleships made 60 winterings east of Point Barrow, and most of the ships carried several native families to act as hunters and seamstresses. At the same time the Brooks Range, the great mountain range that traverses northern Alaska, had been largely abandoned by the interior Eskimos who had moved to Point Barrow and Herschel Island to seek employment. They were deracinated by the famine which had occurred because of a severe cyclical decline in the caribou herds.

In 1893, for instance, with seven ships wintering at Herschel, many of the Point Barrow Eskimos were at the Island. Many Eskimos were able to join whaling crews at Point Barrow, but still the manpower pool was insufficient. Brower was forced to hire men from farther south on the coast, at Icy Cape and Point Belcher. John W. Kelly's crews were almost entirely made up of Point Hope Eskimos who had presumably moved north to avoid starvation in their lands. By this time both men and women were employed as paddlers.

This, of course, left the stations at Point Hope without manpower. The station operators there in turn hired men from farther south, inland Eskimos from the rivers that drain into Kotzebue Sound who were also suffering from the lack of caribou, and Eskimos from the Seward Peninsula.

These southern Eskimos settled around the whaling stations at Point Hope, which were about six miles east of the village. This polyglot community—called "Jabbertown"—in 1900 was made up of two hundred Eskimos from many communities, as well as about two dozen whalemen, including American whites and blacks, Cape Verdeans, Portuguese, Japanese, Germans, and Irish.

Out of this broth of cultures grew a hybrid form of whaling, neither entirely Yankee nor Eskimo. One of the most fundamental and unique differences was in the method of "shipping" a crew. Many whites served as boatheaders. If they were not involved in the ownership of the station, then they were signed on at a lay of one twenty-fifth of the catch, payable at the end of the season in whalebone. Often a bonus was specified in the contract if they caught large whales.

Although some Eskimos signed on a lay, in the intense competition for manpower in the 1890s most Eskimos simply received a salary. Each station signed up the best men and women it could and paid them $100 to $250 apiece in supplies to secure their services. During John Kelly's administration of the Pacific Steam Whaling Company's station, it required from $15,000 to $20,000 to fit out and man twenty or more boats.

Brower, like Kelly, maintained a number of families throughout the year just to have them available in the spring for his whaling crews. The explorer Vilhjalmur Stefansson saw his operation.

> . . . these men get each year as wages about two hundred dollars' worth of supplies. This means that the Point Barrow community leads an easier life than any other community does as a whole in any land where I have ever traveled. The whaling season in the spring is six weeks, and it is six weeks of fairly easy work at that. For all the rest of the year the men have nothing to do,—are their own masters, and can go wherever they like, while their employers must not only pay them a year's wage for six weeks' work, but also furnish them houses to live in, usually, and rations for the entire year. Of course, the men are expected to get their own fresh meat which they do by seal and walrus hunting, and by cutting in the whales,—only the [whalebone] of which goes to their employers. The employer supplies them with cloth for garments, and such suitable provisions as flour, tea, beans, rice, and even condensed milk, canned meats and fruits. Each man each year gets, among other things, a new rifle with loading tools and ammunition.
>
> The pay-day of the Point Barrow Eskimo comes in the spring, and their employer hands them out rifles, ammunition, cloth, provisions, and various things which the people scarcely

Hauling a whaleboat across a pressure ridge of ice toward the lead edge, Point Barrow, 1898

know what to do with. So they load them into their skin boats and take them east along the coast, to sell them at any point in the Colville or at Flaxman Island. To give some idea of the scale of prices it is worth while to say that one of the men whom we met returned with ten deerskins, which was all he had received in the Colville River for a boat-load of supplies consisting of two new rifles, two cases of smokeless powder ammunition for these, twenty-five pounds of powder and a corresponding supply of lead and shot, three bolts of cloth, a case of carpenter tools, some camp gear, three hundred pounds of flour, sixty pounds of good tea, two boxes of tobacco, and various other articles too numerous to mention. The ten caribou skins were of varying quality. The best of them were worth that year about five dollars apiece, and the total value of the ten skins could not have been more than thirty dollars.

Most stations sent out a combination of wooden whaleboats and sealskin covered umiaks. The whaleboats were used for cruising under sail far away along the lead.

A whaleboat cruising in the lead at Point Barrow American Geographical Society

Their disadvantage was that they were comparatively heavy and cumbersome to move out to the lead edge over the pressure ridges, and their thin cedar planks were easily broken on the jagged ice during this transit. Few whaleboats were built in northern Alaska; most were brought from San Francisco or New Bedford.

Umiaks, lacking centerboards, were comparatively poor sailers. They were, however, light, quiet, and fast over short distances under paddles. Because their joints were tied together, not screwed or bolted, the entire frame was flexible—and this feature, combined with the tough bearded-seal cover, allowed them to take a lot of punishment without breakage or leaking.

Umiaks were also easy to build. Brower, for instance, spent much of the winter making boat frames, and he developed a hybrid form of umiak with steamed and bent hardwood ribs. At the end of the whaling season he sent out some of his crews hunting for bearded seals. The boat skins had to be changed every other year, and about six skins were needed for one boat covering. Some of the more marginal operators rented fully equipped boats from others for approximately one half of the whalebone taken.

A whaling crew with their umiak, waiting at the lead edge, Point Barrow, 1898 Private collection

The necessary whaling gear for a properly equipped crew differed from that of a pelagic whaleboat. The boat needed two darting guns, one of them rigged with a toggle iron. A third darting gun, also rigged with an iron was placed at the ice edge in case a whale should surface close by. A shoulder gun was also carried in the boat, and one or two others were kept ready at the ice edge. The toggle iron was usually rigged to about thirty fathoms of line. Two inflated sealskin drag floats ("pokes") were attached about twenty fathoms from the iron, and another, large float was attached at the very end of the line.

Some of the whaleboats were equipped with heavier weapons. Because the water in the leads is often quite calm, hence conducive to aiming fixed firearms, a few men experimented with the Mason and Cunningham swivel gun that fired a bomb nearly two feet long. Others used the three-barrelled Haviside gun that simultaneously

Whaleboats and umiaks returning to Point Barrow, towing a dead whale Glenbow-Alberta Institute

fired two bombs and a toggle iron. These guns were never widely accepted, however, because the power of their recoil strained the boat's timbers, and their extra power made them very dangerous in case of accidents.

At the beginning of April the men prepared for the hunt. First a trail had to be cut through the broken and jumbled shorefast ice out to the lead edge. Only then could the boat be dragged out. Even so it was back-breaking work getting the whaleboats to the lead edge. In 1895, for instance, it took Henry Koenig fourteen hours with eight men and five dogs to drag a whaleboat eight miles to the water's edge.

Once the outfit had reached the lead, the first business at hand was to set up the wall tent on the most stable ice nearby and to light the sheet iron stove, fueling it with blubber for drying clothes and cooking. Next the whaleboat was launched and moored to the ice, or, if the crew was using an umiak, it was set up on ice blocks to prevent it from freezing tightly to the ice. Both procedures were carried out because with the whale's sharp hearing, the foreign sound of a boat's bottom being scraped along the ice as it was launched would gally the whale. If the lead was wide and long enough, a crew with a whaleboat set out, cruising south, against the direction

Beginning to haul out a small whale onto the ice at Point Barrow, 1898. Its left lip has been cut off Private collection

of the whales' movement. Those using umiaks would sit quietly on a sled, protected by a small canvas windbreak, constantly scanning the lead to the south of them. The boatheader tried to place his umiak in a small bay in the ice because the whales often moved along the edge of the shorefast ice, taking shortcuts under the projecting points and rising to breathe in the small bays.

The harpooner often spent much of the time at the lead edge sitting in the bow of the umiak. If a whale surfaced directly in front of him, the crew simply threw the umiak, harpooner and all, directly at the whale.

If a whale surfaced farther out in the lead, they launched the umiak swiftly and silently and paddled noiselessly toward the whale, trying to approach from its blind spot, either directly in front or behind. If they succeeded in reaching the whale before it dove, the harpooner tried to strike it with both darting guns while a man further aft simultaneously tried to bomb it with a shoulder gun. It was the job of the

Chopping the headbone off a very large whale prior to hauling out the carcass, Point Barrow, 1908 U.S. Geological Survey

man directly aft of the harpooner to throw the floats and line overboard a split second after they struck the whale. If the line fouled the boat, the whale could easily pull it under.

The instant it was struck, the whale began its dive. For a moment the floats danced on the water, spinning crazily like tops as the line uncoiled from around them. Then they too were pulled under. Wounded whales usually dive and run back in the direction from which they came, or, if badly wounded, they often double back again. At the sound of the dull whump of the darting gun or shoulder gun, the other boats nearby immediately put off and, paddling furiously, tried to get to the place where they judged the whale would rise next.

First there was a mere ripple on the surface, then one of the floats would pop up and shortly thereafter, with a mighty PAAAAAAAAAAH, the whale would break the surface, exhausting its lungs in a great, V-shaped whoosh of spray and air. The boats charged toward the whale and the harpooner now held the shoulder gun, hoping to get a killing shot at the top of the neck. It was common practice that if a boat killed a whale that was towing another station's floats, the boat that killed the whale was entitled to one quarter of the baleen.

Hauling out a whale at Point Barrow University Museum, Philadelphia

Once the whale was dead the carcass had to be towed to shore-fast ice for butchering. Usually a station's boats assisted each other in towing a kill. They lined themselves along the whaleline to tow the carcass back. Sometimes, however, a crew took a whale far from their own station's boats. If they were also far from the shore-fast ice and in danger of losing the carcass because of the distance or the wind or the current, they were often forced to signal for assistance to another station's crew.

The fee for this aid was variable and negotiable, usually between one-quarter and one-half of the baleen. In 1896, for instance, Henry Koenig killed a big whale at 7 A.M. on May 27, twelve or fourteen miles from the ice. He put some pokes on its flukes to keep it floating horizontally, then lashed its fins together to reduce the drag. After six hours of towing, his crew had made little progress against the wind and current. He hoisted a flag to his masthead and an umiak came out to help. They negotiated the fee at one-third of the whalebone. The pair then towed the carcass together, not reaching the shorefast ice until 4 P.M. the next day. They finished cutting in the whale at 2 A.M. May 29. For Koenig's crew it was more than two days of nonstop work.

Whale out on the ice, Wainwright, Alaska, the line around the whale's rostrum Dartmouth College

When a crew towing a whale found shore-fast ice that was strong enough they began the equally tiring job of cutting in the carcass. While they secured the whale to the ice edge, someone was sent to shore for the station's cutting gear: spades, axes, boarding knives, and massive blocks and tackles. The blocks and tackles were expensive, and if the station did not own a set, it was forced to rent it from another. In one case the fee was sixty pounds of prime whalebone.

As soon as the crew reached the ice with the carcass, they hoisted a flag on a pole and mounted it on a high piece of ice nearby. This signalled that a whale was being cut in, and other crews sent a number of men to help. For this they received shares of the meat, blubber, and blackskin (the Eskimos considered blackskin a delicacy). The baleen was the property of the boat that had first struck the whale, minus, of course, any commitments that had been made in the meantime.

Henry Koenig and his wife Powsanna, Point Hope, 1906 Bancroft Library, University of California, Berkeley

If it was a big whale, the head had to be cut off first, and hauled onto the ice. If the whale was very big, the blubber and skin also had to be cut off while in the water. Surprisingly big carcasses can, however, be hauled out on the ice by a large number of people hauling on intelligently rigged tackle that is firmly anchored to heavy bridges cut in the ice.

It took from twelve to twenty-four hours to cut in a whale and when it was complete, all that remained was a small pile of useless entrails and a vast blood stained plateau of ice. The rest had been taken for trade, for food, for building materials, and for fuel. The meat and blubber were stored, frozen, year round in underground caches that had been dug into the permanently frozen ground.

By the middle of June the whales had all passed the whaling villages and the men then turned first to hunting bearded seals for boat skins, then to scraping and cleaning the whalebone in preparation for the ships' arrival. Whalebone became a currency in the Arctic. The Eskimos and small operators used it to repay debts and to establish credit at the stores run by the larger stations.

The years from 1896 to 1898 marked a clear change in the tenor of Alaska shore whaling. Josiah Knowles died in 1896 and the Pacific Steam Whaling Company decided to discontinue its shore-whaling operations, leaving Charlie Brower's renamed Cape Smythe Whaling and Trading Company as well as a number of small independents at Point Barrow. Nevertheless, the massive shore-whaling operations of the preceding decade had accelerated the bowhead population's decline and never again would the great catches be made. Although the price of whalebone remained relatively high for another decade, the number of shore stations declined gradually.

Significantly, the Eskimos began once again to take control of the shore fishery. A number of Eskimos had accumulated sufficient capital to put out their own crews—a few put out as many as six—and they began going into partnership with whites or simply hiring them outright.

By the turn of the century so many whales had been killed by the ships and the shore whalers that the catches were erratic and generally poor. In 1906 Antone Bett at Point Barrow wrote to his friend Henry Koenig at Point Hope. He complained of his poverty and expressed the hope that in the spring "there might be a blind or sick whale marked Antonio."

In 1907 the European fashion for narrow waists began to change. The whalebone market softened and the agents had trouble selling what stocks they held. The precipitous decline in whalebone prices continued, and by 1914 most of the whalemen, both white and Eskimo, had ceased hunting commercially. Thus, for the first time in nearly half a century, the crews that went to the lead edge in the spring were hunting whales primarily for food.

But by 1914 the Eskimo whalemen were hunting a population of about three thousand whales—between 10 and 20 percent of its size in 1847. Although the whaleships were the primary agent in the bowheads' reduction, the shore fishery was also an important factor. From 1849 to 1914 the whaleships retrieved better than 75 percent of the whales that were struck; however, the shore fishery was far more wasteful. Although it is likely that from 1890 to 1914 shore whalers captured less than six hundred bowheads, it is the nature of shore-based whaling in the western Arctic that whales are only vulnerable for a very short time before they become gallied and dive, seeking the safety of the ice. Consequently most shore whalers made it a practice to fire shoulder guns at most of the whales that were out of range of their harpoons. Some of these wounded whales were captured, but the majority escaped into the ice, where most died—out of reach of the whalemen. The recovery rate for the wounded whales was at best probably 50 percent and in difficult years may have been only 10 percent.

Scraping whalebone at Point Hope California Academy of Sciences

For more than half a century after the collapse of the commercial shore fishery the level of catch and effort remained relatively constant, with the Eskimo shore whalers taking an average of less than twenty whales per year. There was no commercial value to the whaling. The Eskimos were taking them for food and raw materials, solely. But in the early 1970s the situation changed abruptly. The number of crews and the catches began to rise dramatically, increasing roughly three-fold by 1976. Before this time the number of crews had been limited by the high cost of outfitting and maintaining them. With the boom in Alaskan construction projects, particularly the trans-Alaska pipeline, however, it became possible for an ambitious Eskimo to earn the $9,000 or so necessary to buy the darting and shoulder guns, bombs, harpoons, floats, lines, boat, sleds, snow machine, tent, stove, and food, and thus to attain the highly prestigious position of whaling captain.

The changing character of Eskimo whaling did not go unnoticed. As the whale catch began to rise, so did scientists' and conservationists' concern about the impact on the bowheads. In particular, they worried about the unknown, but at least proportionate, increase in the number of wounded whales—those struck and lost by darting or shoulder guns—that were not retrieved (many, if not most, of which are assumed to die).

As the catch rose (it reached 48 whales in 1976 with another 43 known to have been wounded; 28 and 77, respectively, in 1977), the Scientific Committee of the International Whaling Commission issued repeated warnings about the possible consequences of the increasing hunt. In 1976 it underscored its concern by passing a resolution asking the United States to reduce both the catch and the loss, but the United States took no important action regarding this request.

Then in 1977, with international sentiment rising against the uncontrolled bowhead hunt, the IWC voted to rescind the Eskimos' exemption from the otherwise total ban on bowhead whaling that had been in force since 1931. Ray Gambell, the secretary, wrote, "Clearly this was a drastic measure but the evidence presented by the scientists indicated that there was a real risk that the expanded slaughter of the bowhead whales, many of which were going to waste, would lead to the extinction of the stock within the foreseeable future."

The Eskimos' outcry was immediate and sharp, and the United States government was unprepared for the uproar, but it eventually succeeded in securing a quota for the Eskimos—presently a few more than twenty whales per year. The controversy is far from over. One thing is certain, however: the Eskimos and conservationists alike are the recipients of the legacy of the historical commercial whaling industry.

12 / *Herschel Island*

With steam power the whaling fleet was finally able to penetrate to the farthest boundaries of the Beaufort Sea.

In the 1870s the fleet began sailing east of Point Barrow with the intention of searching the waters off the Mackenzie River delta, but these passages were brief and unsuccessful. The whalemen knew that the Arctic bowheads were identical to the Greenland whales and assumed, therefore, that there must be an open-water passage from the eastern Arctic to the western Artic through which the whales could travel. Many had known for twenty-five years that the explorer Sir John Richardson had seen bowheads at Cape Bathurst, 800 miles east of Point Barrow. Thus the westward-traveling whales they met at Point Barrow in August and September were presumed to have come from those waters.

This belief gained support in 1870 when the *Cornelius Howland* was cutting in a bowhead off Wainwright Inlet. In a healed-over wound the men discovered an old whaling iron marked "A. G." It was widely assumed that this iron had come from the *Ansel Gibbs,* an American whaleship that had only cruised in the eastern Arctic, particularly in Hudson Bay. Then in 1881 an Eskimo in Kotzebue Sound produced an English-style whaling iron marked "Scorraos London." He claimed it had been taken from a bowhead the previous autumn. English whaleships never hunted in the western Arctic; they cruised off the west coast of Greenland. Although it remains to be proved whether or not some bowhead whales traverse a northwest passage, it is fact that the easternmost whale kills in the western Arctic and the westernmost kills in the eastern Arctic were separated by only a little more than 500 miles.

So, with the advantage of steam power, all the whalers needed was good weather, little ice, and few whales near Point Barrow to send the auxiliaries eastward. In 1886 the steamers reached Barter Island, 300 miles east of Point Barrow and only 200 from the Mackenzie delta. Two years later Captain George Bauldry took the *Orca* a hundred miles farther and passed the 141st meridian, the boundary between the United States and the British possessions in North America.

It was not the crew of a powerful steam bark, however, but a man in a whaleboat who first reached the Mackenzie whaling grounds. In 1887 a group of Point Barrow Eskimos returned from a trading trip to the Mackenzie delta and told Charlie Brower, the shore-based whaler and trader, that they had seen large numbers of whales playing in the waters of Mackenzie Bay. The following year Brower outfitted one of his boatsteerers, "Little Joe" Tuckfield, "a beachcomber of an adventurous turn of mind," with a whaleboat, gear, and a year's supplies and sent him east with

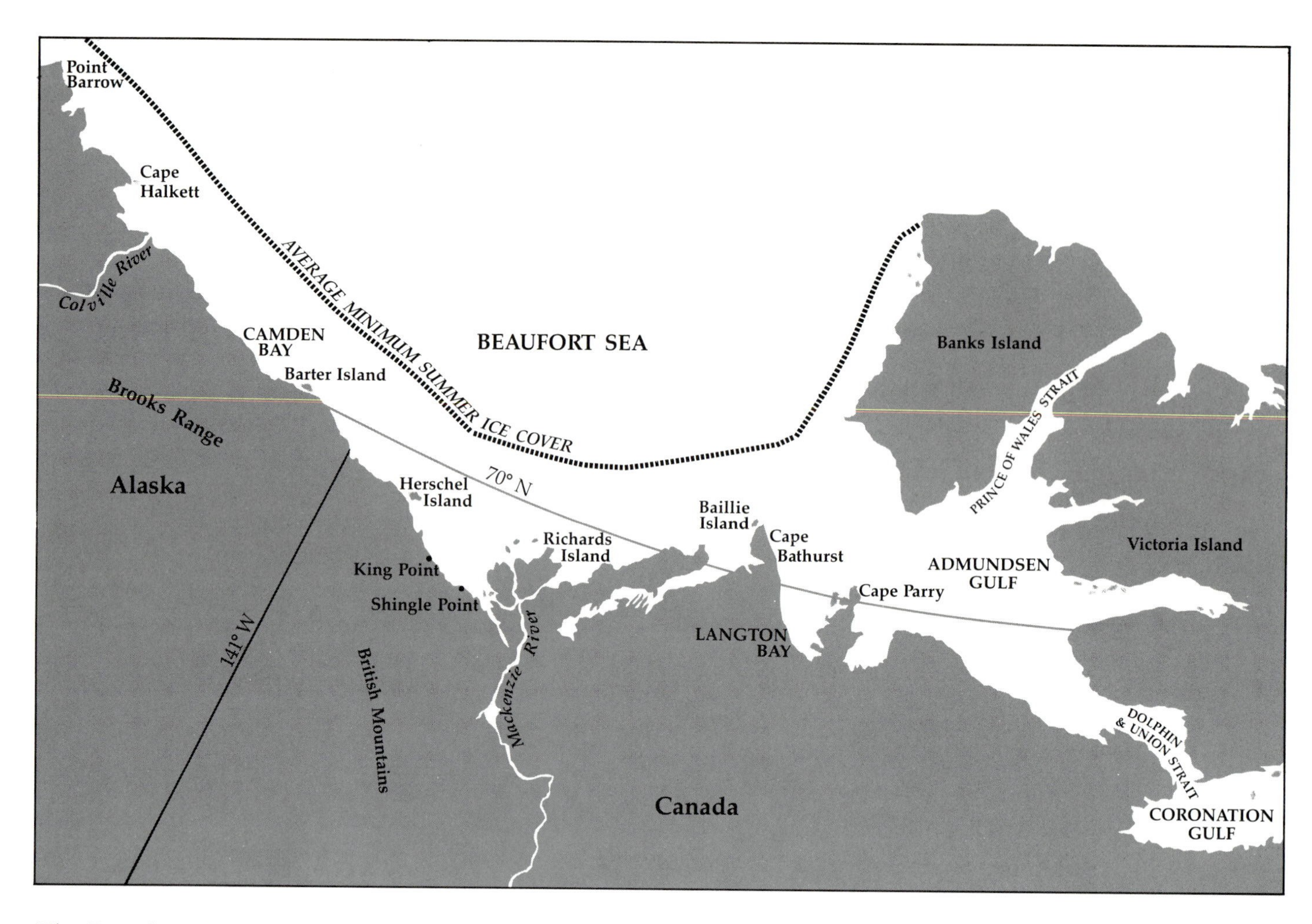

The Beaufort Sea

an Eskimo crew on what was to be one of the most important voyages in the history of the western Arctic.

Tuckfield set out from Point Barrow about July 25, 1888, and returned August 5, 1889, with news that "made about everybody crazy." He had wintered in the delta and told of abundant driftwood there. He had passed the winter without incident and had traded for furs and visited the Hudson's Bay Company's post on the Peel River. He had returned safely, and on the way he had found a good harbor at Herschel Island. But most important the whales were "as thick as bees," and he had struck and saved one—he had the baleen to prove it.

Tuckfield was allowed to stay at Point Barrow only three days. On the eighth of August he was taken aboard USS *Thetis* to act as pilot, accompanying John W. Kelly, who was the interpreter. The *Thetis*, a naval vessel, had been sent north to assist the U.S. Revenue Marine Service in "looking out for the commercial and whaling interests in Bering sea and the Arctic ocean [sic]." Because most of the steamers had

Little Joe Tuckfield, discoverer of the whaling grounds in the eastern Beaufort Sea National Archives

headed into the Beaufort Sea, Captain Charles Stockton saw it as his duty to follow them and to support them if necessary.

That year the whaleships had, in fact, only reached as far as Harrison Bay, about 150 miles east of Point Barrow, when, on August 2, Tuckfield boarded the *Grampus* and told his story, reporting "plenty of open water up there in the forbidden sea." Because their luck had been poor, seven steamers started east. Early on August 12 the *Lucretia, Jesse H. Freeman, Orca, Narwhal, Thrasher, William Lewis,* and *Grampus* anchored at the east side of Herschel Island.

The ships then continued on along the coast in loose formation as far as the west mouth of the Mackenzie River where they dropped back to Shingle Point to find shelter. There the first of the fleet to anchor were boarded by Eskimos who were seeing the first ships in those waters in thirty-five years.

But the *Jesse H. Freeman* arrived at the anchorage after the others with the alarming news that there was a shoal fifteen miles off shore. "This frightened all hands to death," remembered Hartson Bodfish, the first mate on the *Grampus.* "They wouldn't take any chances of getting ashore in that lonesome place." Immediately

Pauline Cove, Herschel Island, 1970. The whalemen's houses are visible on the sandspit Paul Fenimore Cooper, Jr.

the fleet, with the exception of the *Orca* and *Thrasher,* started back toward Point Barrow.

On their way back the ships passed the *Beluga* and USS *Thetis* bound east. Stockton was on his way to look after the *Thrasher* and *Orca,* which had stayed behind to hunt for whales. On August 15 he rounded Herschel Island and found the two steamers safely at anchor. He also anchored and named his anchorage Thetis Bay. A couple of miles to the east the *Thetis*'s men discovered a small cove, protected by a quarter-mile sandspit. Stockton sent a party to sound the harbor, found depths of up to three fathoms, and named the anchorage Pauline Cove in honor of his wife. The sandspit he named Simpson Point after his ice observer, John Simpson. On the seventeenth, after having surveyed the island, he weighed anchor for Point Barrow, leaving the *Orca* and *Thrasher* safe and still looking for whales. Along the way he passed the Pacific Steam Whaling Company's little two-boat sloop *Spy,* which in the excitement had been dispatched on the strength of Tuckfield's report.

The Mary D. Hume *at Herschel Island*

When the *Thetis* reached Point Barrow with the news that, so far, no whales had been taken near the Mackenzie, many thought that Tuckfield was a fraud. George Leavitt, who was in charge of the Pacific Steam Whaling Company's shore station there, was one of the few who believed Tuckfield. Leavitt boarded the *Grampus* at the end of August to go to San Francisco. He was bent on convincing Josiah Knowles, the company's managing agent, to send him back with a ship so that he could winter at Herschel. After Leavitt had left Point Barrow, the *Orca* and *Thrasher* returned with the report that they had seen plenty of whales and that they had taken two apiece. Wind and fog had kept them from catching more. Had they been sufficiently provisioned, the captains announced, they would have tried to winter over at Herschel because the whales seemed to be "sticking on at that place."

George Leavitt was not given a ship until 1893, but his convictions, combined with those of the captains of the *Orca* and *Thrasher,* convinced Knowles of the opportunity in the eastern Beaufort. Knowles may also have been spurred by the knowledge that Captain Louis Herendeen and his brother, Ned, had heard the same reports as Brower and had set out from San Francisco in 1889 in the little schooner *Nicoline.* Provisioned for twenty-seven months, she was to winter north and to explore for new whaling grounds near the Mackenzie delta and Cape Bathurst.

To lead the wintering experiment Knowles chose James Tilton and Albert Norwood, two of the company's first officers who wanted to take the risk of wintering north. Because shallow waters had been reported, Knowles decided to give Norwood the *Grampus,* the lightest of the company's ships, and for Tilton he bought an even smaller one, a tug, the *Mary D. Hume.*

The *Mary D. Hume* became one of the most famous vessels in American whaling history (and is still afloat). Although she was well suited for whaling in the Beaufort Sea, to the whalemen who would have to take her through the mountainous seas of the North Pacific she must have seemed a mixed blessing at best. She was only ninety feet long "smaller than many a pleasure yacht," observed Hartson Bodfish, her first mate.

The *Hume* was too small to carry tryworks—a point which suited Knowles. Knowles had concluded that with the price of baleen approaching four dollars per pound (while whale oil fetched only sixty-five cents a gallon), it made sense to take only whalebone in the short Arctic season—and not to waste time cutting-in and trying-out blubber. The practice of taking only the headbone from a bowhead began about 1880 because of the steep rise in baleen prices. Initially only the smallest vessels limited their take to baleen, but gradually the practice spread until more than half the fleet cast off the carcasses after having taken only the headbone. This waste, in an already wasteful enterprise, was enormous and, regrettably, it continued until the end of the industry.

The *Mary D. Hume*'s voyage began badly. During a hasty outfitting she was sloppily rerigged as a brigantine. Despite the haste, she left San Francisco on April 19, a month late. The shoddy rigging showed up at once. A week after her departure she ran into a heavy gale and both top masts collapsed. The crew replaced the main topmast with a whaleboat's mast, which allowed them to put a lookout aloft, but they could not set a gaff topsail all season. To make matters worse, as the men were working on the foretopmast, it fell and a block crushed two of Hartson Bodfish's toes. "I thought rather fast," he recalled:

> my foot was numb from the accident. I knew that the longer I waited, the more painful the amputation would be, so, with the steward and cabin boy looking on and groaning, I whetted up my knife and cut it off myself. The way it was injured made it necessary for me to unjoint the bone from my foot, too, but I did it, and there was considerable satisfaction in having performed my first surgical operation.

The *Hume* reached Port Clarence early in July and loaded more supplies aboard from the other company ships. Her bad luck still held, however, for she sailed for the Arctic July 10, only to run aground on the Sea Horse Islands near Point Franklin a few days later. When she reached Point Barrow, the plan had been to load some of

Hartson Bodfish at Point Barrow, 1892

her excess supplies aboard the company's little whaling sloop *Spy* and to bring the *Spy* along to Herschel. The *Spy* had been left ashore at Point Barrow and once it was launched, the men found that she leaked so badly that she nearly sank at the anchorage.

So, loaded with yet another infusion of provisions, the *Hume,* in company with the *Grampus,* prepared to set off into the unknown. Most of the other whalemen considered their expedition to be the height of folly, predicting that they would lose their ships and return to Point Barrow in their whaleboats. Only canny old Captain Barney Cogan predicted that they would come back loaded with whalebone. They departed without fanfare.

The *Mary D. Hume* reached the southwest sandspit of Herschel Island on August 10. The men sounded for an anchorage and found five and a half fathoms immediately behind the tip of the spit. The *Hume* was so heavily laden that they could do no whaling until she was freed of her freight. A few days later, having decided to winter there, the men began unloading their lumber and supplies on the sandspit.

Josiah Knowles had sent up 5,000 feet of lumber for a house to store provisions and whalebone ashore during the winter—an intelligent precaution in case of fire aboard the ship. The men set to work at once on the house and had it finished by the eighteenth of the month. They then immediately set out to join the other vessels. The fleet had passed them bound east.

On the thirtieth the other steamers departed for the west, leaving the men of the *Hume* and *Grampus* feeling extremely lonely, but more depressing was the fact that they had taken no whales.

They continued cruising near the island without success, alternating their hunt with collecting driftwood for their winter fuel.

On September 18, while they lay at anchor in Pauline Cove, the ships were unexpectedly frozen in—nearly twenty miles by water from their storehouse at the opposite end of the island. Although they tried to buck their way out a few days later, they surrendered to the inevitable on the twenty-second and began landing their firewood.

The two steamers found themselves in company with the little schooner *Nicoline*. Under the command and ownership of Louis and Ned Herendeen, she had gone north a year before and arrived too late to reach Herschel. Instead she had wintered in Elson Lagoon at Point Barrow. The crew had tried shore whaling there in the spring, but, having no success, the *Nicoline* had sailed east, clean, to the island.

The three vessels were, in fact, extremely lucky to have been frozen in at Pauline Cove. It proved to be a secure, if small, harbor and was protected from the crushing pressures of moving pack ice.

As the weather permitted, the men of the *Hume* and *Grampus* spent most of the next month tearing down the storehouse on the southwest sandspit and sledding the lumber and supplies back to Simpson Point where they rebuilt it and again cached their provisions. At the same time the men were busy sawing up the driftwood and cutting blocks of ice from the freshwater pond they discovered on top of the island a few miles away.

As the cold increased the men began housing their ships by putting frameworks over the decks and covering them with sails so that the men could work in comfort. Before long enough snow had fallen so that they could cut snow blocks and bank the sides of the ship with walls three or more feet thick—effective insulation. Below decks they set up stoves and kept the fires fed with driftwood. On October 6, feeling that they were now securely frozen in for the winter and that no gale, however strong, could break up the ice, the *Hume*'s engineer blew down her boiler and put out the fires.

As the days shortened the cold grew sharper. On November 29 the sun rose only briefly at noon over the jagged outline of the Buckland Mountains, then sank for good, not to be seen again for more than six weeks. By then the men had completed their preparations for winter and could pause to consider their surroundings.

They lay more than 4,000 miles from San Francisco, 200 miles above the Arctic Circle, and 400 miles east of Point Barrow. Their little island was only a mile from the mainland and only about eight miles on its longest axis. To the whalemen, searching for meaningful comparisons, it seemed about the size of Martha's Vineyard, Massachusetts—an error caused no doubt by the island's treeless tundra cover and gentle slope that magnified its size in their eyes.

To the south the flat coastal plain bordered the saw-tooth mountains that are the farthest northern reach of the Rockies, the great cordillera whose southern terminus lies at Tierra del Fuego. This cordillera guides the course of North America's two great arctic river systems. To the south of the mountains lies the 2,300-mile Yukon, and to the east lies the 1,120-mile Mackenzie.

Unknowingly the ships were poised to grope toward the waters east of the delta where the silt from the Mackenzie's vast outflow finally filters out, allowing sunlight to penetrate the formerly murky waters and thus to trigger the plankton bloom on which the bowheads feed.

From Pauline Cove, the whalers' view was framed on the north and west by the gentle slope of the island. To the south was the ragged outline of the mountains. Everywhere else was the vast whiteness of the Polar Sea. And dominating all was the wind.

The whalemen quickly learned that Herschel is a windy place. They watched as the wind, shrieking and ripping at the pack, drove a thirty-foot pressure ridge onto the sandspit, burying their wood pile under hundreds of tons of rubble ice. In the winter at Herschel it blows from the west and northwest almost half the time, and a good portion of the remainder it blows from the east and southeast. It blows a gale an average of four times a month. Coupled with temperatures averaging around –20° F. during much of the winter (occasionally it has fallen to –50°F.) the cold can be severe: the "rule of thirty," an arctic adage, has it that at –30° F., with 30 miles per hour of wind, exposed flesh will freeze in 30 seconds.

Within this environment the men had to consider their food supply. Although they had been amply provided with staples, they needed fresh meat both for variety and as an antiscorbutic. The causes of scurvy were imperfectly understood (one whaleman believed that drinking water melted from snow was a contributor), but it was known that a diet of raw meat would cure it (as we know today, by replenishing vitamin C).

Caribou were the only large source of winter meat and to hunt for them the men had to be dressed to withstand the cold. As the autumn progressed they found that their heavy woolen clothes were inadequate for coping with the temperatures. Heavy and sweaty, woolen clothing restricted movement and kept them perpetually chilly. They soon adopted Eskimo dress, discovering that a double layer of caribou skin clothing is light and totally comfortable even at –30°.

It requires skill to hunt caribou and it is unlikely that the whalemen would have been able to provide for their own needs. Fortunately, the word had gone out that ships intended to winter at Herschel, and the lure of trade goods brought sixty Eskimos to the island for the winter. Most surprisingly, in October a group of Rat Indians from south of the mountains arrived with caribou meat and skins to trade. Theirs was the first of countless trading visits to the island by Eskimos and Indians during the next twenty years. Four days later Hartson Bodfish left on a hunting trip, also the first of many that the ships' officers would make.

By the end of February their supplies of game were large enough to require storage. They settled on the ingenious idea of building a cold storage cellar in the permafrost. At Herschel the ground only thaws about a foot in the summer and below that lies hundreds of feet of permanently frozen mucky silt. The men simply blasted a large hole into the ground, then squared it up to twelve feet on the side with pick axes and roofed it over with driftwood. Later they covered the wood with a layer of blocks of sod four feet thick. They added a small doorway and found that their meat would stay frozen throughout the summer.

Boredom was one of their biggest problems. "Routine for the day month after month," Hartson Bodfish wrote to his mother the following summer: "Get up in the morning for breakfast, lay around till lunch, lay around till supper, lay around till bedtime, varied by playing cards and setting the men to hauling and sawing wood and once in a while going after a load of ice or meat." This boredom and loneliness drove one of the men toward suicide. Thinking that freezing was a comfortable death, he wandered off one morning and was found the next day with his feet frozen. His toes required amputation. Bodfish, who had amputated his own toes at the Aleutians, and George Leavitt—who would often operate on frostbitten men in the next twenty years—were called in to assist the captain with the surgery. The tools were a butcher knife and a hacksaw. "Captain Tilton and I missed our vocation when we started whaling, we ought to have been surgeons," Bodfish wrote to his mother.

> We administer chloroform and off with them, so far everything has turned out all right and none of our patients have died on us. I think there will be a "corner" on toes when we all get back to Frisco, as three of us have lost eight and I think two more will have to go. If Father hadn't bought a livery stable, I think I would have been a doctor.

The monotony and isolation contributed to another type of lunacy that winter: on March 15 Robert Coleman, the *Hume*'s fireman, deserted and on the following day three men from the *Grampus* left, heading for the Yukon gold fields 600 miles to the south. In preparation they pilfered a number of items, and on the night of their departure they broke into the storehouse for more, but with the limited supplies and knowledge they would surely have perished.

Captains Tilton and Norwood, however, could not simply leave the deserters to fate. If an attempt were not made to bring them back, others would surely follow suit. The captains set out at once and quickly picked up the men's trail. Their sled was overloaded and as they tired, they had thrown away stores to lighten their work. The captains soon overtook the men, finding them exhausted in a native's camp. All were badly frostbitten; two were unable to walk.

Robert Coleman confessed that the plot was wider still. He implicated several who had not yet deserted: the *Hume*'s steward, John Meyers, as well as the cook, the cabin boy, and a foremast hand. Coleman's toes were so badly frozen that they had to be amputated; the others, who were in better shape, were placed in irons. The steward and cook were derated and sent before the mast as ordinary sailors.

The other sad event of that first winter took place on November 4 when J. A. Drayton, the first mate of the *Grampus,* died of "dropsy." The next day he was buried ashore in a shallow grave that the men hacked out of the frozen ground. His was the first of several dozen whalemen's graves that would be dug at Herschel over the next two decades.

For someone in his first arctic winter, spring arrives with agonizing slowness, and for the men aboard the ships, suffering from boredom and cabin fever, the thermometer's rise above zero in the long days of April was merely tantalizing. The land and sea were still their uniform bleak whiteness. But soon signs of life began to appear. First they saw a few caribou moving northwest along the coastal plain. They were the vanguard that preceded the arrival of the vast herds that travel north each spring from their winter range south of the mountains to their calving grounds along the arctic coast. And when the first snow buntings reached the island, the Eskimos told them that spring was indeed arriving.

The men aboard all three ships felt the need for action. With the long, hard winter drawing to a close, none of the ships had a whale to show for their pains. At the beginning of May, when a lead of open water appeared at the north end of the island, they began preparations for shore whaling, trying to make something out of what so far had been a dismal failure. Bodfish and George Leavitt, the *Hume*'s second mate, set out with their men and dragged their whaleboats six miles over the ice to the lead edge. They reached the water after six days of back-breaking work and camped there, sleeping in their boats and cooking on the ice. On their first night the temperature dropped to –16° F. "Nobody loved the life," Bodfish remembered, "but . . . we weren't overlooking any chances to get a whale."

They stayed thirty-five days on the ice and didn't see a whale. They could not have known that their efforts would have been fruitless; only in recent years have scientists, aided by aircraft and satellites, learned that in the spring the bowheads move on past Point Barrow, northeastward, in the leads that usually extend to the western side of Banks Island. From there the whales swim south to the rich plankton blooms near Cape Bathurst and Cape Dalhousie. The whales approach Herschel Island only as they travel toward the Chukchi Sea in September.

Disgusted and depressed, the men returned to the island in the middle of June to contemplate their bad fortune. George Leavitt then set out in a whaleboat and cruised along 150 miles of coast, from Icy Reef in Alaska to Tent Island in the delta, again without seeing anything. As the snow melted from the land and the tundra burst into a riot of wildflowers, the men gloomily prepared their ships for the summer's cruise. Bodfish wrote to his mother: "Just as soon as we can get out we are going and are bound to that undiscovered country that lies to the eastward of us. Only hope that it will turn out [to be] the great 'El Dorado' that we expect."

The ships left Herschel July 10 and the *Nicoline* at once headed west to return to San Francisco. Her experiment was a failure. She was clean, after two and a half years in the Arctic. The *Hume* and *Grampus* headed east. One July 24 near Cape Bathurst the men aboard the *Hume* saw their first whale and discovered their "El Dorado."

The *Hume*'s men spent the months of August and September in a flurry of whaling. Before they were frozen in again at Pauline Cove they caught twenty-seven whales. They became so heavily laden with whalebone that they were forced to return to Herschel at the end of August to ship their bone south aboard the *Grampus,* which, having taken twenty-one whales, was heading home to San Francisco.

The second winter at Herschel, without other crews to share the monotony, was harder on the *Hume*'s men. Boredom and loneliness descended and scurvy broke out. The second death occurred when John Meyers, a sixty-year-old black seaman from Baltimore, died of "inflammatory rheumatism" on May 17, 1892. His body was buried in the snow and covered with logs until the ground thawed enough to allow a proper burial.

Having received no news since 1890, the men were anxious for word from the outside world. William Mogg, the third mate, walked 300 miles southward, crossing over the mountains and below the tree line, to Rampart House, a Hudson's Bay Company outpost. When he arrived, the missionaries, also starved for news, were able only to tell him that Queen Victoria was alive and that Lord Salisbury was still prime minister. When so informed, Bodfish wrote that this was "wonderful news for a bunch of New England Yankees who had received no mail for over twenty-six months!" Billy Mogg was reported as saying that when he left the missionaries at Rampart House "he needed their prayers to protect him from damnation for profanity on the return tramp."

The *Hume* made her way out of the ice at Herschel on July 4, 1892, and reached Cape Bathhurst on July 28. Again the crew took whales as fast as they could, then started for home on August 15. When they arrived in San Francisco on September 30, after a twenty-nine-month absence, the news was electrifying: she had taken the baleen of thirty-seven whales and her cargo was valued at $400,000 ($4,296,000 in 1982 dollars). Her voyage was clearly among the most profitable in all of American whaling history. A newspaper reporter provided further details for San Francisco readers:

> The little steamer *Mary D. Hume,* which made the phenomenal take of whales at the mouth of the McKenzie River . . . arrived in the bay at an early hour yesterday morning. The boarding house masters swarmed out to her until the Hume's progress was almost checked by a wall of whitehall boats two to three deep all around her. . . . Several of her crew are suffering from contracted muscles, due to scurvy and rheumatism, contracted during the first season. The Hume had not reached the oil works wharf, however, before nearly every man had left her.
>
> The sailors looked like a lot of wild men when they came ashore; they had long hair, longer

Herschel Island, ca. 1894. At the left is the store house, built in 1890, with the Pacific Steam Whaling Company's house flag flying; in the center is the company's community house Anglican Church of Canada

beards and clothing that was patched and tattered beyond recognition of the original hue and texture of the garments. There were only three pairs of shoes in the party, the remainder of the crew being shod with deerskin and rubbers.

The *Mary D. Hume*'s return merely confirmed what Josiah Knowles and his partners in the Pacific Steam Whaling Company had known for a year. When the *Grampus* returned in 1891 they had immediately begun to outfit four of their steamers—the *Balaena*, the *Grampus*, the *Narwhal*, and the *Newport*—for wintering at Herschel. Knowles had also ordered their other steamers to cruise to Cape Bathurst. With his characteristic foresight Knowles had already conceived of Herschel Island as an Arctic advance base from which his ships could exploit the new whaling grounds at Cape Bathurst. At Herschel his whaleships could winter, make repairs and meet the company's tender that would annually provide them with food, supplies, and fresh crews. The tender would then return to San Francisco with the whalebone and furloughed men, thereby absolving the whaleships of the time wasted shuttling to and from California.

In 1893 the company's 800-ton four-masted steam schooner *Jeanie* made the first of a number of visits to the island. Her supplies were unloaded into the new storehouse on the sandspit that the men had put up the preceding fall, allowing the old one to be used solely for whalebone. The following year Knowles sent up Captain E. C. Murray, a retired whaling master, to act as the company's manager and storekeeper ashore and to serve as summer relief master in the event of the incapacity of one of the company's captains. His presence was necessary to coordinate the operations of the ships and to oversee the large quantities of supplies that the company had placed at the island to cover many contingencies, including the possibility that the ice might prevent the "outside" ships from reaching the island.

Recognizing the destructive effects of boredom during the winter, Knowles also sent up, in 1894, pre-cut lumber for a large community house for his men. Forty feet by sixty feet, it contained a large living room with card tables, a pool table, a billiard table, and an office for Captain Murray. To keep his ships as efficient as possible he had a blacksmith shop built on shore, complete with a turning lathe, and stationed a blacksmith there.

But Knowles's plans did not include the necessity for the ships to winter solely at Herschel; he planned to establish satellite bases. The Beaufort Sea's "El Dorado" was more than two hundred miles from Herschel Island; hence his ships wasted valuable time breaking out from the island and dodging ice on their way to Cape Bathurst. As early as 1892 he ordered his ships to search for suitable wintering sites east of Herschel. Their principal discovery of that summer was at Baillie Island, the tiny, low, eroding, silt island off the tip of Cape Bathurst. In 1893, his men sounded the channel between the island and the cape and concluded that, although it was by no means an ideal harbor, whaleships could winter there behind the southwest sandspit.

Thus, quickly and efficiently, Knowles had once again set his company's ships on a profitable, if brief, course. His preparations allowed his ships the maximum time to hunt whales in the ten-week Beaufort Sea season—and the *Mary D. Hume* and the *Newport* would prove this by remaining in the Arctic for six consecutive seasons. His plans also showed a glimmer of change in the otherwise static American whaling industry: with his centralized operations at Herschel he foreshadowed the fully mechanized procedures of the twentieth-century British and Norwegian whaling stations in the antarctic islands.

So in August 1892, when the fleet rounded Point Barrow, it was "Cape Bathurst or bust the boiler." The entire fleet, including Knowles's ships, was in a mad race to reach the "El Dorado." At first there were so many ships cruising near Cape Bathurst that the whales were constantly gallied, but the Pacific Steam Whaling Company won the day: Knowles was the only manager who had had the foresight to prepare his vessels for wintering. One by one the other ships bore away for Point Barrow, worried about being trapped by the pack ice. When the winter began, only the *Balaena*, the *Grampus*, the *Narwhal*, and the *Newport* were in Pauline Cove.

The next summer, 1893, the ships enjoyed fantastic successes. They had discovered the bowheads' last refuge, and in that season, with comparatively little ice, they pursued them mercilessly. In all, 286 whales were taken east of Herschel. The *Narwhal* and the *Balaena* each took more than fifty whales, the greatest seasonal catches in the history of the fishery.

When freeze-up arrived in 1893 seven ships were in Pauline Cove. Knowles's four had been joined by the *Mary D. Hume*, and the Roth, Blum Company, meat packers of San Francisco, had sent up two steam brigs, the *Jeanette* and the *Karluk*.

The success of 1893 was the last hurrah of the American whale fishery. As with the other discoveries of new grounds in the western Arctic only those ships that exploited the grounds in the first year or two did well; those that acted more slowly enjoyed only modest returns. There were no further discoveries; the ships had penetrated to the last refuge of the bowheads and had scoured almost all of their known range.

The next summer, 1894, was a poor one. Heavy ice retarded the ships' progress and the catch was small. Nevertheless, fifteen vessels arrived to winter at Herschel, accompanied by a number of others trying a short summer cruise. The *Jeanie* visited the island briefly, bringing supplies for Knowles's ships. The steamer *Lakme* supplied the rest of the fleet.

With so many ships wintering at Herschel, life there soon developed a unique character and rhythm. By about 1890 many of the ships had abandoned the old routine of sailing from San Francisco late in the year with the first mate in charge on an easy winter shakedown cruise to Hawaii. Instead of sending the captain by steamer to Hawaii to join the ship, the owners simply laid the ships up in the autumn when they returned from the north. Many ships were towed across the Bay to the sheltered waters of Oakland Creek, and left with only a shipkeeper aboard. Then in February the ships were overhauled and prepared for a March departure directly to the Arctic.

In preparing for a wintering voyage the ship was put in first-class shape, for she would be gone for more than two and a half years. She was loaded with tons of supplies, food, and clothing. Because the crew might have to carry out major repairs on her, she carried extra tools and spars and a set of cofferdams to clamp over the stern, should it be necessary to change the propeller blades.

The ships usually headed directly to Unalaska in the Aleutian Islands. There, at the town that had grown up around the old Russian-American Company trading post, they topped up their coal bunkers and took on water. In the safety of the spacious harbor they often sent down their fore topgallant and royal yards, put up their crow's nest, rigged their cutting stage, and generally prepared the ship for whaling. They also sent several boats codfishing and often salted down several thousand pounds for the winter.

The ships then departed on their spring whaling cruise. As they traveled along the Siberian shore, they usually traded for large quantities of skin clothing, principally reindeer parkas and sealskin coats, "Masinker boots" (native boots),

Deck scene on a steam whaleship bound to Herschel Island. A coffer dam is visible at the upper left of the photograph. At the upper right a man's clothes hang drying from the cutting stage. Codfish are hanging from the house, and the ship's dogs are being fed. In the center the blacksmith is at work

vests, and trousers. Boots from the Diomede Islands were particularly sought because the women there had the reputation as being the best bootmakers. It was not unusual for a steam whaler to take on five hundred pairs of Masinker boots on its way to Herschel Island.

At the major native villages—Plover Bay, Indian Point, the northwest cape of Saint Lawrence Island, East Cape, the Diomedes, Point Hope, and Point Barrow—they also signed on Eskimos, usually with their entire families, to go along as "ship's natives"—hunters and seamstresses for the voyage. Each ship carried from four to eight native hunters. In addition to the more than forty whalemen there were at least fifteen natives aboard for the passage to Herschel and often as many as fifty huskies that had been brought to pull the dog sleds. Thus, approaching Herschel, the decks of a steam whaler were a jumble of supplies, equipment, whalemen, Eskimos, dogs, and a few pigs as well.

The ships reached Port Clarence in early July. There they smoked ship to kill rats, carried out repairs, topped up their water tanks, and usually took on from the tender about two hundred tons of coal, as well as fresh fruit and vegetables. Often further supplies were loaded at Point Barrow. In 1893, when the *Mary D. Hume* was headed to Herschel on her second voyage, already laden, she took from the tender at Point Barrow:

> 65 tons of coal, 50 bbls. flour, 5 bbls. butter, 10½ bbls. sugar, 6 bbls. beef, 6 pork, 1 bbl. molasses, 1 [case] onions, 10 box black tobacco, 12 box chewing tobacco, 2 [cases] peaches, 2 [cases] green peas, 2 [cases] bacon, 1 [case] hams, 8 boxes bread, 20 boxes potatoes, 1 box onions, 200 gals. coal oil.

The whaleships and the tenders usually reached Herschel around the middle of August. The newly arrived whalers frantically discharged their stores and started on their cruise. Hurrying to Cape Bathurst they often met the "inside" fleet on its way to the island to exchange crews and to ship out their whalebone. Once the crews and cargo were exchanged these inside vessels returned at once to the whaling grounds for another three or four weeks of cruising as the fleet gradually worked its way west with the whales.

By the middle of September most of the wintering fleet was staying close to Herschel in case of an early freeze. At the same time most of the bowheads had migrated west past the island. As the whaling tapered off, the ships went to the mainland to collect firewood, for the island's supply had long since been exhausted. Depending on the requirements of the ship, they gathered anywhere from 100 to 300 cords of driftwood.

By the beginning of October all the ships were in Pauline Cove and ready for freeze-up. One last duty remained: securing sufficient dog feed. Usually a few whale carcasses had washed ashore near the island and the ships towed a couple to the outside of the sand spit, where great chunks of meat could be chopped out during the winter for the dogs.

The crew was often detailed ashore in the last days before freeze-up to cut sod blocks for covering the decks of the ships and to build a house for the ship's natives. This house was usually constructed in the local Eskimo style: a simple, central-four-post construction with wall timbers leaning onto it and walls and roof thickly covered with sod blocks. It was an extremely warm structure. At the same time other crewmen might be at work blasting and digging an ice cellar.

As young ice began to form in Pauline Cove, any heavy ice floes that had blown into the cove were towed out to clear the harbor before the ships positioned themselves for the winter. At this point everyone was back aboard ship because for a few days it would be impossible to reach the shore through the thickening slush.

When the ice was a couple of inches thick the ships hove up their anchors and drove southeast into the young ice. On the southeasterly axis they could minimize their surface exposure to the brutal northwesterlies of winter. Deeply fearing fire, the captains also took care to keep their vessels as well separated as possible and kept a "fire hole" chopped open through the ice to provide quick access to water.

The ship itself had then to be prepared for the winter. Then men unbent several of the sails and sent down some of the yards. They placed the yards running fore and aft from the poop to the forecastle, then stretched the sails over them and battened them down. The main deck was thus completely enclosed, creating a large room called the "bull room," where the carpenter could work—building sleds, among other things. The room was heated by a wood stove made from a 100-gallon metal oil drum. Next to the stove was another drum for melting fresh-water ice.

The men also chopped open a hole through the thickening ice around the rudder and propeller and cleared it daily to save the gear from being torn off should the ice move suddenly.

When the ice was judged strong enough, they blew down the boiler, drained the water from all pipes and joints, greased or painted the joints, and drove wooden plugs into the discharge pipe holes. They installed a stove in the engine room to maintain the temperature. To heat other parts of the ship the men often dismantled the donkey engine and rigged heating pipes from its boiler. Other ships simply put stoves in the crew's quarters and after cabin.

By the middle of November the wind had usually compacted the snow drifts sufficiently so that blocks could be sawed from them. The men then built a wall of snow blocks, four to five feet thick, completely around the ship, up to the main deck and around the afterhouse. They also built a small house over the fire hole and rudder hole to retard the ice formation there. Even with this housing, four to six inches of ice would form on the water in twenty-four hours.

The men also had to attend to the ship's supply of drinking water. As soon as about six inches of ice had formed on Herschel Island's fresh-water pond, which was several miles away at the top of the island, crews were sent there to saw out blocks. The blocks were stacked nearby. Each ship had its own cache, and it was one of the winter's chores to haul blocks of ice from the top of the island to the sandspit where they were placed on stages to keep them above the drifting snow and away from the dogs. The ice blocks could then be fed into the melting tank as needed. October, before the ice became too thick, was the easiest month to saw ice, but ice-cutting went on all winter—the crew of a steam bark used four tons per week.

Once the hectic preparations for winter had been completed, the pressing concern of the captains was to provide sufficient fresh meat for the men. Late autumn is a barren time at Herschel. The Rat Indians who lived 200 miles from the island in the forests south of the mountains, were able to work this to their advantage. Game was relatively more plentiful there in the autumn. Groups of Indians arrived at the island two or three times a winter with as many as twenty-two toboggans carrying caribou, moose, and fish. In the autumn their arrival usually touched off a scramble among the captains to get their trade. The meat was bought for six to seven cents per pound in trade goods (principally tea, fixed ammunition, and tobacco) valued at the San Francisco cost price.

The Narwhal *housed in at Herschel Island* Dartmouth College

Apart from this trade with the "Itkillicks," as the whalers and Eskimos called the Indians (*Itkillick* is the Eskimo word for Indian), in the first year the only other trade food the whalemen obtained in the autumn was fish from the Kogmullicks. Kogmullicks were the indigenous Eskimos who inhabited the coastal areas around the Mackenzie River delta. A large part of their winter diet was fish, and the whalemen never found them to be very energetic or effective caribou hunters. The whalemen valued meat far more than fish, consequently their contact with the Kogmullicks was less intense than it was with other native groups.

By far the most effective hunters were the Nunatarmiuts, inland Eskimos who originated in the mountains of the Brooks Range in Alaska. These Eskimos were primarily caribou hunters. In the 1870s they lived in the western and central drainages of the Brooks Range, but a cyclical decline in the caribou herds of those regions, among other factors, forced them to move east, where the caribou were

Indian traders visiting Herschel island, 1895 Mystic Seaport

more plentiful. It is likely that these deracinated Eskimos reached the Arctic coast near the international boundary only a few years before the whalemen began wintering at Herschel, and once they perceived the market for their skills, they ranged throughout the mountains on both sides of the boundary and provided large quantities of meat to the whalemen, often inspiring enmity from the Kogmullicks.

The whalemen soon developed a system for acquiring their fresh game. They traded directly with the natives at the ship and sent out hunting parties of officers and ship's natives, who set up satellite camps inland. The whalemen also grubstaked Eskimos from Herschel to go on their own hunting trips.

The whalers found the ship's Siberian natives (the "Masinkers"), who had been signed aboard at their villages, to be good dog drivers but indifferent "deer" hunters; consequently, they often used the Masinkers to operate their teams traveling back and forth to the hunting camps. The ship's natives from the Alaskan coastal villages were usually the hunters that they stationed in the field. The need for hunters was so great that in 1894 and 1895 most of the Point Barrow natives were at Herschel, as were nearly one hundred from Point Hope.

Lured by the trade goods at the island, a number of Eskimos from Point Barrow—as well as Nunatarmiuts and Kogmullicks—simply moved to Herschel to live, thus raising the "resident" population of the island to between fifty and one hundred natives in any year.

With as many as fifteen ships wintering at Herschel in the 1890s, food consumption was prodigious and the hunting activity intense. Hunting and trading parties traveled as far as 150 miles from the island—from Camden Bay in the west to Richards Island in the east and throughout the mountains to the south. They took a wide variety of game: moose, mountain sheep, polar bears, seals, ptarmigan, ducks, geese, swans, eggs, and fish, but it was caribou that they found to be the most plentiful and palatable.

For the winters of 1894-1895 and 1895-96 it is likely that more than two thousand caribou were taken annually. Fortunately, the herd from which they were primarily taking their caribou was undergoing a natural increase and may well have numbered more than a hundred thousand head. Several writers have claimed that this hunting pressure decimated the herd, but at the end of the whaling era the herd was larger than at the beginning, and the whaler's kill, averaged from 1890 to 1908, probably amounted to little more than 1 percent of the herd's size annually.

With the Nunatarmiuts, Kogmullicks, Itkillicks, Alaskan and Siberian ship's natives, whaling crews, and "beachcombers" (the few whalemen who had chosen to stay at the island after their tour of duty was over), the little settlement at its height approached a thousand persons. It was a polyglot, cosmopolitan community, its members living amid "a mixture of wooden and canvas buildings, native huts, spare casks, boats, wood and all spare stuff put on shore to make room . . . on the ships." In this broth of humanity a trade jargon grew up which allowed all to communicate basic ideas more or less effectively in a mixture of words drawn from Western Eskimo dialects, Polynesian, Danish, French, and English, among other languages. Some of the natives wore equally eclectic clothing: "One of the men wore a new sombrero with a very broad brim. Others had miscellaneous odds and ends combined with their native costumes. . . . Several wore tight-fitting red flannel drawers over their deerskin trousers."

The natives were drawn to the island because of the ready access to manufactured goods: gunpowder, primers, lead, rifles, fixed ammunition, flour, sugar, tea, calico, tobacco, soap, knives, combs, files, whaleboats, small stoves, clothing, and many other items, including alcohol. Throughout Herschel's whaling heyday (1890 to approximately 1908) the price of meat seems to have been steady. The whalemen bought it for six or seven cents a pound, paying in trade goods that were valued at their San Francisco wholesale prices. Because the whalers did not add the cost of transportation or profit to these trade goods, the prices they charged were only 20 or 30 percent of those charged at the Hudson's Bay Company store at Fort McPherson on the Peel River, less than two hundred miles from the island. The Hudson's Bay Company, of course, had to make a profit and accrued substantial costs in shipping their goods overland from eastern Canada.

This point was made dramatically to John Firth, the Hudson's Bay Company's factor at Fort McPherson. In the summer of 1892 a group of Eskimos arrived at Fort McPherson from Herschel Island in a whaleboat loaded with trade goods. At the whalemen's prices, the Eskimos were becoming traders themselves, bartering with others and then returning to Herschel with the furs. The number of fox and beaver skins offered for trade at the Hudson's Bay store declined significantly.

The company then played its only strong card. In the winter of 1894-95 Firth apparently sent word to the whalers that unless they desisted from the fur trade, the company would no longer forward mail to them via its string of trading posts on the Mackenzie. In their intense isolation the whalemen eagerly awaited the two overland mails that reached them each winter. News from home was more important to the whalemen than the minor profits from fur trading. When John Firth visited the island in the autumn of 1896 he was able to report that the whalemen had largely ceased encouraging the Eskimos to bring them furs. The whalers continued to trade for meat, of course.

The fact that the Hudson's Bay Company did not establish any posts on the western Arctic coast until after the whaling industry had collapsed seems to have been one result of the presence of the whaling fleet at Herschel. Even after the whalers had begun paying customs duty on their trade goods, they were able to sell 100 pound bags of flour to the Eskimos for $2 while the Hudson's Bay post at Fort McPherson was asking $30. Many natives simply abandoned the fur trade to become contract hunters for the whalemen and thus to acquire proportionally more trade goods than if they had been trapping.

There was, also, a darker side to the whalemen's relations with the natives. A number of writers have claimed—mostly after the fact—that the whalemen "converted Herschel Island and indeed most of the Delta into a hive of debauchery: drunkenness and immorality prevailed everywhere." "Down the gangplanks," another wrote, "surged a motley horde of mixed humanity till the sandspit was overrun with a drunken mob of dark-visaged Kanakas, bearded Russians, ebony-faced Negroes, and the off-scouring of the Barbary Coast. Rum flowed like water. Fighting, drinking and debauchery became the order of the day." Had Herschel been like this few whaling masters would have wintered there or brought their wives and children along. According to Inspector D. M. Howard of the Royal Northwest Mounted Police, the wild reports appearing in newspapers were greatly exaggerated: "the Esquimaux greeting at the ships arrival belies stories of abuse and mistreatment; the women would certainly stay away."

But there was a germ of truth to these stories. By the mid-1890s there were more than twenty sod houses on the sandspit, many of which were inhabited by ships' officers who chose to live ashore with Eskimo women during the slow days of the long winter. They returned to their ships only to carry out their few duties. Most of these shore-based officers, who were derisively called "squaw men," kept their Eskimo mistresses ("coonies") by giving alcohol and trade goods to the women and to their husbands. Other women apparently made themselves available on a less permanent basis, causing Frank Russell, a naturalist who visited the island in 1894, to identify them simply as "prostitutes." This was a relatively common practice; by 1895 "many" mixed-race babies were seen at the island.

A few captains also carried their mistresses with them aboard ship. On their way to the island they often picked up women or young girls at villages on the Alaska coast. Captain E. W. Newth of the brig *Jeanette* was notorious as the "kindergarten captain" because of his penchant for native girls eleven to fifteen years old. He usually carried five aboard his ship, having more or less rented them from their parents in Alaska with trade goods and alcohol. The police at Herschel, who also frequently kept native mistresses, were unable to press charges against Newth because he had signed the girls on the ship's articles.

Alcohol caused its familiar problems. Although the Pacific Steam Whaling Company forbade its men to trade alcohol to the natives, a few bottles always found their way ashore from the officers' private stocks. Some of the officers of other ships, who were under no such restrictions, traded alcohol freely, and white men and natives alike made liquor in gun-barrel stills. Some natives visited Herschel with furs and meat for the expressed purpose of obtaining whiskey.

As was to be expected, whiskey not only caused the usual drunken brawling but also accounted for a few deaths. On September 1, 1896, for instance, during a drunken row the third mate of the *Balaena* shot the second mate in the head with a .38 caliber revolver.

Reports of drinking and debauchery began to filter up the river from Herschel almost as soon as the whaleships began wintering there. In the summer of 1892 the stories reached the ears of an energetic young Anglican missionary, the Reverend Isaac O. Stringer, who had just arrived at Fort McPherson on his first posting. In the spring of 1893 he set out on a long dog sled journey to the island. He arrived on the first of May and was confronted by the presence of the *Balaena*, the *Grampus*, the *Newport*, and the *Narwhal* and their attendant natives on shore.

To his surprise Stringer was warmly received by the whalemen. Captain Horace P. Smith immediately invited him to stay aboard the *Narwhal*. The captains he found to be "jolly fellows" and—he wrote to his wife, Sadie—"as kind and considerate as they could be. I had the best of everything and they helped me in every way they could." Stringer, who was later to become known as "the Bishop who ate his boots" on a survival march, must have appealed to the captains through his self-reliance and forthright manner. And Stringer consequently found himself highly ambivalent about his experiences there.

The captains genuinely welcomed his presence; not only did they like him and appreciate his pastoral visit, but he had a mollifying influence on the tensions that always build among men in isolation. "I did not *see* [his italics] much liquor or drinking," he wrote to Sadie, "but I cant have my ears open without knowing that there is a good deal given to the Eskimos."

Stringer returned to Herschel in November 1893 on the second of his many visits. He remained there two weeks and during that time conducted his first service at the island—in the Pacific Steam Whaling Company's house, using the billiard table as an

The Reverend Isaac O. Stringer and his wife Sadie Anglican Church of Canada

altar. He returned to Herschel frequently thereafter to campaign among the whalemen against the whiskey trade and to proselytize among the Eskimos. In 1895 he began preparations to establish a permanent mission and received donations from thirty-one captains and officers (including E. W. Newth). The following year he succeeded in persuading all the captains to sign an agreement that they would allow no whiskey to be traded to natives whose original home was in Canada (Alaskan and Siberian immigrants were, therefore, not covered by this document). In 1897, with the mission established, he and his wife Sadie began their first full year of residence at the island. They were always to enjoy good relations with the whalemen.

That Stringer was partially successful in suppressing the whiskey trade was probably beneficial for the Eskimos, but his presence at the island reinforced a European code of morality, and this posed powerful problems for the Eskimos, to whom many of the concepts were foreign. By far the most painful expression of these new elements in the Eskimos' lives concerned the concatenation of tragedies that surrounded Pysha, a Point Hope Eskimo.

In a drunken frenzy on November 30, 1895, Pysha beat his wife and then killed his little daughter by beating her head against the wall of his house. On December 3, an indignant crowd of whalemen seized him, handcuffed him to an upright log, stripped him to the waist, gave him one hundred lashes, and ordered him to leave the island that day, never to return. C. E. Whittaker, the missionary who was temporarily replacing Stringer, laid on the first dozen lashes. The punishment deranged Pysha, who set off to the west and murdered eight Eskimos at Flaxman Island. He was later executed by the Eskimos at Point Barrow.

The same reports that brought the church to Herschel also militated for the arrival of the police. These reports to the government were repeatedly reinforced by the Hudson's Bay Company, which was suffering the invasion of its trading monopoly. The company's reports emphasized that untaxed trade goods were flowing into Canada from the whaling fleet at Herschel. Despite these warnings, the government took no really active role in the Arctic until after 1903, when two members of the Royal Northwest Mounted Police were sent to the Island.

When Sergeant Fitzgerald and Corporal Munroe arrived at Herschel, like Isaac Stringer they were welcomed by the whaling captains. The captains were glad to cooperate with the police because, among other things, the presence of officers of the law reinforced their control over their increasingly unruly crews. The presence of the police also further reduced the trade in alcohol, but it was of little consequence. By 1903 the whaling industry was in steep decline and only a few more whaleships would winter at the island. When Fitzgerald and Munroe first reached the island they were told they were six years too late.

Alcohol merely aggravated a more fundamental problem for the whaling captains. Once the ships had been housed in for the winter there was very little work to do until May, and boredom, tedium, and isolation took their toll on the men's spirits.

There was no use in standing watches during the winter, the ships merely made do with a pair of nightwatchmen who tended the fires during the night. During the day the captains tried to keep the men as busy as possible by dividing the crew into groups that sawed and split firewood, kept the fire and rudder holes chopped free of ice, hauled fresh-water ice from the pond, and the like. The men usually worked only from 8 A.M. to noon and from 1 P.M. to 4 P.M. Saturdays were spent washing and bathing; there was no work on Sundays. It was inevitable that with so much time on their hands some would think of deserting the ship.

Several men deserted the ships during the first winter at Herschel—and a few jumped ship every year after that. Always the captains went after them, for although the weather would usually either kill the men or force them to return, the captains knew that if these departures were allowed to occur unchecked, soon the forecastles would be sparsely inhabited. When the runaways returned—often with appalling frostbite—those that were healthy enough to stand punishment were always put in irons.

Sergeant Fitzgerald (left) of the Royal Northwest Mounted Police chopping wood in front of his sod house Mariners Museum, Newport News, Virginia

Throughout the 1890s, as the rumors of gold in the Yukon grew, so did the desire of the men to reach the gold fields—and in the winter of 1895-96, with thirteen ships in Pauline Cove, the rumors were flying. The first mass desertion took place on January 21 when Dan Sweeney, a boatsteerer on the *John and Winthrop* and a well-known hard character, lead away six others. The captains immediately sent a search party after them. The party returned after only a few days, empty-handed—it was thought, out of fear of Sweeney. The party started out again, better armed, and after some hard traveling caught up with Sweeney and three others. One of the deserters had already returned frostbitten, but Joe Carroll and Dick Martin became two of the handful of runaways to make good their escape from Herschel.

The captains realized that Sweeney was both a dangerous man and a ring leader. Captain Abe Simmons had him put in a "spanish buckle" as punishment. He was handcuffed in a sitting position with his legs drawn up and with his arms around his legs. A rod was then pushed through the gap between his elbows and the back of his knees, keeping him painfully immobile. The captains thought more desertions were likely, so they formed a patrol of officers to guard the storehouses at night and allowed none of the men off the ships after 10 P.M.

But it was impossible to prevent the men from jumping ship. On March 16, twelve men broke into the storehouses and stole a large supply of food, guns, and ammunition. After stealing dogs and sleds as well, they departed. The small search party that followed almost at once found that the deserters had held up two hunting parties, taking all their supplies. The next day the deserters got the drop on the search party and drove the men off with gunfire. The search party returned to Herschel for reinforcement and again set off, this time better armed.

By coincidence the deserters came across a hunting camp in the foothills of the mountains where Harry Huffman, the fifth mate of the bark *Wanderer,* and Peter Peterson, a boatsteerer on the *Alexander,* were camped with a number of natives. All were away hunting at the time. The deserters destroyed the camp, took what they wanted, and moved on, thus virtually assuring the death of the hunters who were then without food, sufficient clothing, sleeping bags, tents, or dogs.

The hunting party returned to the camp just as John Bertonccini, returning from Fort McPherson with the mail, arrived. Bertonccini remembered the scene:

> Every item of the stores was scattered about, the huts burned, the snow igloos destroyed, clothing cut into ribbons and sleds broken and the remaining dogs they did not want were killed. . . . After a consultation we decided to take up the chase as it would be suicide to try to get back to the ships. . . . Our little party numbered thirty natives and four white men.

The hunting party was traveling light, carrying only guns and ammunition and after three hours of hard traveling, overtook the deserters in their camp. Although the deserters had five sleds loaded with supplies and a dozen dogs pulling each, they lacked snow shoes and the soft snow had retarded their progress. Huffman and Bertonccini approached within shouting distance, demanding that the deserters return their supplies.

One of the leaders, a sailor from the bark *Northern Light* named Kennedy, who had been on the January escape, replied that they would start shooting if the hunting party was not out of sight in ten minutes. Huffman apparently fired first and soon both sides were shooting. Kennedy was killed and another runaway wounded before the deserters surrendered. They left Kennedy there and returned with six men. Three of the runaways had been out of the camp at the time and two others escaped in the fracas. They were left to fend for themselves. One or two may have reached the gold fields.

In response, the captains established boundaries around Pauline Cove, beyond which the sailors were not allowed to go, and advanced the curfew to 8 P.M. There were no more desertions that winter. But Dan Sweeney made good his escape at the beginning of September 1896, inexplicably stealing a whaleboat and going to Shingle Point just before his ship, the *John and Winthrop,* left Pauline Cove for San Francisco. He returned to Pauline Cove to live insolently ashore in a sod hut, sharing it with another deserter and hard case, Big George Madison. Both stirred up trouble among the ships' crews. Big George was slashed across the stomach in a drunken brawl that winter.

There was also a pleasant side to life at Herschel. As we have seen, seven ships wintered in Pauline Cove in 1893-94, fifteen in 1894-95, and thirteen in 1895-96. The captains tried to keep themselves and their crews happy and busy. In their foreign surroundings they tried to make life as much like home as possible.

The men began sledding and skiing in the small ravine on the north side of the harbor, and sometimes as many as a hundred people could be seen on the hill. At first the sailors were shaky on their skis, and one reported that he "capsized" regularly on his runs. A few teeth were loosened in sledding accidents.

Their sport soon became more organized. In 1893 as soon as the ice was thick enough in the cove, then men put up goal posts and laid the perimeter of a soccer field. Throughout the winter the crews of all the ships played there—even at –25° F. On bad days they occasionally held boxing matches on the main decks of the ships.

On February 19, 1894, the men played what Hartson Bodfish considered to be the first baseball game north of the Arctic Circle. The boatsteerers beat the officers by one run, and the game was such a success that four teams were immediately formed: the Herschels, Arctics, Northern Lights, and Pick-ups. The captains of the four teams met on March 5 to form a league and to draw up the rules and regulations. A diamond was laid out using ashes to mark the lines and bases and a ship's sail for the back stop. During the spring the teams entertained one another socially. On April 9, 1894, for instance, the Herschels gave a supper for the other clubs. "The bill of fare: ham sandwiches, popcorn, ice cream, and hop beer."

Baseball at Herschel could also be dangerous. On Saturday afternoon, March 7, 1897, the men from three ships were in the midst of a game in beautiful weather, unseasonably warm at 20° F. Shortly after 1:30 P.M., in the bottom of the second inning, a dark billowing cloud suddenly loomed over the island. Within minutes it was blowing the worst gale some of the men could remember. They ran for the ships. Almost at once they could see only a few feet and the temperature plunged toward –20° F. Scrambling and unable to see their way, the men simply ran to the first ship or building they found, staying all night while the wind shrieked around them.

At eight o'clock the next morning the wind stopped as suddenly as it had started. The ships' crews quickly fanned out and found five frozen corpses, three whalemen and two Eskimos. One native was found on the sandspit, only a hundred yards from his house.

Some captains brought their families with them. They had to pay the owners $1,000 for this privilege but it made their lives—and the lives of others at Pauline Cove—far more pleasant. The winter of 1894-95 was the first with women and children in the fleet. Aboard the *Jesse H. Freeman* Captain W. P. S. Porter had his attractive twenty-nine-year-old wife, Sophie, and daughter, Dorothy. Mrs. John A. Cook was aboard the *Navarch;* Mrs. Charles E. Weeks on the *Thrasher;* Mrs. Albert Sherman and her son, Bertie, on the *Beluga;* and Mrs. F. M. Green and her niece, Lucy McGuire, on the *Alexander.*

The Navarch *at Herschel in the spring of 1896*

The presence of women and children softened the hard edge of life that grows among men in isolation. At once the women began organizing parties, using every conceivable excuse for celebrations to leaven the gloom of winter. They gave small card parties, elaborate dinners, birthday parties, dances—as well as celebrating Leap Year Day, Washington's Birthday, Valentine's Day, the Fourth of July, St. Patrick's Day, Thanksgiving, Christmas, New Year's Day, and so on.

Shortly after freeze-up, when all the ships' decks had been housed over, Sophie Porter described the first party of the season, a "deck house warming" on October 4, 1894, aboard the *Beluga*.

We all went on deck, where we found a fine comfortable room entirely covering the poop deck, and lit by two or three dozen light-lanterns and colored side lights. There was an

Fancy dress party at Herschel Island, 1895-96. Left to right, back row: *Sophie Porter, Captain George Leavitt, Captain James Wing, Captain Hartson Bodfish, Captain James McKenna;* front row: *Mrs. F. M. Green, Lucy McGuire, Viola Cook, Bertie Sherman, Mrs. Joseph Whiteside, Dorothy Porter*

excellent band of three pieces—violin, banjo & accordian—and we were treated to all the latest (up to March 1894) airs, and not withstanding [the] unpolished floor and heavy boots, we had some jolly dances. . . . Ice cream and cake, also beer and cigars were served during the evening.

A week later Mrs. Green gave a tea party.

At 6 o'clock ten of us sat down to high tea, and as soon as we cast our eyes over the tastefully

Mrs. Albert Sherman with Helen Herschel Sherman aboard the Beluga *at Herschel Island, late June 1895*

arranged table we ladies mentally decided that Mrs. Green was blessed with the 'boss' steward of the fleet. The menu consisted of, as far as I can remember:

Lobster salad & olives
Oyster Paté with French Peas
Veal loaf with Jelly
Chops a la français with Saratoga chips
Sea Bisquits
Bartlett Pears, with citron & sponge cake
and most delicious tea.

On Christmas day Dorothy Porter received a special treat:

Dorothy woke bright and early (9 A.M.) to see what Santa Claus had brought her. About a week ago she sent him a letter which she told one of the officers to leave in a hole out on the hill. Well, she got all she asked for and a great deal more in the way of toys, books, aprons, and a lovely stocking stuffed with everything good & pretty. A delightful letter from Santa Claus came tucked between the bedclothes of a dear little cradle. I don't think the children at home ever had a nicer one from their Xmas friend.

The men responded energetically to the ladies' entertaining. The captains frequently gave parties on their own and many of the men formed minstrel and theatrical groups. The "Herschel Island Snow Flakes" and "Fry's Theatrical Company" were groups that performed regularly on the large main deck of the *Beluga*. For these performances whaleman-artist John Bertonccini was persuaded to paint the backdrops. The captains hung swings from the ships' bowsprits for the children. The women added an exotic aspect to the formerly all-male sport of skiing by appearing on the slope dressed in native costume with trousers "cut with legs quite loose and very full aft," as Captain McInnis noted.

The dances and dinners were often elaborate and dressy; on January 15, 1896, Captain George Leavitt wrote in his journal, that this was "the first time I have had a boiled shirt on for thirty-three months." But with only five wives among the fleet, dance partners were scarce. Captain McInnis reported that at Captain Weeks's birthday party, "for a partner in the Virginia Reel I had Miss Dorothy Porter, aged 5 years."

Those masters who were accompanied by their wives soon became known as "the Four Hundred"; those without them organized themselves into "the Hoodlums" and "Dry Throats." Although the groups entertained one another, a certain amount of badinage was passed between them. The Hoodlums, for instance, gave a party aboard the *Karluk* and invited, among others, Captain Joseph Whiteside and his young wife. Captain and Mrs. Whiteside were considered extremely stingy, not having shared their abundant supplies of fresh pork with the others. The Hoodlums, among them Hartson Bodfish, in consequence stole half of one of the Whiteside hogs and served it to them without their knowledge. After dessert they read this doggerel, written by Sophie Porter:

This little pig went to market on the good ship
  Belvedere
Until he came to the Arctic, where he ended his
  career.
Of a roving disposition, even while he was pork
He got into a sack one night and thought he would
  take a walk
He wandered among the ships, and thought he would
  have some fun
Until he came to the Karluk, and then his days were
  done.

Joaquin Gonsalves, cook aboard the Beluga. *Herschel island, late June 1895*

Certainly one of the happiest events of those years took place on May 8, 1895, when Mrs. Albert Sherman gave birth to a 7½ pound girl, Helen Herschel. One of the saddest was on March 29, 1895, when Captain Charles Weeks was killed by a fall down the main hatch of the *Thrasher.* His body was frozen, then, in the spring, salted and taken by his wife to San Francisco.

The Fourth of July, 1896, at Herschel Island. Sailors watching an obstacle race

In the lengthening days of April the men began to turn to the tasks of preparing the ship for the summer whaling cruise. First they removed the snow banking from the ship, then sawed a trench around the hull through the six-foot-thick ice. They did this because sometime in early May the ship, lightened of the winter's supplies, would usually "jump"—rise a foot or more. Unless it were allowed to jump evenly it would list, making life aboard unnecessarily uncomfortable.

Next they removed the housing from the decks and sledded the coal out to the ships to fill its bunkers. Cofferdams had to be put over the stern of the ship and pumped dry to allow the men to inspect the propeller blades and shaft gland.

At times the whalemen had to devise ingenious methods to carry out repairs so far from home. In 1895 the *Newport* twisted its propeller blades so badly that the cofferdams were useless. In San Francisco the ship would have been hauled out on a marine railway, but at Herschel the men had to work things out for themselves. With spars they rigged a huge pair of shears on the ice over the *Newport*'s stern. Then they hung three or four cutting falls on the spars and hooked on to the stern. They were able to winch the *Newport*'s stern high enough out of the water to change the propeller.

During the spring the crews scraped the masts and hull, painted them, tarred the rigging, painted the interior of the ship, scraped last season's whalebone, dried the sails, rigged the boats and put them onto the cranes, sent the crow's nest aloft, overhauled the darting guns and bombs, checked the main engine and the donkey engine and the steam tryworks and the launch's engine, shackled on their anchor cables, and put an anchor through the ice. In the middle of June the ice began to break up in the cove and at the beginning of July the ships were ready for sea.

The celebration of the Fourth of July concluded the events of winter. The men dressed the ships with all their flags and at eight A.M. they fired a salute to begin a day of games and contests: tug-of-war, jumping, foot races, wheelbarrow races, sack races, tub races, three-legged races, obstacle races, a whaleboat race, shooting contests for whalemen, Eskimo men, and "squaws," and, of course, baseball. In 1895, $300 in prizes were divided among the two or three leaders of the contests.

The ships were usually able to buck their way out of the harbor and into clearer water by the tenth of July. The ships, having spent ten months in the ice, then started on their whaling cruise.

13 / *Federal Support and the Disaster of 1897*

The loss of the steam bark *Navarch* in August 1897 foreshadowed the wrecks of September. In July of that year the nineteen vessels of the whaling fleet worked their way north through Bering Strait and as far as Icy Cape, following the melting ice. Late in the month ten days of light northerlies forced many small pieces of ice from the edge of the pack, driving them toward shore. The *Navarch*, in advance of the fleet, was moored to grounded ice about twenty miles beyond the Cape, and there, for more than a week, the ice slowly packed in around her.

Captain Joseph Whiteside, fifty-eight years old, had taken *Navarch* past Icy Cape and into a dangerously exposed position apparently for the sake of his twenty-year-old wife, a former New Bedford shop girl. Mrs. Whiteside thoroughly disliked the middle-aged wife of the captain of the steam brig *Karluk*. The enmity was mutual. The two could not stand to be anywhere near one another. When the *Karluk* reached the *Navarch*'s anchorage south of Icy Cape, Whiteside moved on to avoid a marital squabble.

On July 27 Charlie Brower, the manager of the whaling and trading station at Point Barrow boarded the ship. He had come down the coast from Point Barrow by open boat in the narrow leads. He saw that the ice was heavy and dangerous that year, and once aboard the *Navarch*, he watched with alarm as the ice closed in:

> the morning of the twenty-eighth the ice came all around the ship; instead of coming back south, we took the anchor, tying to a large field of apparently grounded [ice]. While there, the smaller drifting ice came in, completely surrounding the ship. The first thing we knew, the piece we were tied to started drifting; even then, we might have, by hard work, pushed our way out from the ice. Instead we just drifted, the ice coming more inside of us all the while, until we were fairly adrift in the pack with no chance to move.

While the rest of the fleet lay in comparative safety to the south near Icy Cape, the *Navarch*, with Brower aboard, was swiftly carried northeastward, ever farther from shore. By August 3, she was twenty-five miles northwest of Point Belcher.

Seeing that their situation was very dangerous, with little chance of escape for the ship, Captain Whiteside suddenly decided to abandon her. He intended to try to reach the shore by crossing the moving ice floes with whaleboats. Brower had spent many whaling seasons hauling whaleboats over the broken and piled floes. He advised Whiteside to add a four-inch false keel to the boats to help keep their thin cedar planks above the ice and to carry canvas for quick repairs to the holes that would surely be stove in the hulls. Whiteside ignored Brower's advice. He added only one inch to the keels and decided to take no canvas.

All but eight of the crew abandoned the *Navarch* on August 3. Those who left, including Mrs. Whiteside, made slow progress dragging their three boats to the southeast, against the direction of the drift. They reached heavier ice the next day and were soon exhausted. Two of the boats, now badly broken, were abandoned. When they reached the edge of the pack, they launched the third boat, against Brower's advice, in dangerously milling ice. The ice smashed it to pieces and the party had no choice but to return to the ship, by then ten miles away across broken ice. According to Brower, Captain Whiteside broke down and began to drink heavily from a flask, declaring that it was everyone for himself or—including his wife—herself.

The group finally regained the ship five days after abandoning it. Brower knew there was no time to lose before starting out again. The other ships south of Icy Cape, knew nothing of their peril and by this time the *Navarch* had drifted nearly to Point Barrow. Once past the Point, Brower knew that the strong alongshore current forked, with one branch swinging northwest into the Arctic Ocean, the other turning east into the Beaufort Sea. Their only real hope of safety lay in reaching land before they were swept past the Point.

Brower quickly began building a light canvas boat along the lines of an Eskimo umiak. His experience in hauling boats over the ice to the lead edge for spring whaling had taught him to prefer the umiak, a lighter and more resilient craft than the heavy, fragile whaleboat. His boat, made with a hickory frame and oiled canvas covering, was large enough to hold ten men and light enough to be carried by two.

The group started out again on August 10. This time nine men chose to remain with the ship. Brower agreed to go ahead, breaking trail for the others, with the captain and his wife following with the boat. Late in the afternoon the vanguard reached the edge of the pack, only six miles from shore and a few miles south of the Point. They could see the U.S. revenue cutter *Bear* and the steam bark *William Baylies* at anchor inshore. When the men straggled up to Brower, he asked where the captain and the boat were. The last arrival told Brower that Whiteside had taken the boat, compass, food, and ammunition back to the ship, leaving all the others stranded on the ice. "We were in a fix," Brower remembered,

> trying every way we could think of to make the ships see us, to no use. In an hour it shut in foggy; then we did not know which way to turn. The ice was drifting fast to the north. The nearer we came abreast of the point, the faster we travelled. Then, the piece we were on separated from the pack, and we were marooned. Everyone was hungry; no one had anything to eat. . . . Not more than two besides myself had taken spare boots with them.

Brower realized that the only hope for the party of thirty-two men was to move into the current of ice flowing east around Point Barrow into the Beaufort Sea. To do so, he knew that they must leave the solid ice and reach the smaller drifting pieces of young ice, farther out at the pack's edge. Once in the Beaufort Sea there was a chance—a small chance—that they might be sighted by a whaler.

The men had to move fast, but the ice was heavy and rough. The only food they had was the few pieces of hardtack that Brower had stuffed into one of his spare boots. They rested frequently, but the ice was too wet to allow sleep. On the second day out, the first man died, shooting himself. A day later, two more dropped dead, and on August 13, the chief engineer died and two of the firemen stayed behind with him. The survivors, without food or sleep, became weaker and weaker. During the next two days six more men perished. Others fell back until only sixteen remained. Brower kept the remnant walking south as fast as possible, knowing that every day the summer pack ice retreated further from the coast, lessening their chances of being spotted by a passing ship.

Eight days after leaving the *Navarch,* the group reached the edge of the pack somewhere east of Point Barrow. Most of the men wanted to wait where they were, on relatively solid ice, in hopes of being spotted by the ships. Brower persuaded them to join him on a small piece which might be blown south to shore. He clinched the argument by pointing out that they might just as well die on a small piece of ice as on a larger one. Boarding a cake only twenty yards square, they drifted southeast for four days, until they were within two miles of Cape Halkett, about one hundred miles east of Point Barrow. Here, on the twelfth day of their ordeal they saw the steam bark *Thrasher* coming toward them along shore. Providentially, a Siberian Eskimo at the *Thrasher*'s masthead thought he saw a group of walrus on the ice, but, knowing that walrus were sparse east of Point Barrow, he pointed them out to the officers on deck. They turned their glasses on the small party and Brower and his men were rescued.

Sixteen of thirty-two survived the march; eight of them were unconscious when rescued. Brower saved the lives of fifteen men despite the cowardice of Captain Whiteside. Whiteside, Brower learned, had reached Cooper's Island near Point Barrow after little more than a day's travel with the umiak. Brower's march and drift surely rank with mankind's most punishing tests of survival: his journey over rough and moving sea ice with little sleep and only ice and boot soles for food, and without assurance of rescue at the end, is testimony to a man of iron will and body.

While the *Navarch* fell victim to the ice, the rest of the fleet, proceeded more cautiously, rounded Point Barrow without difficulty and reached the whaling grounds near Cape Bathurst. But the whaling season of 1897 was poor. By August 20 several of the ships gave up and began to work their way west along the north coast of Alaska. On August 27, near Cape Halkett, the steamers *Belvedere, Orca,* and *Jesse H. Freeman* ran into thickly scattered ice only a mile from shore. By September 2, they had covered only half the distance to Point Barrow and they found the ice hard on the land. Slowly they worked their way west during the next two days, but they were only able to get within ten miles of the Point. There they were joined by the steam bark *Alexander* and the schooner *Rosario,* also from Herschel.

One can only speculate whether or not at the end of August the captains of these vessels sensed their impending ordeal. Until 1897 the general rule was that steamers could safely pass Point Barrow as late as September 10. In fact, the *Orca* had rounded the Point as late as September 20, 1888, and most of the steamers had power enough

Captain and Mrs. Joseph Whiteside and others aboard the U.S. revenue cutter Bear *shortly after their rescue in August 1897*

to hammer their way through as much as five inches of young, thickening ice. But no doubt some of the captains recalled the only real rule of Arctic ice pilotage: that the movements of the ice were unpredictable. Certainly all of them must have hoped for a south or east wind. Either of these winds would have lifted the ice off the point.

But the day the ships reached Point Barrow, a stiff westerly sprang up. It was September 6. By midnight it reached gale force and drove the ice down hard onto the Point and within a half mile of shore elsewhere. During the next three days light winds allowed the pack to ease slightly, letting the ships sneak around to the west side of the Point, but they could go no farther. Their anxiety increased as word came from the east that the Pacific Steam Whaling Company's four-masted steam

Left to right: Tom Gordon, Fred Hopson, and Charlie Brower at Point Barrow in September 1898

schooner *Jeanie* was three days overdue on her return from the company's base at Herschel Island.

The dirty weather of the arctic autumn was upon them. On September 10, the wind shifted into the north. The temperature dropped and ice began to form in the patches of open water. Jim Allen, an engineer aboard the *Jesse H. Freeman,* claimed that the *Orca, Belvedere,* and *Freeman* wasted several crucial days anchored at Point Barrow while the captains drunkenly caroused together. Captain Benjamin Tilton worked the *Alexander* close inshore on the west side of the Point, getting inside a long ridge of grounded ice, the remnant of a massive winter pressure ridge. Behind the protection of the ridge, Tilton quickly unloaded a year's supplies at Brower's whaling station, then left hurriedly, ramming his way ahead and breaking through the ridge as soon as the ice outside slacked off slightly. The *Alexander*'s engineer described the fight to reach open water:

For eighteen hours it was 'full speed astern,' then 'stop her' and then 'full speed ahead,' followed by the crash as we struck the icefield. Back and forth we went and every succeeding crash seemed to us down in the engine-room as though it would be our last. It did not seem possible that wood and iron could stand the strain much longer. After getting through the pack we had to fight our way through . . . miles of young ice an inch and a half thick. I can tell you when we reached Sea Horse Islands and saw open water before us we were a happy set of men.

The *Alexander* was the last vessel to leave the Arctic from Point Barrow that year. It is puzzling that the other ships did not try to break out with her. One observer wrote, "I suppose they think the ice will go off from the shore, and they can get out without bucking."

Opportunity for escape slipped by. The *Orca, Jesse H. Freeman, Belvedere,* and *Rosario* were anchored in a small patch of open water southwest of the Point, hemmed in on the east by the land, on the south by the pressure ridge, and on the north and west by the pack. Young ice was forming quickly. The captains now fully realized the danger; they faced the prospect of being forced to winter on an exposed coast. The ice would almost surely destroy the vessels where they lay, as it had in 1871 and 1876, and none of the ships had sufficient supplies to feed the crew for the ten months until spring breakup.

The schooner *Rosario,* lacking auxiliary power, had to stay where she was in the dangerous roadstead west of Point Barrow. She never escaped; she was crushed the following July. But on September 17, a week after the normally safe departure date, the other vessels started to fight their way out. The *Belvedere* and the *Orca,* more powerful than the *Jesse H. Freeman,* took the lead, ramming their way through the thickening ice. Brought to a standstill by the heavy floes, they put their crews on the ice to blast a channel with gunpowder while the ships rammed ahead. In two days they made six miles and finally lay off Brower's station, but the venture had cost them dearly: the *Orca* had lost her rudder, and the *Belvedere* had sprung hers badly. They were forced to spend a day making repairs.

On September 20 more bad news reached Brower's station. An Eskimo arriving from the east reported that three more vessels were trapped by the ice along the north coast. The steamers *Fearless* and *Newport* were seventy miles away at Pitt Point and the supply schooner *Jeanie* was thirty miles farther east at Cape Halkett. Of the vessels at Herschel Island, the bark *Wanderer* was the last to leave. She had been stopped by ice and had returned to winter in Pauline Cove. All four vessels survived, probably because their late departure from Herschel kept them from reaching the menacing waters west of Point Barrow.

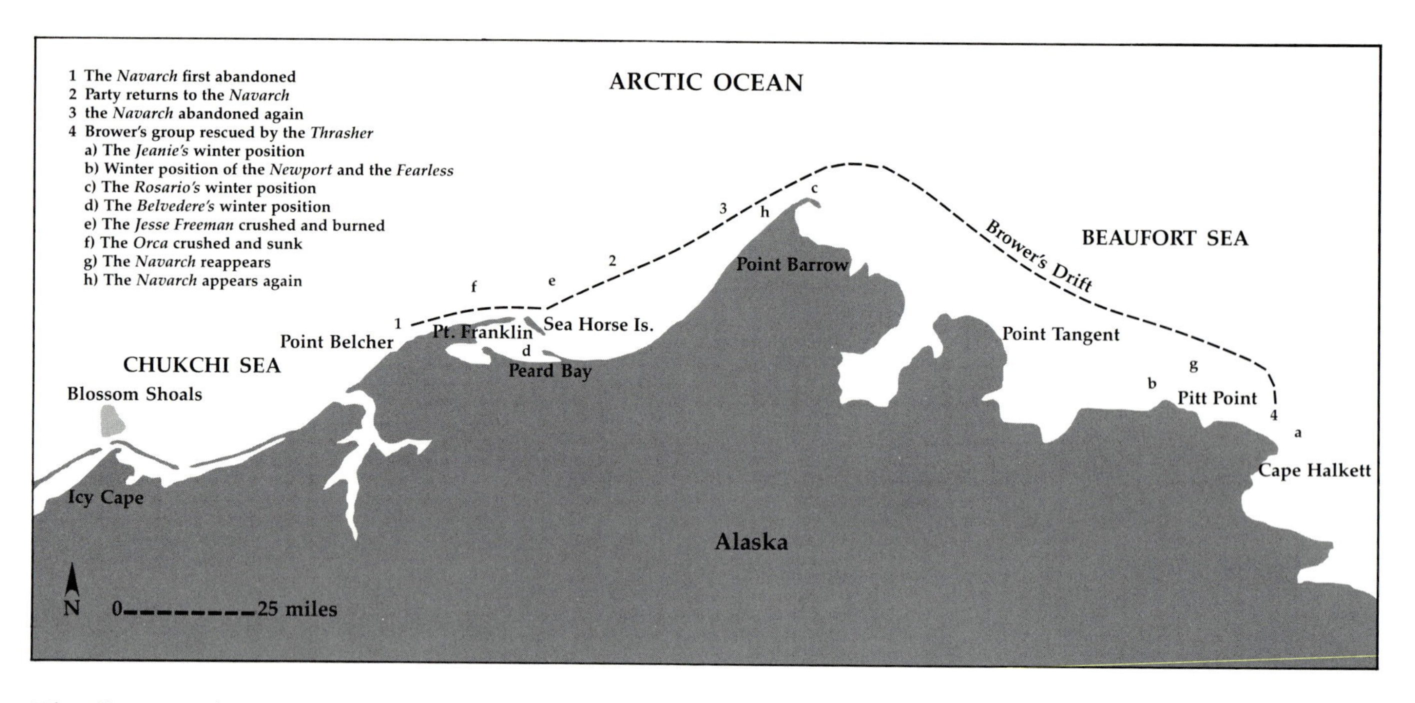

The disaster of 1897

Despite the increasing danger and necessity for speed, when Captain Martin Van Buren Millard of the *Belvedere* heard of the ships trapped east of the Point, he paused long enough to unload supplies of flour, molasses, coffee, and beans for them—a gesture displaying not only the generosity of northern people but also Millard's belief that his release from the ice was imminent.

As soon as the supplies were unloaded, the *Belvedere, Orca,* and *Jesse H. Freeman* resumed their effort to escape, slowly cutting through the young ice. By the evening of September 21, they reached Peard Bay near Point Franklin and close to open water, but before the night was out, their fate was sealed. A stiff northwest wind sprang up at 2 A.M., sending the ice in and catching all three in a totally exposed position as they worked their way toward Point Belcher and the open water beyond. The *Orca* was "caught between two immense ice floes and crushed with such force as to take the stern post and steering gear completely out of her and hurl the wheel through the pilot house." The *Belvedere* and *Freeman* were able to get within three-quarters of a mile of the *Orca* and rescued all her men, but by then escape from the closing ice was impossible. The two vessels turned northeast to seek safety behind Point Franklin. The *Belvedere,* under steam and sail, was able to ram her way through the thickening ice. But before the night was over, the *Freeman* was caught and crushed. The *Belvedere,* now safe behind heavy ground ice at the Sea Horse Islands, took her crew aboard as well.

By the time the *Freeman*'s men reached the *Belvedere,* the moving pack, driven by the freshening northwesterly, had met the young shore ice, piling and rafting it into pressure ridges and cutting off any hope of escape. The *Belvedere*'s crew and her 102

The Orca *(left) and* Jesse H. Freeman *at Point Barrow in September 1897, shortly before their loss* Private collection

passengers could see open water only a few miles away. But the odds that the ice would slack off so late in the season were slim. Captain Martin Van Buren Millard took the only prudent course and continued on to the northeast, bringing the *Belvedere* around behind the Sea Horse Islands into the shallow waters of Peard Bay. It was a wise move. The northerly gale continued for two days, forcing the pack, grinding and shearing, harder on shore.

The *Belvedere* was clearly trapped and had to be made ready to winter. There was no time to lose. Parties were sent to salvage supplies from the two trapped ships, both held afloat by the pressure of the ice. One group reached the *Orca,* but the *Freeman* was out of reach beyond dangerously thin ice. Before the whalemen could get to her, a party of Eskimos boarded the vessel. Hastily ransacking the cargo, they knocked over a lamp and the *Freeman* burned to the water's edge. Jim Allen watched his ship burn: "All the yards were lit up with the fire, and once in a while a 100 gallon drum of coal oil would explode and the flames would shoot up. We could hear the cracking of the bombs that were to have been used for killing the whales. It sounded like a New Year celebration in Chinatown."

Captain Millard organized the men to prepare the *Belvedere* for winter. Her supplies and moveable equipment were taken ashore, and her bulkheads were torn out for materials to build a storage house for supplies on one of the islands. Water was jettisoned, and the hull, now sufficiently lightened, was drawn through a channel cut in the young ice to a position close under the islands.

While the *Belvedere* was being secured, the *Orca* remained in sight. The *Orca* was full of supplies and two of her crew, Charles Walker, the fifth mate, and James Lee, a sailor, were sent out to her to see what could be saved. They were told that if any provisions could be pulled from the wreck they were to set one of the ship's topsails as a signal to bring a crew from the *Belvedere* to help with the salvage. For several days there was no sign; then a native arrived at the *Belvedere* and described how Walker and Lee had mobilized the natives living at Point Belcher and how, with their help, they had dragged ashore three heads of whalebone (worth about $20,000) as well as plenty of supplies. The two had, furthermore, built a canvas house for themselves on shore so that they could live there for the winter and await their ships in the following spring to sell their baleen.

The three captains aboard the *Belvedere* realized that they must get the *Orca*'s supplies. None of their ships had more than two months' supplies aboard and even with the *Orca*'s it would be a hard winter. The next day a party of thirteen armed men set out to retrieve the supplies. They were under the command of George Fred ("Big Foot") Tilton, third mate of the *Belvedere.* They surrounded the *Orca* and managed to get 300 eider ducks that were strung in the rigging—just as water was coming over the main deck. They then headed ashore and took possession of the camp. To their pleasant surprise, they found that Walker and Lee had managed to salvage, among other things, 400 sacks of flour. The party then returned to the *Belvedere,* leaving a guard with the stores. Soon the *Orca* slid beneath the ice.

With the sinking of the *Orca,* the forecast of events for the next six months was written. Most of the crew of *Belvedere* remained aboard their vessel on short rations for the winter, as did the crews of the ships trapped east of Point Barrow: the *Rosario, Newport, Fearless,* and *Jeanie.* But nearly one hundred men from the *Orca* and *Jesse H. Freeman* had to make their way on foot to Point Barrow, sixty miles away, to seek shelter with Brower.

If the disaster had taken place two years earlier, the men would have passed the winter in far better circumstances: until the summer of 1896, the United States government had maintained a well-provisioned refuge station at Point Barrow. Although the Point Barrow Refuge Station was not built until 1889, the first call for assistance to the Arctic whaling industry had come four decades earlier from Captain Thomas Roys. In the report of his discovery of the Arctic whaling grounds he warned:

On account of powerful currents, thick fogs, the near vicinity of land and ice, combined with the imperfection of charts, and want of information respecting this region, I found it both difficult and dangerous to get oil, although there are plenty of whale. . . . Some provision ought to be made to save the lives of those who go there, should they be cast away.

Roys's report—and more importantly, the oil rush it set off—caught the attention of politicians in Washington. With more than six hundred vessels in the American whaling fleet at that time, and with one quarter of these cruising in the North Pacific, there were powerful incentives, both commercial and political, to support the industry. And Lieutenant Matthew Fontaine Maury, the keeper of charts and instruments of the Naval Observatory, who was charged with developing sea charts for the navy and merchant fleet, called attention to the urgent need of whalemen for accurate charts of the North Pacific.

Maury was right. In the Bering and Chukchi seas, only the coast of Alaska between Bering Strait and Point Barrow had been accurately charted—by Captain Frederick William Beechey of the Royal Navy in 1826 and 1827. When the results of the phenomenal season of 1849 were abroad, the firm of E. and G. W. Blunt of New York published the first comprehensive chart of those waters that was readily available to American mariners. They combined Beechey's accurate chart of the Alaska coast with other Russian charts of the Bering Strait region. The Blunts unknowingly reproduced a number of errors from the Russian charts, but their composite was timely and useful.

When the news of the disastrous Arctic season of 1851 reached the East Coast—ten whaleships had been lost in the Aleutian Islands and the Bering and Chukchi seas—it provided a strong argument to those wishing to sponsor the American commercial and political expansion throughout the North Pacific. Accordingly, on March 22, 1852, the United States Senate adopted a resolution asking the secretary of the navy to advise them on the usefulness of sending a surveying expedition to the North Pacific.

Lieutenant Maury supplied the secretary with a detailed (and inaccurate) description of the great investment of American ships, men, and capital operating in the western Arctic fishery. This reply was submitted to the Senate in April 1852. When the bill was taken up in July, Senator William Henry Seward (who would become known as the "purchaser" of Alaska in 1867) gave it his full support in a flowery and bombastic speech with strong expansionist undertones. At the end of August, after a great deal of agitation by both whaling and shipping merchants, Congress provided $125,000 to survey the Bering Strait region and northwest Pacific.

The five ships of the North Pacific Surveying Expedition left Hampton Roads, Virginia, in June 1853. After various adventures, the flagship, USS *Vincennes* reached Petropavlovsk, Kamchatka, in July 1855. The *Vincennes* was to begin her northern explorations by surveying the bay and harbor there and by learning from the Russians about the seas to the north. To their surprise the Americans found Petropavlovsk largely deserted and many of the houses burned, the result of the recent Anglo-French assault—the easternmost action of the Crimean War.

The *Vincennes* reached Saint Lawrence Island on August 1 and then paused in Seniavin Strait on the Chukchi Peninsula to deposit Lieutenant John Brooke and several assistants on Big (Arakamchechen) Island to carry out surveys while the ship continued on into the Chukchi Sea. The *Vincennes* reached Herald Island on the thirteenth and surveyed for the most part in the northwestern part of the sea. The ship then returned south, collected Lieutenant Brooke's party and continued on, reaching San Francisco on October 13, having become the first American federal ship to enter the Arctic Ocean. (During the same summer the USS *Fenimore Cooper* surveyed the Aleutians and the USS *John Hancock,* the Sea of Okhotsk.) The charts of the Bering and Chukchi seas, published soon after the *Vincennes'* return and periodically revised throughout the century, marked a substantial improvement over previous publications.

Federal support for the industry essentially ceased during the Civil War and was not renewed until after the purchase of Alaska from Russia. From July 27, 1868, however, when a customs district was created for Alaska, Alaska's maritime affairs were controlled by the United States Revenue Marine Service. The Revenue Marine controlled trade in contraband (liquor, breech-loading firearms, and fixed ammunition) and provided support for the safety of commercial vessels.

In the first years after the purchase the Revenue Marine was too poorly funded, equipped, and staffed to carry out all its duties on the 33,000 miles of Alaskan coastline. Although the revenue cutter *Reliance* passed north of Bering Strait in 1870, not until 1879 did regular patrols begin, first with the northern cruise of the *Richard Rush,* then from 1880 to 1885 the *Thomas Corwin* carried out the work, and from 1886 to 1925 the *Bear* patrolled northern waters.

The whalemen at first resented the surveillance of the cutter, which curtailed the trade of alcohol and breech-loaders to the Alaskan Eskimos. By the end of the 1884 season Captain Michael A. Healy, USRM, reported "the whiskey traffic in northern Alaska almost entirely ceased."

Despite his revenue duties, "Hell Roaring Mike" Healy won the whalers' respect and support in the 1880s and 1890s. As the number of bowhead whales declined in the Arctic, the ships were forced to take greater and greater risks to secure their catches. The presence of the cutter in the Arctic until the end of August allowed the whalemen to push the season to the limit, knowing that the revenue cutter was never far from the cruising grounds and was always ready to give aid. Healy could be counted on to take risks beyond the call of duty.

In 1885, for instance, while the fleet was waiting at the ice edge near shore between Point Belcher and Point Franklin, a sudden strong gale blew up from the south-southwest. The bark *George and Susan* parted one chain cable and began to drag her other anchor. She ran afoul of the *Mabel* and the collision carried away the *George and Susan*'s jibboom and headgear and broke the *Mabel*'s mainyard. Soon the *George and Susan* was ashore, bilged, and the *Mabel* began to drag.

Seeing the *Mabel* in grave danger, Healy took the *Corwin* right into the breakers, with seas pouring over her foredeck and with only four-and-a-quarter fathoms under her keel. Healy could not save the *Mabel,* but he provided medical aid and

Captain Michael Healy aboard the Bear *with his parrot*

shelter for the crews of both vessels until they found berths aboard other ships in the fleet.

In 1887 when the *Bear* reached Port Clarence to inspect the fleet for contraband, Captain Barney Cogan of the bark *Hunter* gave Healy an old chip of wood with crude letters carved on it.

On the front: 1887 J. B. V. /TOB
BK. NAP. /BACO
/GIVE
On the back: S. W. C. NAV
M 10
HELP COME

A native had given the chip to Cogan near Cape Bering on the Chukchi Peninsula nearly a month before.

Healy concluded that this was a cry from a survivor of the bark *Napoleon* that had struck a piece of ice and sunk in fifteen minutes two years before in a terrible gale, fifteen miles south of Cape Navarin on the night of May 3, 1885. Only fourteen of

J. B. Vincent's wood chip. This cry for help was carried hundreds of miles in Siberia and resulted in his rescue by the Bear.

the *Napoleon*'s thirty-six men had been rescued. They translated the message as follows: the survivor, J.B.V., was ten miles southwest of Cape Navarin and the message had been carried by hand northward along the coast; the survivor wanted the message carrier given tobacco in return for his efforts.

Healy took a native interpreter, Jake Rainbow, from the steam bark *Beluga* and set off for Cape Navarin. He arrived there July 14, and after three days of searching, located James B. Vincent of Martha's Vineyard, a boatsteerer from the *Napoleon,* in a Koryak village about forty-five miles west of Cape Navarin.

Vincent told Healy that the *Napoleon*'s boats had been separated in a gale and that his boat and another had remained at sea on the ice for twenty-nine days. During the time they had only two seals for food. Gradually the men died of starvation and exposure. Only Vincent and three others lived long after reaching shore. The other three died of scurvy in March 1886, leaving Vincent alone among the Koryaks on a coast infrequently visited by ships. Vincent had almost given up hope of rescue.

These stories merely illlustrate Healy's dedication and cool head. He and the Revenue Cutter Service were, in fact, of much more fundamental help to the whaling fleet. The cutters carried the only qualified physicians into the Arctic. They

J. B. Vincent aboard the Bear *after two years in Siberia* Smithsonian Institution

lent equipment, such as anchors and caulking materials, when none could be obtained elsewhere. They carried "sick and destitute" whalemen south. They rescued a number of ships from tight spots on lee shores or in the pack. They also carried "spring catches" of whalebone south when the whaleships were heading into risky situations and could find no other means of transport.

Healy was also deeply interested in the welfare of the natives and complained frequently to his superiors that the contraband laws against selling breech-loading firearms to the Eskimos (a relic of the Indian Wars) prevented them from hunting effectively. For the same reason he was diligent in examining whaleships and trading vessels for alcohol and in sending parties ashore at native villages to break up stills.

Healy and only a few others stand out strongly against the background of America's western Arctic of the latter nineteenth century, a time and place exceptionally rich in strong, colorful men. Healy admired the captains and officers of the whaling fleet, and they in turn loved him; he was as tough and brave and capable and profane and bibulous as they were—if not more so.

But these same qualities ultimately led to Healey's downfall. During the last half of the nineteenth century, as the profits from whaling voyages declined, so did the quality of the foremast hands. By the 1880s, the *San Francisco Chronicle* described these sailors as "drunkards . . . jail birds and thieves." Healy was very firm in reinforcing the officers' authority, and he was called upon to suppress a number of mutinies. His strong measures—his antagonists called them brutal—collided with the growing sentiment for the maintenance of seamen's rights. This, combined with his chronic alcoholism, caused his demotion after the 1895 cruise.

One of Healy's projects was the establishment of an Arctic refuge station for shipwrecked whalemen. Thirty years earlier during the searches for Sir John Franklin's lost expedition, the British had demonstrated the value of placing supplies in the Arctic for shipwrecked men. In 1880 the whaling merchants petitioned Congress—unsuccessfully—for an appropriation to place provisions at several points in the western Arctic.

In 1881 the loss of the *Daniel Webster* supported their petition. The *Webster*'s captain, David L. Gifford, lacked northern experience and was on only his second Arctic cruise. When he reached the pack ice at Icy Cape and saw two ships entering a strip of open water that ran northward along the coast, he followed, assuming that their greater northern experience would keep him from danger. During the night the two leaders outdistanced him, and somewhere southwest of Point Barrow they noticed that the lead was closing quickly. They turned and beat frantically to the southwest, only narrowly escaping the ice. The two ships passed the *Daniel Webster* in a fog. Gifford only discovered his mistake when he reached the end of the lead near Point Barrow—and by then it was too late. He turned back, but the closing lead allowed him very little room to work his ship, and the strong northerly current sealed his fate. On July 2, hoping that the pack would soon ease, he anchored his ship to the shorefast ice about five miles from the Point. Within an hour the pack came shearing in and took the bottom off the *Daniel Webster.* Twenty minutes later she rolled over on her beam ends and sank. The crew had only time to throw some provisions onto the ice before escaping with little more than the clothes on their backs.

The Eskimos confiscated almost everything. According to John Muir, the naturalist who visited Point Barrow six weeks later, "at the village a general division was made in so masterly a manner that by the time the officers and crew reached the place their goods had vanished into a hundred-odd dens and holes; and when, hungry, they asked for some of their own bisquits, the native complacently offered to sell them at the rate of so much tobacco apiece."

Most of the *Daniel Webster*'s men immediately started south on foot toward the safety of the fleet, but about half soon tired and returned to the Point "to pick up a living of oil and seal meat until relieved." The rest, weak and hungry, arrived at Icy Cape and boarded the bark *Coral* less than a week later. The whaling fleet, with the revenue cutter *Corwin,* reached the wrecked men at Point Barrow on August 16.

But not until the following year, 1882, when the steam bark *North Star* was lost at almost the same place and date was the idea of a manned refuge station considered seriously. The *North Star* "was ground as fine as matches" two miles from Point Barrow. Fortunately for the crew, as part of the first International Polar Year the government had established the U.S. Army's Signal Service station near the Point for meteorological observation. The *Webster*'s crew was given shelter there until the fleet arrived later in the month. The meteorological station was closed in 1883, but the Pacific Steam Whaling Company of San Francisco rented it the following year to use as a trading post and as a base for springtime shore whaling at the leads in the ice. The government leased the station with the proviso that, should any shipwrecks occur nearby, the company would take care of the men until help arrived.

In that year, 1884, government surveys in the Noatak and Kobuk river drainages showed that there was no practical overland escape route for the men, should the whaling fleet be trapped again in northern Alaska. Michael Healy wrote to the general superintendent of the Life Saving Service suggesting the establishment of a lifesaving station at Point Barrow. He pointed out that it would be inadvisable for the *Corwin* to remain in the Arctic until the whaling fleet had left these dangerous waters late in the autumn. Healy reinforced these remarks in his annual report to the secretary of the treasury by proposing that such a station could be used to control the illicit whiskey trade. Shortly thereafter a group of ninety-six men—owners, agents, captains, and officers of the whaling fleet—petitioned the secretary in support of Healy's proposal.

With backing for the station growing, Josiah N. Knowles, the managing partner of the Pacific Steam Whaling Company, proposed to the general superintendent of the Life Saving Service that his company build a station and staff it with six men for $3,000 per year. Knowles estimated that a year's provisions for fifty men (roughly the complement of a steam whaler) would cost $9,000. He added that his company would provide the freighting of supplies free of charge. Knowles's offer was probably not made solely in the light of humanitarian interests. The Pacific Steam Whaling Company had just established their shore station in the old Signal Service building at Point Barrow and Knowles had already led the company into several highly profitable ventures—steam whaling vessels, an oil works, and salmon packing. The potential benefits of reinforcing its Arctic operations must have crossed his mind.

Even though the industry advocated the station, it took a crisis before the government acted on the proposal. The crisis occured in August 1888. On August 2, about thirty whaleships lay at anchor off the west side of Point Barrow waiting for the ice pack to open just enough to let them cruise into the Beaufort Sea. About 6 P.M. a southwest breeze sprang up, increasing to a gale in three hours. The fleet scurried to the east side of the Point for shelter, some of them anchoring between the Point and the two-and-one-half fathom shoal lying less than half a mile to the east.

By 6 P.M. the next day the wind veered into the west and the gale increased, driving a tremendous sea into the shallow anchorage. In the midst of the screaming wind and flying scud the bark *Eliza,* anchored bow and stern, began to labor heavily. Finally, she parted her stern cable and swung down on the nearest vessel, the schooner *Jane Gray.* To avoid the collision both vessels quickly slipped their cables and ran off down wind.

Nearby to leeward, the *Bounding Billow* held her ground as the two ships bore down on her. The *Jane Gray* shot by, narrowly missing, but the *Eliza* struck her a glancing blow, stoving in the *Bounding Billow*'s bulwarks and losing her own head gear.

The *Jane Gray* "was going at a fearful rate when she crashed into an iceberg and a great hole was stove in her starboard side into which the water poured in a great stream." As the *Jane Gray*'s crew tried frantically to stop the hole with canvas, Captain William Kelley, seeing it was useless, steered for the bark *Andrew Hicks,* a quarter mile away. As he neared the *Hicks,* the schooner settled in the water and the crew took to their whaleboats, reaching the bark safely.

Meanwhile, the bark *Young Phoenix,* farthest to the west and in the most exposed position, tried to take in her two anchors and run before the seas. The strain on the cables was enormous, and as the crew worked on the windlass one of the cables parted. The other held, but its anchor dug in so deeply the men on the windlass could not break out the anchor. With the remaining cable drum-tight, the nose of the bark was down into the full force of the oncoming seas "until a huge wave lifted her and dashed her down with a terrible shock. The anchor refused to let go and she was again and again dashed on the bottom." Finally the anchor broke loose and Captain Martin Van Buren Millard tried to beach his ship, now mortally wounded with her rudder sprung and seams open. As she "floundered about among the breakers," she smashed into the bark *Triton.* Captain Millard gave up trying to reach shore and headed her before the wind. At 4 A.M. the *Young Phoenix* was thirty miles east of Point Barrow. Her crew abandoned the vessel and was picked up by the steam bark *Beluga.* The *Young Phoenix,* with her back broken, did not sink; she was driven into the ice. A year later she was seen drifting in the pack, an Arctic ghost ship.

The barks *Fleetwing* and *Mary and Susan,* like *Young Phoenix,* were lying on short scope near the point. As the night grew wilder, the seas repeatedly pounded them on the bottom. The *Fleetwing*'s anchor chains finally broke and she drifted onto the shoal, fully exposed to the breakers. The storm eased in the morning but by then both vessels were battered wrecks and their crews abandoned them. The *Fleetwing* eventually washed over the shoal and drifted into the pack. She was sold at auction for ten dollars. Charlie Brower later reached the hulk by boat, jury-rigged some sails

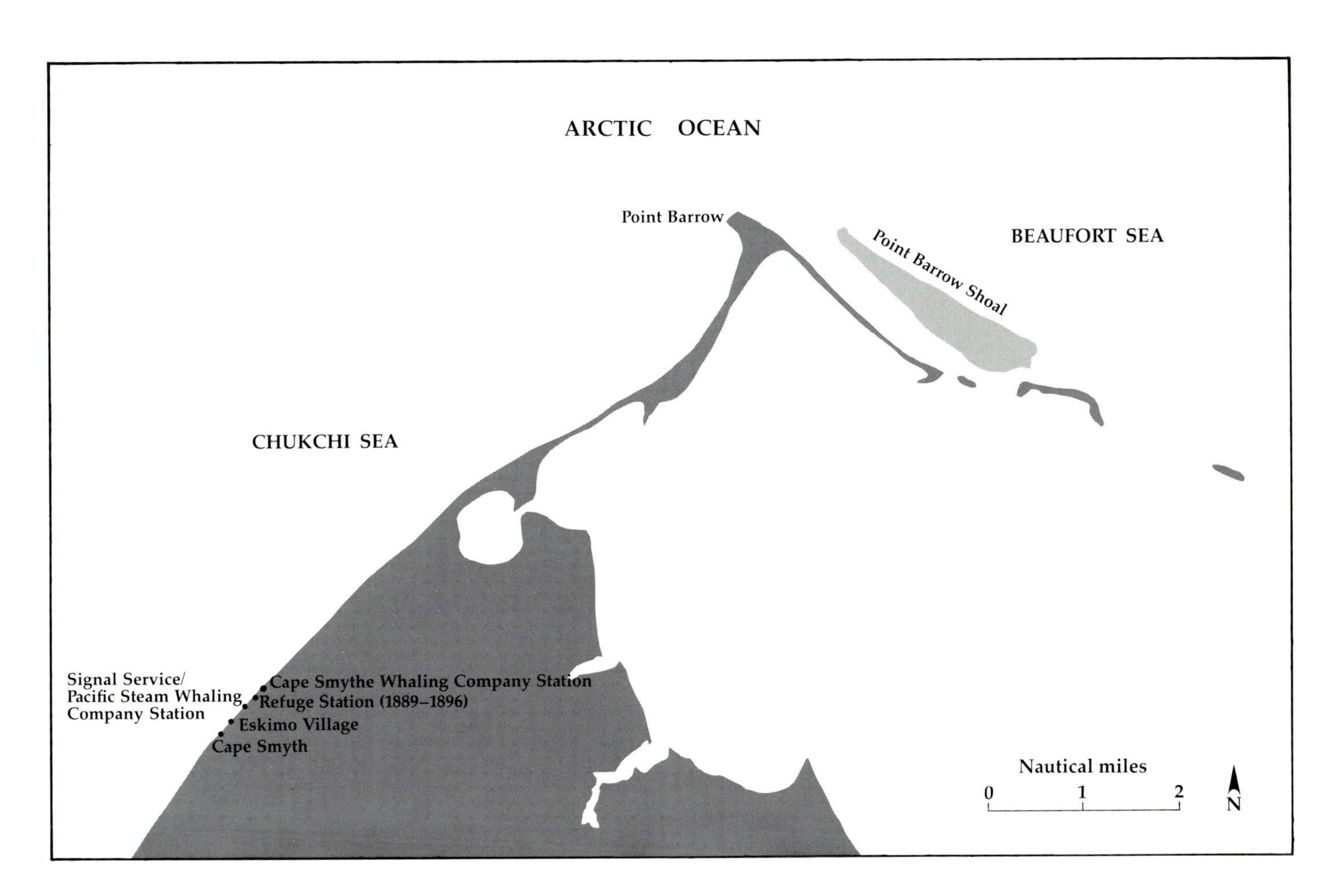

Point Barrow

and steered her, half full of water, to shore where he salvaged her gear.

"It was," Thomas P. Warren wrote aboard the *Triton,* "a night to be remembered, in fact a regular picnic."

Ironically, the little schooner *Ino* survived the gale, only to be lost a few days later. She was anchored at the west end of the shoal when the storm began, and as it increased, the seas "made a clean sweep over her." Being a light vessel, the crew was able to slip her cables and stand off shore, running before the storm. She returned to Point Barrow on the eighth of August, but there she was caught unprepared by a sudden squall and quickly driven ashore. She was declared a total loss.

Other vessels, too, fared poorly in the storm of August 2. Both the *Reindeer* and the *John Howland* lost their anchors and drifted onto the shoal, but pounding hard, they were swept across the shoal into clear water. The *Rosario* and the *Ohio* each lost both of their anchors. Most of the rest of the fleet suffered damage to their ground tackle, windlasses, hawse pipes, and boats.

It was fortunate for the shipwrecked men, cast ashore without food or shelter, that

The Jane Gray *capsized east of Point Barrow, 1888* Alaska Historical Library

both the revenue cutter *Bear* and the USS *Thetis* were nearby when the gale struck. When the *Bear* arrived in San Francisco with 110 shipwrecked whalemen aboard, the news caused a sensation in the presses of both coasts. The loss to insurers and owners was more than $100,000 ($1,074,000 in 1982 dollars) and only the fortuitous presence of the *Bear* and *Thetis* had prevented a greater tragedy.

But before the season of 1888 was finished, another disaster menaced the whaleships. In late September, after the *Bear* and the *Thetis* had gone south, the fleet moved west as usual to cruise on the autumn whaling grounds near Herald Island, deep in "the Hole," the great cleft in the pack ice kept open late in the season by a warm current passing north from the Bering Sea. The whalemen could count on finding bowheads there—but not without risk, for it was a dangerous cul-de-sac where a combination of prolonged freezing temperatures and calms could put six inches of ice on the sea in twenty-four hours.

Equally hazardous were the movements of the pack ice. As the autumn wore on, the pack ice began a steady march south, expanding as a huge pincer, with one mass reaching down along the coast of Chukchi Peninsula and the other, down the Alaskan side. The whalemen knew that if these jaws closed around the ships, only with luck could they escape. The New Bedford barks *Mount Wollaston* and *Vigilant* had suffered this fate and were lost with all hands in 1879, while the bark *Helen Mar* escaped to tell the tale.

History repeated itself on September 23, 1888. Captain Frederick A. Barker, master of the schooner *Rosario*, sensed the danger and sailed south in company with the steam bark *Lucretia* only to find the passage closed. The *Lucretia*, with auxiliary power, rammed her way out but the *Rosario* could not escape. Barker turned back to warn the ten vessels still in the Hole.

Once out of the ice the *Lucretia* quickly spread the word, and four steamers turned back north to try to hammer their way in to the trapped fleet; two others took on supplies at Unalaska and started north as well. Meanwhile, the trapped ships were frantically searching for a way out. "I'm not going to stay here and die like a rat in a hole," Captain Edmund Kelley declared.

By sheer luck a strong gale blew up on September 29. By October 2 it had forced the pack apart, allowing the ships to escape. The *Lucretia*, however, reached San Francisco far in advance of the fleet. The report of the apparent tragedy, following the wrecks at Point Barrow, consolidated support for the refuge station. (It is doubtful, however, that a station anywhere in the Arctic could have helped the fleet in the Hole, had the ships been trapped for the winter.)

As newspapers throughout the nation carried stories about the perils of the whaling fleet, on November 22, 1888, the new Bedford Board of Trade (the city's Chamber of Commerce) petitioned the president of the United States, Grover Cleveland, to establish a station. On December 1, New Bedford's "citizens, merchants, and others" also petitioned Congress. These voices, added to those of San Francisco led by Senator Hearst, forced the government into action: within a week requests went out for advice on the location, staffing, and provisioning of Arctic refuge stations. The consensus of the whaling captains was that these stations should be built near the Arctic's most dangerous and frequently traveled areas; their suggestions included stations at Cape Prince of Wales, East Cape, Point Hope, Cape Lisburne, and Point Barrow, and a cache on Herald Island. But on January 15, 1889, the Commerce Committee of the House of Representatives sent a bill to the floor to appropriate $15,000 to build and equip one station only, at Point Barrow.

Even before the bill had passed, applicants began seeking the station's superintendency. Gilbert Bennett Borden, a retired whaling captain who was then night watchman at the New Bedford Five Cents Savings Bank, was chosen for the position although he was by far the weakest candidate. He obtained the position through the political influence of his Masonic brothers and friends who were members of the New Bedford Board of Trade.

The refuge station was built near Cape Smyth, eleven miles south of Point Barrow, near the old Signal Service station that was then being rented by the Pacific Steam

Building the Point Barrow Refuge Station, 1889 National Archives

Whaling Company. Work began on July 27, 1889. Pre-cut lumber had been brought north on the revenue cutter *Bear.* Under the supervision of Captain Michael Healy and Lieutenant David Jarvis, the building was put together *gratis* by carpenters from many of the whaling vessels. The station, 30 by 48 feet with a twenty-ton coal bunker on one side, had a large central room with bunks for fifty men along the walls, and its storeroom was stocked with provisions for them for more than a year.

Borden took command on August 16. He had appointed two men to his staff: Frank Goetsche, first assistant, and John Cuba, a Japanese, second assistant and steward. By good fortune no shipwrecks were to occur during his three-year incumbency and no doubt things should have gone smoothly at the station, but for Borden—suspicious, contentious and dishonest—trouble started immediately.

Before leaving San Francisco Borden had joined in partnership with Captain James McKenna, a whaling merchant and whiskey trader, and received a trade outfit to barter with the Eskimos for furs and whalebone. Borden began his enterprise at once—although it was in violation of the station's regulations—and quickly annoyed the local whalers and traders, none of whom had the benefit of government subsidy.

Borden's relations with Charlie Brower were no better because Borden also planned to outfit a shore-whaling crew from the station. "Borden was a mean old skin," according to Brower,

> Every time we came in contact with him he was all for whaling. He had nothing in the shape of gear. Thought he might borrow a bomb gun from us and a few bombs; if we would let him have them, he could walk out to the edge of the ice and maybe he could get a shot at one. No doubt he was a good enough whaleman in his time. Now all he wanted to do was to have us pick up any whales he might shoot, tow them to the ice, and cut them up for him; all for his benefit.

Borden wrote to the secretary of the treasury and exculpated himself by claiming ignorance of the station's rules, but in the summer of 1890 he entangled himself in a naked fraud that brought about his downfall.

It all began in August, 1890, when the *Bear* arrived off the station. To Borden's surprise and anger he learned that Captain Healy had appointed Leander Stevenson, a missionary school teacher, as an assistant keeper of the station. Stevenson was to conduct classes for the Eskimos at the station until a schoolhouse was built.

Borden immediately wrote to the secretary of the treasury, protesting the appointment. He claimed that another assistant was unnecessary and that if he was to be responsible for the conduct of his subordinates, then he should choose them himself. Stevenson was under the protection of Sheldon Jackson, the politically powerful Presbyterian minister who was general agent of education in Alaska. For Borden, whose letters display a martinet's love of autocracy, it must have been galling to have no real control over Stevenson. From the start sparks flew between Borden, garrulous and obscene, and Stevenson, prudish and pious.

The next link in the chain of events that led to Borden's ouster was forged in the autumn of 1890 when Borden met Peter Bayne, who was master of the little twenty-eight-ton schooner *Silver Wave*. Bayne, the man who had taught the Point Hope Eskimos how to distill alcohol, was as thoroughly disagreeable as Borden. He reached the refuge station in November after he had run the *Silver Wave* aground at the Sea Horse Islands, about seventy-five miles southwest of Point Barrow. Although the schooner was still structurally sound and was only lightly grounded in soft silt, Bayne and Borden soon devised a scheme to defraud the *Silver Wave*'s owner, James McKenna (Borden's former trading partner). They planned to declare the schooner a derelict, condemn it, auction it to themselves, and then use it and the refuge station's supplies to support a shore-whaling operation in the spring.

The usual procedure for condemning a wrecked vessel in a remote region was to convene several ships' captains to judge the vessel's seaworthiness. If they decided that the vessel was beyond repair where she lay, they would condemn her. Then someone would auction the ship and cargo on behalf of the owners and insurers. Bayne and Borden probably concluded that, with the exigencies of the time and place, they could organize a credible sale if Bayne were to judge the vessel unseaworthy and Borden were to auction it to a cooperative third party.

The staff of the Refuge Station. Left to right: Leander Stevenson, Gilbert Borden, Frank Goetsche, John Cuba National Archives

The white residents of Point Barrow, among them John W. Kelly of the Pacific Steam Whaling Company, immediately saw through this flimsy charade and refused to have anything to do with the sale, calling it a "fraud and a farce." Nevertheless, Borden, acting as auctioneer, sold the schooner, with its supplies and gear, to himself for $100.

Leander Stevenson immediately protested the conspiracy. This challenge to his authority enraged Borden. On Christmas Day he entered Stevenson's room at the back of the station, where Stevenson both lived and conducted classes for the Eskimo children, and announced that if Bayne should arrive at the station and require shelter, then the school would have to be closed. Stevenson objected, saying that Bayne's presence would be no grounds for closing the school, that the building would hardly be crowded. Borden replied that it was up to himself to decide when the station would become too crowded for classes to continue. He added, "I have heard a noise in your room last night. I had a notion to turn out and see if it was someone breaking in, and it would be only a word and a shot."

Later, after Stevenson had stated that it would be unlawful to supply government provisions to Bayne, Borden replied, "There is no law in Alaska," and "I don't care a damn for regulations. *I* am superintendent and you can take your appointment and wipe your backside with it."

Borden was clearly unable to deal with Stevenson, possessing as he did Healy's and Jackson's protection. Borden's only rejoinder of the winter was to write to the secretary of the treasury, giving several reasons why Stevenson should be fired, claiming that he had acted insubordinately and had "alienated the allegiance of the other assistants from their duties to the Superintendent."

The rest of the winter and spring was relatively uneventful, but everyone understood that an inquiry into the events surrounding the sale of the *Silver Wave* would take place when Healy and the *Bear* reached Point Barrow. But the summer of 1891 was exceptionally icy. The pack never retreated far from the coast north of Point Belcher, and in early August only a small shore lead of melt water opened from there to Point Barrow.

When it became likely that the *Bear* would not reach Point Barrow, Borden, who no doubt suspected the worst and may have heard that Healy disapproved of his actions, wrote to the secretary of the treasury, setting out his case for helping Bayne. He had already written to the secretary in April, to notify him of the sale of *Silver Wave,* but in that letter he had neglected to mention that he himself had bought the vessel.

Meanwhile, Bayne had been unable to free the vessel from the sand bar and had gone south from the Sea Horse Islands in a whaleboat to meet the whaling fleet. Somewhere near Icy Cape he sold the schooner for $300 to Captain Albert C. Sherman. When Sherman and Bayne returned north they found *Silver Wave* swinging at anchor. She had worked herself clear of the strand. Bayne moved her south to winter quarters in a lagoon near Icy Cape where he planned to use her for a shore-whaling station.

The resale of the *Silver Wave* and her repositioning apparently ended the association between Borden and Bayne, but Stevenson's presence at the station and the prospect of Healy's return in 1892 must have weighed heavily on Borden's mind. His correspondence took on an increasingly strident irrational tone. On October 1, without authority, he suspended Stevenson as assistant superintendent and wrote to the secretary of the treasury, justifying his action on the grounds of Stevenson's insubordination, claiming that a "powerful conspiracy" existed between Stevenson, Healy, and the Pacific Steam Whaling Company to have him fired. Trying to discredit this group, he concocted ridiculous charges: Healy he accused of smuggling two boxes of untaxed tobacco to Antone Bett, the Azorean shore whaler and trader at Point Barrow; and he claimed "by reliable authority" that Josiah Knowles, the managing partner of the Pacific Steam Whaling Company, had sold twelve barrels of molasses (for distilling alcohol) to Dr. John Driggs, the Episcopal missionary physician at Point Hope.

Certainly no conspiracy existed against Borden, but his meddlesome, combative ways had earned him powerful enemies. On February 29, 1892, a group including the Pacific Steam Whaling Company, the Wright, Bowne Company, and James McKenna, the rightful owner of the *Silver Wave,* petitioned the secretary of the treasury for Borden's ouster, stating that they constantly received complaints about Borden's activities, claiming he was using the station for his personal trading, and protesting his conduct in the sale of the *Silver Wave*. They added that "a government officer's action in such a position should be to suppress rather than encourage any collusion . . . for defrauding owners."

During that winter the secretary of the treasury also heard complaints about Borden from Healy and Sheldon Jackson. He decided to fire Borden sometime in the spring. Borden learned of the decision July 29, 1892, when Lieutenant David Jarvis arrived at the station on the steam bark *Thrasher,* ahead of the *Bear.* Three weeks later Jarvis arrested Bayne aboard the *Rosario* and, pending Healy's investigation, confiscated the 1,300 pounds of baleen from Bayne's 1891 shore-whaling operation at Icy Cape.

Healy reached Point Barrow aboard the *Bear* in mid-August and immediately heard testimony about Borden's conduct. He reported to the secretary of the treasury that Borden had allowed the building to deteriorate, that he had left the stores in disorder and kept poor records of the expended provisions, and that he seemed to have "mixed himself more into other people's business than his own and to have made reckless and indiscriminate charges against everyone about him without being mindful of the property under his charge or his own respect for the law. . . . He used his every endeavor to convert the place into a whaling and trading place of his own." Although the testimonies strongly indicated Borden's and Bayne's guilt, Healy, gruff as ever and clearly irritated by the matter, was glad to be rid of "the old scoundrel" and content to return the schooner to McKenna, concluding that "the whole affair was not worth the time it took to investigate it."

Healy appointed Leander Stevenson as interim superintendent and the following year installed another old whaleman, Captain Edward Akin of New Bedford. The station ran smoothly under both Stevenson and Akin, punctuated only by the loss of the bark *Reindeer* and the maintenance of part of her crew for a few days in August 1894.

Perhaps it was Borden's chaotic tenure, followed by three uneventful years, that caused the station's supporters to lose interest. Discussion about whether to close the station began in late 1895. In January 1896 Calvin Leighton Hooper, the Revenue Cutter Service's superintendent of construction and repair, wrote to the secretary of the treasury, stating that the refuge station was a useless expense, citing Healy's opinion that it should be closed and the whaling stations allowed to care for any wrecked crews. Shortly thereafter, Josiah Knowles, previously one of the station's strongest supporters, wrote to Senator George Perkins:

> from the first I never could see the real benefit that it would be to any of the whaling fleet to have a station there, as ships are seldom lost in that vicinity. There has never been a wreck there to my knowledge when the station would have been of any use . . . and should a wreck occur in that locality there are other stations that would provide for the crews, there being two stations at Point Barrow and another at Point Hope.

He added, accurately, that ships lost late in the season, when the fleet was near

Herald Island, were too far west from Point Barrow for the station to be of use, and those lost early in the season were too far south. He suggested that a sole station would be better placed at Point Hope.

Knowles's comments were a mixture of accuracies and inaccuracies. The station at Point Barrow certainly had the potential to save lives. As the wrecks of 1876, 1881, 1882, and 1888 indicated and the disaster of 1897 would prove, the area was definitely a ships' graveyard. But he was correct in stating that the station would be of no use for losses in early spring or late autumn, remarks that point to the carelessness with which the refuge station project was planned in 1889. Nevertheless, Knowles's comments may have sprung from another motive: he was certainly aware that his own station was dilapidated and that the refuge station, if sold, could probably be bought cheap.

It is less easy to understand Healy's apostasy. Four years earlier he had written:

> The encouragement and assurance given [the fleet] by the Refuge Station with its supply of provisions near by in case of accident, where heretofore they would in each case be at the mercy of the elements, has enabled them to go farther and to more profitable places in pursuit of their business and contributed greatly to their success.

That winter they decided to abandon the station. In March 1896 the government called for bids on the sale of the station and its stores. Knowles, on behalf of the Pacific Steam Whaling Company, bid $4,059.86, exceeding by $1,000 the offer made by the H. Liebes Company, a furrier in partnership with Charlie Brower's Cape Smythe Whaling Company. Knowles's bid was accepted in April, but payment was not required until the bidder had been able to survey the station. In December 1896 Knowles wrote to Calvin Leighton Hooper to report that sixty-eight barrels of pork were found to have spoiled. The company ultimately had to pay only $3,013.51 for the station and its contents, a property whose stores alone had been valued to $8,373.74 in June 1895. On December 29, 1896, the deed was transferred to the Pacific Steam Whaling Company, ending government administration of the refuge station.

Knowles died shortly thereafter, and the company decided to abandon its Point Barrow operations. Some of the equipment was sold and some taken to Herschel Island. The station was placed under a caretaker for a year. In the spring of 1897 the company leased the station to Edward Avery McIlhenny, a twenty-four-year-old naturalist who planned to winter at Point Barrow with two assistants to make an ethnological collection for the University Museum in Philadelphia and to collect ornithological specimens for his own study.

Then came the disasters of 1897—the losses of the *Navarch, Orca,* and *Jesse H. Freeman.* At the end of September 1897 the men aboard the trapped vessels—the *Belvedere, Rosario, Newport, Fearless,* and *Jeanie*—had plenty of time to reflect on the government's abandonment of the refuge station, for they had no one to turn to with responsibility for their welfare. As always, arctic people helped one another, but the destitute whalers were lucky to have Charlie Brower at Point Barrow.

E. A. McIlhenny, 1898 Private collection

Brower had sensed early in the autumn that the ships might not escape the ice and he started to store as much game as possible in his ice cellars. Then on September 20 word came from the east that the *Newport* and *Fearless* were trapped at Pitt Point with only two months' supplies aboard. Brower figured that with the provisions left earlier by the *Belvedere* and with his own stores, he could easily supply the *Newport* and the *Fearless* for the winter. He was not worried about the men aboard the *Jeanie;* she was farther to the east at Cape Halkett with many of the refuge station's supplies aboard. Brower took charge and sent Ned McIlhenny and Tom Gordon, one of Brower's associates, by dog sled to Pitt Point to notify the men aboard the ships there that they would be cared for.

The crew of the Rosario, *1898* Private collection

Amazingly, the morning after McIlhenny and Gordon reached the ships they were greeted by the spectral sight of the *Navarch,* far out in the pack, slowly drifting eastward.

No sooner were supplies for the *Newport* and the *Fearless* arranged than more bad news arrived. A messenger from Point Barrow told McIlhenny of the loss of the *Orca* and the *Freeman* and of the 148 men aboard the *Belvedere* with only two weeks' provisions. Leaving Tom Gordon to organize a salvage attempt for the *Navarch*'s supplies and to notify the *Jeanie,* McIlhenny started back to Point Barrow. There he found that Brower had mobilized his Eskimos; most had been sent inland to hunt for caribou, and others, under the leadership of Fred Hopson, another of Brower's men, had been sent south to bring the crews of the *Orca* and the *Freeman* to Point Barrow.

McIlhenny, unwilling to discard entirely his scientific plans for the year, agreed to take only the sixteen ships' officers into his quarters. To house the rest of the men, he and Brower decided to repair John W. Kelly's dilapidated old Pacific Steam

The wreck of the Rosario, *July 1898*

Whaling Company house, which had been unused for a year. With Brower's lumber they made it tight and built bunks for 144 men along the walls. On October 5 the house was ready and on that day the first party from the ships straggled in. They had been two days en route. The sleds were heavily laden. Many of the whalers were unfit for the journey and were unable to walk. In all, 94 men arrived at the station. In the following days McIlhenny organized their feeding and issued clothes and medicine from his own supplies while Brower set the men to making bedding from his stocks.

Brower was immediately forced to take control of the men because the ships' officers had abrogated responsibility for them as soon as the ships were lost. Brower disarmed all of the men and quickly set up a routine for the winter. He forced them to exercise, making them haul ice for their drinking water. He issued each man his weekly ration of food and Richmond Marsh, the Presbyterian missionary doctor, took charge of health problems.

Meanwhile the men aboard the *Newport* and the *Fearless* had been busy building an ice road to the *Navarch* to salvage her supplies. Brower reached the ship a few days later. To the surprise of the salvage crew, they found all well with the nine men who had refused to leave the *Navarch* the second time the ship was abandoned. Eight of them immediately left the *Navarch*. But one man remained. He claimed salvage rights and protested the removal of the ship's supplies. Brower had foreseen the possibility. The man was immediately put in irons and sent on his way to the *Newport*. The foodstuffs from the *Navarch* were taken to the *Newport* and the *Fearless*, making them self-sufficient for the winter, as was the *Jeanie*. They lacked only fresh meat. Brower then took a few of the men to Point Barrow to ease their crowding.

When Brower returned to Point Barrow, Captains Porter and Sherman, as well as McIlhenny, decided that messengers should be sent south to report that there was no cause for undue worry. At a cost of about a thousand dollars apiece, the captains and McIlhenny fitted out two men for the journey south: George Fred Tilton, third mate of the *Belvedere*, and Charles Walker, who had tried to salvage the *Orca*'s whalebone for himself. Walker was to go via Herschel Island and the Mackenzie River—a well-traveled, relatively easy route. Tilton was to cut a new and difficult trail down the west coast of Alaska to the Pacific Ocean. They set out in October—each on a journey of more than 1,700 miles on foot—and both arrived in the United States only a few days apart.

It was thought that Walker, had he pressed on, should have beaten Tilton by a month. Later the men learned that Walker had spent valuable weeks drinking at Herschel Island.

By the middle of October, Brower and McIlhenny could see that apart from maintaining discipline, their most serious problem would be getting enough fuel, as well as fresh meat and fish to stave off scurvy. McIlhenny, one of America's most brilliant wing shots, had already stocked more than 2,300 ducks and geese in his ice cellar. He and Brower then turned to organizing Eskimo hunters to go inland for caribou and fish. Fortunately, there were more caribou near Point Barrow that winter than any of the residents could remember; more than 1,200 carcasses and 30,000 pounds of whitefish were brought in, enough to supply all the men at Point Barrow and all the ships as well. Later in the winter they took more than a thousand beluga whales—some weighing more than a thousand pounds apiece—at a small patch of open water where they had been trapped by moving ice.

Fuel was a more serious problem. Although McIlhenny and Brower were rationing

Whalemen at Point Barrow waiting for the ships to arrive, July 1898

coal, neither had enough for the winter, and driftwood, the only substitute short of blubber, became so scarce that the sailors began to pilfer wood from the graves in the Eskimo burial ground. Providentially, on January 10 the *Navarch* again drifted into sight, this time near Point Barrow. With eighty tons of coal aboard, her supplies would be sufficient for everyone throughout the winter. Brower immediately turned his Eskimo workers to making an ice road out to the ship, which was found in perfect condition, frozen in an ice floe.

As the ice moved the ship up and down the coast about six miles from shore, the men worked quickly to salvage the coal. By February 5, McIlhenny calculated that with one more day's work they would have enough coal to last the winter. But the next morning McIlhenny awoke to the sight of a "pillar of fire" above the ship. Two men from the *Orca* had set the *Navarch* on fire, evidently because they were sick of unloading coal.

While Brower and McIlhenny were busy maintaining the crews, the United States government prepared to send help. Word of the trapped ships reached San Francisco with the arrival of the *Alexander* on November 3, 1897. As soon as the *Bear* returned from her 1897 Arctic patrol she was ordered back to the Bering Sea. She tried to

The ships at Point Barrow waiting for escape from the ice.
Left to right: The Jeanie, Fearless, *and* Newport *inside the pressure ridge;*
Jeanette *and* Bear *outside the pressure ridge*

reach the Seward Peninsula with a relief expedition, but ran into heavy ice in the late autumn and was forced to land the expedition on the mainland north of Nunivak Island. Under the command of Lieutenant David Jarvis, first officer of the *Bear,* and with Surgeon Samuel Call and Lieutenant E. P. Bertholf, the expedition made its way north along the coast collecting domesticated reindeer and slowly driving them toward Point Barrow for men who, they assumed, were starving. However, when the reindeer arrived at Point Barrow in March, having been driven nearly seven hundred miles in fifty-five days, they were too skinny to be of any use and, thanks to Brower, McIlhenny, and Marsh, the stranded whalers were in far better shape than Jarvis and Call expected. Only two serious cases of scurvy were found, and only three deaths had occurred during the winter—none due to the living conditions.

The Bear *jammed in the ice* Private collection

Jarvis's presence helped to maintain order. The whalers had grown increasingly restive from enforced idleness and became more and more resentful of Brower's and McIlhenny's command. Jarvis stopped their thefts from the Eskimos, supported the captains' authority over those aboard the ships, and enforced standards of hygiene and regular exercise. As the spring wore on and the ducks, seals, and whales returned, he was able to increase their allocation of fresh meat. To his credit he persuaded McIlhenny and Marsh each to take twenty-five men into their houses. With that, Brower took twenty-eight and they were able to tear down the now filthy Pacific Steam Whaling Company station for firewood.

With Jarvis's authority controlling the men, the spring passed reasonably smoothly, eased by the additional medical attentions of Surgeon Call and a successful shore-whaling season, and marred only by a sudden ice movement on July 2 that pushed the *Rosario* ashore at the Point, crushing her starboard side.

At the end of July the *Bear* and steam brig *Jeanette* reached Point Barrow. Ninety-seven of the shipwrecked men went aboard the *Bear,* hoping to leave right away. But the ice, capricious as always, was not yet through with them. It packed in closer and closer until on August 2 it squeezed the *Bear* with such pressure that the iron plates in her engine room deck buckled. The two vessels survived only because they sought refuge in a small indentation in the grounded pressure ridge. The *Bear* and the *Jeanette* were held captive until mid-August when finally the pressure eased. The *Newport,* the *Fearless,* and the *Jeanie* broke out and appeared on the other side of the Point, making their way from the east by blasting a channel through the pressure ridge. Once released from the ice, the *Bear* and the *Jeanette* were able to pass coal and supplies to the other ships. The *Bear* then departed for San Francisco, carrying the officers of the expedition, the shipwrecked men, and Brower, who was going to Washington, D.C, to petition Congress to repay him for the supplies he had used to keep the whalemen alive. He was successful, but the refuge station was never reopened. It stands today as Brower's Restaurant, one of the oldest frame structures in northern Alaska.

14 / *Decline of the Fishery*

By 1895 Herschel Island's heyday was over. After the concentrated assault on their last refuge, only a few bowheads were left. In a vain effort to find whales, the ships scoured waters even farther east—along the coasts of Banks and Victoria islands.

Josiah Knowles, the manager of the Pacific Steam Whaling Company (and always the leader in the industry), must have sensed that the whales would become more difficult to catch and that Herschel Island would become too remote from the whaling grounds. Almost as soon as he had established a company base at Herschel Island in 1890 he tried new methods.

First, he tried to reduce the expenses of a voyage by reducing the master's share of the profits. In one of his rare errors of judgment he reasoned that because the captains in the Arctic fishery did not command whaleboats and thus did not directly participate in killing whales, their share from a voyage, which could run as high as $10,000 or more, was exorbitant. Because the whales in the eastern Beaufort Sea were found in only a few well-known areas, any competent navigator, whether a whaleman or a merchant skipper should be able to take the vessel to the whales—or so Knowles thought.

He hired Captain W. P. S. Porter, who had commanded tenders that resupplied his Alaskan salmon canneries. "Alphabetical" Porter—as he was called, according to Hartson Bodfish, "because of the string of initials he carried astern of his name"—was paid $175 monthly. Bodfish and others thought "he lacked the instinct for whaling." He took no risks, never ventured away on his own, and always stayed with the fleet. He made stupid mistakes as well. He enraged his boatsteerers by being incapable of letting the ship lie quietly while his boats were down chasing whales. Instead he often engaged his propeller to follow the boats and, as a result, gallied the whales. His lasting contribution to the Arctic fishery was in bringing his handsome and talented wife, Sophie Porter, to Herschel Island. Her parties helped leaven the gloom of winter.

Knowles's second innovation was in beginning a shore-whaling operation east of Herschel. In 1893 the company outfitted a group of Point Barrow Eskimos and established them at Cape Bathurst. They were to go whaling in the spring before the ships were free of the ice at Herschel.

The group was unlucky in the spring of 1894, but their failure no doubt convinced Knowles that his ships had to be closer to the Cape Bathurst whaling grounds in early summer. In any case, he had already instructed his captains to look for suitable wintering sites east of Herschel.

In 1892 Knowles sent W. P. S. Porter searching for winter harbors. Porter sounded the mouths of the Mackenzie River and found them unnavigable. The following year, 1893, the *Narwhal, Newport,* and *Balaena* reached Cape Bathurst and sounded the small harbor there at Baillie Island.

Knowles must have suspected that the richness of the Cape Bathurst grounds would not last. He instructed his captains to push as far into the eastern Beaufort Sea as they could—to go through the Northwest Passage if possible. His ships went past Cape Lyon in the east, then swung north and reached the northwest corner of Banks Island. After these voyages, there remained very little water to explore for whales in the western Arctic.

The search for new wintering sites paid off in 1895. The season was cold, icy, and a "dismal failure." When an early autumn sent the pack ice down on the coast, the *Balaena* and *Grampus* found themselves trapped far east of Herschel. But they were able to retreat to Cape Parry where they used the safe, if small, harbor that they named Balaena Bay.

The failure of 1895 was only the beginning of the misfortunes. The summer of 1896 was exceptionally icy as well. The ships on their way north from San Francisco first discovered this state of affairs when they found heavy ice south of Saint Matthew Island. At the beginning of July they found Port Clarence completely frozen over with shore-fast ice protruding eight miles at sea. They had to resupply from their tenders at sea, using what little protection they could find under the lee of King Island. In the eastern Beaufort sea the ice was so heavy that the "inside" fleet was unable to get past the Mackenzie delta until the very end of July.

The ships took only a handful of whales. To make matters worse the ice quickly slammed down on the coast again and sent the fleet running for Herschel. Only three ships made it to the island. The ice trapped the *Fearless* and *Mary D. Hume* near Shingle Point and the *Jesse H. Freeman* near King Point. The crews of the ships that were left exposed on the coast spent a very nervous winter, living in constant expectation of being crushed by moving ice.

The following year, 1897, the ice claimed the *Narvarch, Jesse H. Freeman,* and *Orca,* and trapped the *Belvedere, Fearless, Newport, Jeanie,* and *Rosario,* and eventually crushed the *Rosario.*

That same winter the Pacific Steam Whaling Company faced the fact that Herschel Island was now too far from Cape Bathurst for the ships to make paying catches among the remaining whales. They had to winter closer to be able to hunt on those grounds as long as possible in the summer. Leaving only the little *Mary D. Hume* at Herschel, the company sent four of its ships to winter in the east. The fleet tried to reach Balaena Bay at Cape Parry but was blocked by ice and instead had to go to the bottom of Franklin Bay to winter in the harbor at Langton Bay.

The men discovered the hard way that Langton Bay usually does not lose its ice until much later than Herschel. When the ships broke out of the Bay and reached Cape Bathurst in mid-July, they found the old bark *Wanderer* already cruising there. She had no auxiliary power and had sailed from Herschel.

At the same time the industry as a whole was facing a serious problem. The calibre of the foremast hands had never been high, but by the last years of the nineteenth century it was abysmal. With little chance of making a profit, intelligent and energetic men worked elsewhere. By 1891 the Reverend Dr. Sheldon Jackson, the general agent of education for Alaska, described the sailors in the arctic fleet as "an inferior grade—some of them being, emphatically, hard cases." Those that were not hard cases were either drunkards or inexperienced. In 1901, for example, on setting out from San Francisco Captain George Leavitt found that the landsharks had provided him with a crew that "you could see the grass growing on them, they were so green." Within a few days of putting to sea he had a knifing in the forecastle and a man down with a severe case of gonorrhea. The cook and steward were so incompetent that he had to derate them and he found that the carpenter was also useless: "He can't even sharpen a jack-knife." At Unalaska one of the sailors tried to jump ship and another pulled out a .38-calibre revolver and tried to murder him.

Drinking had become such a problem that on sailing day in March 1899, Leavitt took the trouble to note in his journal that to his surprise all hands were *sober.* Nevertheless, seven weeks later one of the mates was off duty for three days. "I threw overboard part of a barrel of beer, ten gallons of whiskey and five gallons of alcohol belonging to Mr. Mulligan." The next day he wrote, "Asst. Engineer and one fireman drunk. Threw 8 gals. liquor overboard belonging to Asst. Engineer, also 5 gals. belonging to fireman."

In this weakened condition the industry staggered into its final years. As the turn of the century approached it was clear that the fishery had no bright prospects. Although the price of whalebone remained high profits were very erratic. It was difficult for the shipowners to sell their vessels, so they simply kept them operating in hope of some return. The Pacific Steam Whaling Company withdrew from shore whaling in 1896 and did not replace its ships after the 1897 disaster. After that, no large vessels entered the fleet, except the *Bowhead* [II], which was a latecomer by all standards and a senseless investment by her owners.

As the new century approached, the only sensible solution for those that remained in the fishery was to winter in the Arctic, getting as close as possible to the Cape Bathurst whaling grounds. This meant wintering in the harbor at Baillie Island off the tip of Cape Bathurst.

Baillie Island was "a nasty place to winter," according to George Leavitt. The only protection was the low sandspit off the southwest corner of the island. Only four large ships could be wedged in there with their bows almost on the beach and spaced only one hundred feet apart—far too close in the event of fire. They had no protection from the southwest and there was no water at the island. The whalemen had to get water from drift ice.

Herschel Island's comforts seemed luxurious by contrast. Nevertheless, as Captain John A. Cook wrote, Baillie Island was "so much nearer the whaling grounds that it offers an opportunity not to be ignored, as in the short summer being within a few days of where the whales feed means sometimes a year's profit."

In 1898 Captain Hartson Bodfish put the *Beluga* in at Baillie Island for the winter, pioneering the wintering venture there as he had helped to pioneer the first winter at Herschel Island eight years before. The *Beluga* went to Baillie accompanied by another Pacific Steam Whaling Company ship, the *Mary D. Hume.* Her captain, William Hegarty, saw that the anchorage was exposed and dangerous, offering only partial protection fom the moving ice. He refused to winter there and returned to Herschel.

Alone, Bodfish set about the usual autumn routine for the *Beluga,* preparing the ship for the winter as at Herschel Island. He built a house ashore for the ship's natives and dug an ice cellar to store their fresh meat. To his delight it turned out that the area was full of game. The men enriched their diets with ducks, geese, seals, ptarmigan, caribou, fish, and bears. On November 27 Bodfish recorded in his journal, "killed Mrs. Collins. She weighed 250 pounds." Mrs. Collins was a pig. The trapping was also good and Bodfish did well both on his own trap lines and through trading for furs.

The *Beluga* had already made a good catch in the 1898 season, and Bodfish contrived to increase it by mounting a shore-whaling expedition. The local natives had previously taken whales at Cape Bathurst in the spring. Bodfish reasoned that he could do as well. As summer approached he sent two boats to the sandspit at the tip of Cape Bathurst. At the end of June the boats caught their first whale, and by the eleventh of July, when the *Beluga* broke out, they had three more. Several Alaskan Eskimos also outfitted their own crews and took at least three whales there that spring.

The *Beluga* reached San Francisco in 1899 with 106,000 pounds of whalebone aboard. It was the last good catch of the fishery.

The following winter three ships were in that insecure harbor, and the winter went much the same as at Herschel. A group of Kogmullicks set up camp nearby and a number of Nunatarmiuts also arrived. These Eskimos, plus the ships' officers, boatsteerers, and ships' natives, kept the crews well supplied with fresh meat, sending hunting parties as far as one hundred miles from the island.

The men engaged in theatricals and played games of all sorts. On Dewey Day, May 1, they celebrated the defeat of the Spanish fleet at Manilla two years before—dressing the ships, playing baseball, and holding target shooting contests.

As it had at Herschel, alcohol became a problem. The men traded small quantities to the Eskimo men in return for the loan of their wives. When the Reverend I. O. Stringer reached Cape Bathurst in August 1900, he reproached the Eskimos for their drinking. In reply, one Eskimo "disposed of the whole matter by asking why the whitemen had brought liquor to them & taught them how to drink. I am completely discouraged," wrote Stringer.

One great disappointment of the spring was the failure of the shore whaling despite a concentrated effort. Heavily packed ice kept the men from reaching the whales. They ceased shore whaling on July 27, when the ships were able to get out from behind the southwest sandspit and work around the island to Cape Bathurst.

That summer, 1900, Hartson Bodfish returned to Baillie Island in the *Beluga.* She was towing the *Sophia Sutherland,* the three-masted schooner that Jack London had earlier sailed aboard while poaching fur seals in the Bering Sea. Bodfish had bought the schooner for the Eskimos with three heads of bone they had caught while shore whaling at Cape Bathurst in 1899. The *Sophia Sutherland* carried not only two whaleboats but also Captain E. C. Murray. He had once been in charge of the Pacific Steam Whaling Company's warehouse at Herschel and had agreed to come north for the summer to teach the Eskimos how to handle the schooner and how to hunt whales "American style." The Eskimos' venture came to naught, however. In a strong northeast gale on August 31 the schooner rolled over and went ashore, a total loss.

Near the wreck that winter there were four ships in the harbor, Bodfish's *Beluga*, the *Narwhal*, the *Bowhead* and the little forty-three-ton schooner *Penelope*. The *Penelope* had been built as a yacht on the Pacific coast. At seventy-two feet she only carried two boats and was there solely for head hunting. Her presence was telling. With catches erratic and generally poor, a vessel the size of a steam bark was a distinct disadvantage, requiring as it did fifty men or more for a wintering voyage. Small schooners like the *Penelope* required sixteen men, and one or two whales made the voyage highly profitable—whereas four whales were needed to break even for a sailing bark and more than that for a steamer. Small schooners were by no means rare in the fishery before the turn of the century, but they were the only new entrants into the fleet after that and they came to characterize the industry's closing years.

The *Penelope* was, in fact, moderately successful, having taken four whales. Her owners, Stabens and Friedman, San Francisco outfitters, were luckier still when in 1902 they sold her at Baillie Island to four Alaskan Eskimos. The new owners sailed her behind the whaleships, picking up carcasses. They also did a little whaling—until 1907. Unfamiliar with the ways of a ship, they periodically took the lead pigs (ballast) from her hold to melt down into bullets. As *Penelope* steadily lost ballast she grew crank. At last in 1907 she rolled over in an autumn gale and was wrecked.

The winter of 1900–1 passed relatively uneventfully at Baillie Island, punctuated only by a few deaths and the usual expeditions to Fort McPherson for mail. With Herschel's importance as a wintering site on the wane, there were camped at Baillie Island, in addition to the local natives, more than a hundred Alaskan Eskimos who had moved to the area for the trapping, shore whaling, and possibilities of employment. These immigrants, as well as the ships' natives made up a large heterogeneous Eskimo population.

The year 1900 coincided with a rise in the price of furs. Trapping and trading became extremely important, drawing natives from as far as Richards Island in the Mackenzie delta. Hartson Bodfish, through his own trapping and energetic trading, took 1,500 fox skins that winter.

One sad aspect of the winterings at Langton Bay and Baillie Island was the natives' near extermination of the musk oxen in that area. The Eskimos found that there was a market for their skins with the whalemen. Because these gentle beasts form a defensive ring when threatened—an effective defense against wolves—it was a simple matter to slaughter them with rifles.

On March 19, 1901, Bodfish was surprised to see a group of Eskimos arriving at the

Captain Hartson Bodfish with a mammoth tusk at the stern of the Beluga, *Baillie Island, 1901*

island with two musk ox calves harnessed to the sled. He bought them both. Although the dogs killed one of the calves, the other, named "Olive Jones," he took aboard the *Beluga* to San Francisco. He sold her to the Bronx Zoo for $1,600.

When the ships broke out of Baillie Island in 1901 things were bleak. There had been no luck in shore whaling and the ships were confronted by an icy and barren

"Olive Jones," Baillie Island, 1901

season. To the men's disgust, they had still taken no whales when the ships arrived there from San Francisco. Thus the entire winter's effort had been to no purpose (with the exception of any fur trading they had done on their own).

Into this unpredictable quest for bowheads—by then called "the greatest of marine lotteries"—came the industry's last major innovation. Small schooners like the *Penelope* had proved economical, and William Lewis, who in 1880 had introduced steam to the Arctic fishery, now saw the potential of the internal combustion engine. He outfitted the 126-ton schooner *Monterey* with a small gasoline engine—despite the protests of the proponents of steam power who claimed, as the sailing whalemen had before them, that the engine would scare the whales.

The *Monterey*'s voyage was moderately successful. "I found it easier this season to take the few whales we got than under the old methods," said the *Monterey*'s master. Because "of her light draught and comparatively small size, [she] could work to better advantage, and the chugging of the engine was no drawback to the work." Sailing skippers told him, he said, that they could hear the steam whalers farther off than they could hear us. "I was able," he continued, "to get a whale that had sounded near an ice pack by simply running through the small openings and meeting him soon after he arose on the opposite side." His opinion was "that the gasoline engine has come to stay in the Arctic whaling business." The *Monterey*'s greater maneuverability capitalized on the handicaps of the less agile steamers, just as the first *Mary and Helen* had outmaneuvered the sailing vessels twenty years earlier. Three other gasoline auxiliaries were quickly fitted out—*Barbara Hernster, Charles Hanson,* and *Olga.*

Wintering voyages by large whaleships might well have ended in 1901. No large ships were in the North for the next two winters. The arctic fleet had shrunk to only thirteen vessels. There were few whales to be had and the price of whalebone had stagnated below three dollars per pound.

As it turned out, however, the introduction of the gasoline-powered auxiliaries in 1903 coincided with the last important rise in the whalebone market. Suddenly it went sky-high, reaching five dollars a pound, and a few ships returned to the Arctic to winter. Because Baillie Island had not proved much of a head start on the season, the ships returned to the greater comforts of Herschel Island—with the exception of the winter of 1904-5, when two ships used Langton Bay again.

The summer of 1905 was also exceptionally cold and the pack ice never retreated far from the Mackenzie delta. The ice conditions looked so bad that Hartson Bodfish, who had only reached Herschel with the *William Baylies* from San Francisco on August 4, decided to leave on the tenth. He wormed his way along shore in company with the *Monterey*, and even then almost did not make it out. Still, he reached San Francisco with sixteen whales, a handsome profit.

The other ships did badly. But what was worse, the ice trapped most of the fleet. Few of the ships were planning to winter in the Arctic and none wanted to winter on an exposed coast. Six ships were eventually able to work their way to Herschel Island, but none could get past there. Four others were forced to winter at Baillie Island. In greater danger were the *Charles Hanson* at Toker Point and the *Gjoa* off shore at King Point. The *Gjoa* was Roald Amundsen's exploration sloop that had just completed the first traverse by water of the Northwest Passage.

No one knew where the whaling schooner *Olga* was. She was owned by a porcine whiskey trader and failed small-time whaling merchant, Captain James McKenna. McKenna had been Gilbert Borden's partner in the latter's attempt to use the government refuge station at Point Barrow as a trading post. In sleazy dealings McKenna met his match in a duplicitous and dangerous Danish cook, Charlie Klengenberg. Klengenberg had served under Peter Bayne aboard the *Emily Schroeder* at Point Hope in the winter of 1893-94, and he seems to have been an apt understudy for the larcenous and corrupt Bayne.

The Olga *at Herschel Island, ca. 1906* Royal Canadian Mounted Police Archives

In the summer of 1905 McKenna cruised east of Herschel aboard the *Charles Hanson*, keeping his other schooner, the *Olga*, close by. McKenna found that he distrusted the officer that he had placed in charge of the *Olga*; consequently, he replaced him with Klengenberg. As the explorer Vilhjalmur Stefansson recounted, Klengenberg "had acquired at least two kinds of reputation; one for enterprise . . . and the other for a character not very different from that of the buccaneers of old, or the Sea Wolf of Jack London's story."

McKenna, who knew larceny when he saw it, was suspicious of Klengenberg and suspected that he might try to make off with the *Olga*. To keep Klengenberg close by the *Charles Hanson*, McKenna removed all the stores from the *Olga*, leaving only two weeks' provisions aboard. He assumed that Klengenberg, being so short of food, would not dare to desert him. "This showed how little he knew [Klengenberg]," said Stefansson.

East of Herschel a dense fog enveloped the two schooners. When it lifted, the *Charles Hanson* was alone. Klengenberg had vanished. During the winter speculation ran high, generally focusing on whether Klengenberg was dead, whether he had taken the *Olga* east in search of "Eskimos who did not know the present high price of fox skins," or whether he had headed out, to sell the *Olga* in a foreign port.

When the *Olga* reached Herschel on August 10, 1906, the story gradually emerged. Klengenberg went directly to the police. He admitted intentionally taking McKenna's schooner and said that he had gone to Langton Bay, where he had robbed one of the whaleship's warehouses of its stores, then torn it down and put it aboard the *Olga*. From there he had sailed to what he thought was the coast of Banks Island and had wintered, trading with primitive Eskimos. (Later it was discovered that his navigation was so poor that he was in fact nearly 200 miles from the nearest point of Banks Island and had actually wintered behind Bell Island, off Cape Kendall on the southwestern corner of Victoria Island.)

When Klengenberg gave McKenna the jump, he had nine men aboard the *Olga*. When he returned to Herschel, he had only five. He claimed that during the winter two men had fallen through the ice and drowned, another had died of natural causes, and a fourth he had been forced to shoot and kill to defend himself from murder.

The whaling masters at Herschel were extremely skeptical of the whole story and wanted the police to arrest Klengenberg. But at the time McKenna was off on a cruise. The police had no complaint to act upon. They merely told Klengenberg not to leave the island with the *Olga*.

Klengenberg was as good as his word. He did not leave the island in the *Olga*. That night he took a whaleboat, which may or may not have been his, and with his family sailed into United States waters, headed for Point Barrow.

As soon as Klengenberg left the island an entirely different story emerged. The four sailors on the *Olga* claimed that Klengenberg had arbitrarily killed the engineer, that the man who had died of "natural causes" had, in fact, starved and frozen to death in chains in the forehold, and that the two sailors who "fell through the ice" had been the only witnesses to the murder of the engineer. Stefansson recalled that as the *Olga* approached Herschel, Klengenberg addressed the survivors, saying something like, "Boys, you know the penalty for killing five men is the same as killing four." The sailors claimed that while Klengenberg was at Herschel they had been afraid to tell their story.

Eventually a warrant was put out for Klengenberg's arrest. He was tried and acquitted for lack of evidence. He returned to the western Arctic to run a number of trading posts. Because of his devious behavior, he became a thorn in the side of

Charlie Klengenberg and family at Herschel Island Dartmouth College

many of the traders in Alaska and Canada. Joseph Bernard, a pioneering maritime trader, recalled that when the police were investigating the murder of an Eskimo woman on Victoria Island, Klengenberg told them that Bernard had taken the body and pickled it because of his interest in collecting curios. The police sought Bernard to question him about the allegation. His reply: "I think it would be pretty strange for me to kidnap a dead woman."

The winter of 1905-6 was hard on everyone. The ships, few of which had planned to winter in the Arctic, were on very short rations; the crews aboard some of the ships were on the verge of mutiny. To make matters worse, the shipping articles expired for the crews of the *Bowhead* and *Karluk*. When the articles expired, the officers had no legal authority over the men—they were free to leave the ships.

After much wrangling and ill feeling the seamen agreed to work the vessels back to their home ports. This, in addition to the brutal and possibly sadistic behavior of at least one captain, John A. Cook, made the winter a nightmare at Herschel. Among other tragedies, Mrs. Cook suffered a nervous breakdown.

The summer of 1906 was no better. Heavy ice blocked the passage east from Cape Parry to Nelson Head on Banks Island. The last event of a tragic year occurred on August 13 when the old steam bark *Alexander* was charging along in a dense fog under both sail and steam. Just as the officers were sitting down to breakfast at 7 A.M. the ship struck so hard on Cape Parry that she drove her bows into only nine feet of water.

The men were in a panic. They thought that no ships were going to winter north that year. All the ships in the eastern Beaufort Sea were short of supplies and expected to leave the Arctic early. The men knew they would have to race to Herschel in their boats to catch one of the ships before it sailed.

In little more than an hour the entire crew was in the boats, heading for Herschel, 450 miles away. On August 26 the first boats reached Herschel, just hours before the last ship, the *Charles Hanson*, was to leave.

In essence the Arctic fishery ended that year. The catches had become so poor throughout the North Pacific that the New Bedford owners began returning their vessels to the Atlantic. "The whaling grounds of the far north need a rest," said William R. Wing, one of the owners. By then the ships had scoured every navigable part of the Beaufort Sea, having reached into Coronation Gulf in the east and been on the west and north coasts of Banks Island, to the north end of Prince of Wales Strait (the strait that separates Banks from Victoria Island). In the Chukchi Sea they had reached seventy-four degrees north and been along the north coast of Wrangel Island. At their easternmost the San Francisco vessels had reached within 500 miles of the farthest western penetration of the British whaleships in the Greenland fishery.

There were few whales left to catch. By 1908 it had become clear to scientists and many whalemen that unless whaling ceased quickly the bowhead would soon be extinct. Curiously enough, however, it was the bowheads' very scarcity that helped to save them. The high price of whalebone had stimulated the use of artificial substitutes—spring steel, Celluloid, and "featherbone," an amalgam made from processed goose quills.

But the end came in 1907. Although the price of whalebone stood at five dollars per pound that year, corset staying remained the last significant market for baleen; substitutes had taken over almost everywhere else, and indeed they had penetrated the less expensive end of the corset market. Whalebone was only used in the volatile *haute couture* market—in the finest, most expensive creations—and suddenly there was no demand for it. In 1907 Paul Poiret, a rising young Parisian couturier, introduced a slim figure in his collections, banishing at one stroke the exaggerated narrow waists and S-shaped figures that had characterized late Victorian and Edwardian fashion. According to Elizabeth Ewing, a historian of costume, this change was "nothing less than the start of modern fashion." The modern straight-

Edgar Lewis in his office in New Bedford, ca. 1912

sided corsets did not require boning and the bottom dropped out of whalebone sales.

By the spring of 1908 the market was soft. The nine vessels in the Arctic fleet found one of the iciest summers of the last quarter century. It was such a bad year that few even tried to pass east of Point Barrow. As it was, the fleet took only twenty-four bowheads, and when they reached San Francisco, they found the price of whalebone well below the cost of its acquisition.

At the end of the 1908 season the leading whaling operators—the Pacific Steam Whaling Company, William Lewis and Son, and John A. Cook—laid up their remaining steam vessels. In 1909 the market was so low that only three ships went north, primarily for the fur trade. By 1910 there were no buyers for whalebone.

Ironically, the last convulsive lurch of the bowhead fishery took place in New Bedford in 1908, where five years earlier the Old Dartmouth Historical Society had been founded to commemorate the history of the whaling industry. In that year, with the market for baleen collapsed, the remaining whaling companies organized the "whalebone trust" in an effort to force the price of baleen upward. This cartel, run by William Lewis's son, Edgar, received all the whalebone brought into the United States and stored it in New Bedford until it could be sold. Lewis was able to control the market, such as it was, but for only six years. His last sales were made in 1913 at $1.35 per pound. Then he, too, went out of business.

On December 29, 1914, the *Whalemen's Shipping List*, the industry's trade journal since 1843, ceased publication. It closed with this epitaph to the whalebone market:

> Whalebone: We are unable to quote any sales. At the beginning of the year it was reported that a small quantity of Arctic had been sold for export but particulars could not be learned. There does not seem to be any demand for the large stock on hand, the principal reason, probably, being the foreign war.

But the *Shipping List* had, in fact, proclaimed the industry's own epitaph in only its third issue, March 28, 1843, when describing the decline in the Pacific sperm whale fishery it said, "A business *overdone*, must lead to the same results, whether upon land or ocean."

Epilogue

When the whalebone market collapsed in 1908 the few whalemen that remained active in the fishery were not greatly affected. Their trading activities had been increasing since virtually the opening of the fishery. By 1912 the handful of remaining whaling vessels had given up any pretense of serious whaling. That year, for example, the bark *John and Winthrop* returned from the Arctic with a total catch of only thirty barrels of sperm oil (taken in the North Pacific). She had spent most of the month of August anchored at Point Barrow, trading. At the same time Captain Stephen Cottle of the *Belvedere* faced the inevitable. "It is no use," he reported, "to pay out three or four dollars for every pound of bone you catch and then receive $1.50 or $2.00 a pound."

Cottle decided from then on to concentrate almost exclusively on the fur trade. Outfitting the *Belvedere* for a trading voyage would allow him to staff his ship with only half as many crew. Only at the end of the trading cruise, when he had completed his circuit in the Beaufort Sea, would his vessel and others go whaling, spending a few weeks cruising near Banks Island and then near Herald Island before returning south. This genuflection to a whaling voyage made sense. It allowed the ships to clear port as whalers—whereon the crew worked for shares of the whaling profits. The profits from fur trading were a matter of negotiation among the captain, owners, and crew.

For about ten years after the collapse of the whalebone market Stephen Cottle's *Belvedere* and the H. Liebes Company's *Herman* were the only large vessels going north, although they were joined by a few small schooners. However, with the end of the First World War in 1918, the market for furs, especially for white foxes, accelerated sharply. In this environment emerged the maritime fur trade's central figure and most skillful mariner and trader, Captain Christian Theodore Pedersen.

C. T. Pedersen's fortunes mirrored the fall of the whaling industry and the rise of the maritime fur trade. Pedersen arrived in the Arctic as a boatsteerer aboard the steam bark *Fearless* in 1894. The bark wintered at Herschel Island, where Pedersen became a proficient traveler and dog team driver. He continued in the fishery in various capacities and eventually commanded the small schooner *Challenge*, wintering at Pont Barrow in 1908-9.

Then, with whalebone nearly valueless he turned his attention to sea otter hunting in the Aleutians for a few years. This adventure coincided with the beginning of the rise in the market for white fox skins: in 1900 the average price was a dollar; in 1910 it was ten dollars. In 1912, however, he was able to arrange for backers to purchase a Japanese schooner that had been seized for poaching fur seals.

The Nanuk, *ca. 1920* Private collection

He renamed it the *Elvira* and set out on a whaling and trading cruise. In 1913, on her second Arctic voyage, the *Elvira* was crushed in the ice on the north coast of Alaska.

The following year, in 1914, Pedersen was hired by the H. Liebes Company to command the *Herman* to resupply their trading posts at Point Hope and Point Barrow and to carry out a whaling and trading voyage in the eastern Beaufort Sea. That year the ice trapped him east of Point Barrow again but he blasted his way out with black powder. In 1915 he was hired by the Alaska Packers Association to command their ship, the *Santa Clara*. Aboard her he set a record for a sailing vessel,

Captain and Mrs. C. T. Pedersen Private collection

racing from San Francisco lightship to Cape Flattery in 87 hours and 30 minutes. But he found life too tame with the Alaska Packers. In 1916 he returned to the Arctic fur trade as captain of the *Herman*, a position he held until 1922. In 1921 Pederson, aboard the *Herman*, became the last captain to take a bowhead commercially.

During his years with the *Herman* Pedersen became universally recognized as the most skillful ice pilot ever to sail in the western Arctic. He never lost a vessel after the *Elvira* and the ice never trapped him, although there were many close calls. He achieved his extraordinary success through his uncommon intelligence and energy. He constantly studied ice movements and his exceptional stamina allowed him to stay in the crow's nest without sleep for as long as forty-eight hours while he wormed his way into and out of the Arctic.

The Patterson *in northern Alaska, 1935. She is carrying the* North Star *on her deck for delivery to native trappers who lived on Banks Island*
Private collection

In early 1923 Pedersen left the H. Liebes Company in a bitter dispute over his salary. He immediately arranged for financing, and purchased the motor schooner *Ottillie Fiord*, which he renamed the *Nanuk*. In 1923 he left San Francisco six weeks after the *Herman* but because of his exceptionally skillful ice work, he managed to pass his former ship and nearly completely denude the coast of furs by the time the *Herman* arrived.

He was able to do this, not only because of his skill as an ice pilot, but also because of the high regard in which virtually all Arctic residents—native and white alike—held him. His honesty, his reliability, his prices, and the quality of his trade goods are remarked upon to this day. It was not long before he essentialy drove the H. Liebes Company out of the maritime fur trade.

Pederson not only traded on his own, but also supplied a number of independent traders as well. Thus, almost single-handedly he prevented the Hudson's Bay Company from establishing a monopoly in western Arctic Canada. The Eskimos were the beneficiaries of this competitive atmosphere and were able to sell their furs

The Nigalik *at Baillie Island*

to the highest bidder—a situation entirely different from that in the eastern Arctic, where either the Hudson's Bay Company or Revillon Frères maintained a monopoly in most settlements.

By 1925 Pedersen's business had expanded to the point where the 261-ton *Nanuk* was far too small. That year he bought the 580-ton former Coast Survey vessel *Patterson*. He reinforced his operations with two small schooners, the *Nigalik* and *Emma* and established trading posts for his newly formed Canalaska Company as far east as Gjoa Haven on King Willliam Island, more than a thousand miles from Herschel. There his agent, George Washington Porter, Jr., received furs from as far east as Somerset Island in the eastern Arctic. Each year the *Nigalik* and *Emma* met the *Patterson* at Herschel Island to exchange furs and trade goods. There, too, Pedersen met countless natives who traveled to the island from all over the western Canadian Arctic in their whaleboats and small schooners to trade with him.

But the fur business was cyclical. In 1929 white fox skins had fetched $50 to $55 apiece, but prices dropped steeply after the stock market crash. Prices again

End of the industry. The Patterson *at Oakland at the end of her last Arctic voyage* Private collection

declined dramatically in 1933 when the United States recognized the Soviet Union and thereby opened its market to Soviet furs. By 1935 Pedersen had concluded that his business was not profitable and that he could do better in fur farming. In 1936 he sold the Canalaska Company to the Hudson's Bay Company. On the fourth of September 1936 the *Patterson* left Herschel for the last time.

September 25, 1936, was a Janus date. Pedersen, the last of the Arctic whalemen, was concluding what Thomas Roys had begun almost a century before. As the worn and ice-scarred *Patterson* entered San Francisco Bay, this relic of the nineteenth century encountered one of the enduring symbols of the twentieth—spanning the entrance to the Bay were the first cables of the Golden Gate Bridge.

Appendix 1

Chronology of Major Events in the Western Arctic Whale Fishery

1750s	American whaling fleet begins expansion into the South Atlantic.
1790s	Whaleships enter the Pacific Ocean.
1819	Whaleships reach Hawaii.
1840s	Whaleships on the rim of the North Pacific.
1848	The *Superior,* first whaleship to Bering Strait.
1849	Fifty whaleships at Bering Strait.
1852	More than 220 whaleships near Bering Strait. One quarter of the total pelagic bowhead whale catch in the western Arctic reached by this date.
1852–54	Poor catches in western Arctic; North Pacific fleet shifts to Sea of Okhotsk.
1855–57	Western Arctic deserted by whaling fleet; peak of activity in Sea of Okhotsk.
1858–60	Catches decline in Okhotsk Sea; fleet begins to return to western Arctic.
1859–63	Wintering voyages to the Chukchi Peninsula.
1865	CSS *Shenandoah* cruises to Bering Strait. One half of total pelagic bowhead whale catch reached by this date.
late 1860s	Beginning of intensive walrus harvest.
1867	Southern shore of Wrangel Island delineated.
1868	Fleet begins to probe northeastern Chukchi Sea.
1871	Thirty-two whaleships abandoned near Point Belcher and Wainwright Inlet without loss of life.
1872	One whaleship salvaged from abandoned fleet.
1873	Fleet begins to probe western Beaufort Sea.
1876	Twelve whaleships lost near Point Barrow; fifty men die.
1878–79	Massive starvation on St. Lawrence Island.
1879	First annual patrol by U.S. revenue cutters.
1880	The *Mary and Helen,* first steam auxiliary whaleship to cruise in western Arctic.
early 1880s	End of intensive walrus harvest. Center of the American whaling industry shifts from New Bedford to San Francisco.
1883	Incorporation of Pacific Steam Whaling Company.
1884	Shore-whaling station established at Point Barrow.
1887	Shore-whaling station established at Point Hope.
1888	Four ships lost at Point Barrow; eleven ships nearly lost in northwestern Chukchi Sea.
1888–89	Joe Tuckfield winters in Mackenzie River delta and discovers eastern Beaufort Sea whaling ground.
1889	Government Refuge Station established at Point Barrow. Whaleships reach Herschel Island.

1890–91	First ships wintering at Herschel Island.
1891	Wholesale price of whalebone reaches $5.38 per pound.
1894–95	Fifteen whaleships wintering at Herschel Island.
1896	Government Refuge Station closed. Pacific Steam Whaling Company withdraws from shore whaling.
1897–98	Four whaleships lost near Point Barrow; four whaleships trapped by ice for winter; crews of lost vessels winter at Point Barrow. Government relief expedition drives reindeer herd to Point Barrow for food.
1903	First gasoline auxiliary whaleship in western Arctic.
1905–06	Ten whaleships trapped by ice in eastern Beaufort Sea.
1907	Price of whalebone collapses.
1909	Three whaleships in western Arctic fleet.
1914	An estimated 16,600 bowhead whales were caught out of an estimated 18,650 killed by whaleships in the western Arctic to this date.
1921	Last bowhead taken by a whaleship.

Appendix 2

Estimated Harvest of Bowhead Whales and Walruses in the Western Arctic Fishery

Sources: Bockstoce and Botkin, 1982 and 1983

	BOWHEAD WHALES				WALRUSES		
Year	Estimated Annual Catch	Estimated Annual Kill	Estimated Cumulative Catch	Estimated Cumulative Kill	Estimated Annual Catch	Estimated Cumulative Catch	Year
1849	507	571	507	571	28	28	1849
1850	1719	2067	2226	2639	180	208	1850
1851	758	896	2984	3535	107	315	1851
1852	2188	2682	5172	6217	109	424	1852
1853	628	796	5800	7013	68	492	1853
1854	105	130	5905	7143	110	602	1854
1855	0	2	5905	7146	3	605	1855
1856	0	0	5905	7146	0	605	1856
1857	72	78	5977	7224	174	779	1857
1858	424	459	6401	7683	551	1330	1858
1859	335	366	6736	8049	946	2276	1859
1860	211	221	6947	8269	108	2384	1860
1861	293	306	7240	8575	1395	3779	1861
1862	150	157	7390	8732	130	3909	1862
1863	288	303	7677	9035	58	3967	1863
1864	396	434	8073	9469	602	4569	1864
1865	455	588	8529	10057	239	4808	1865
1866	503	540	9031	10597	273	5081	1866
1867	566	599	9598	11196	1144	6226	1867
1868	456	516	10054	11712	2300	8526	1868
1869	340	370	10393	12082	5998	14524	1869
1870	594	620	10987	12702	14443	28967	1870
1871	125	133	11112	12835	6674	35641	1871
1872	163	194	11275	13029	5775	41416	1872
1873	147	147	11422	13176	4515	45931	1873
1874	95	95	11517	13271	9215	55146	1874
1875	200	200	11717	13471	13080	68226	1875
1876	57	76	11774	13547	35663	103889	1876

	BOWHEAD WHALES				WALRUSES		
Year	Estimated Annual Catch	Estimated Annual Kill	Estimated Cumulative Catch	Estimated Cumulative Kill	Estimated Annual Catch	Estimated Cumulative Catch	Year
1877	244	262	12018	13810	13294	117183	1877
1878	72	80	12090	13890	13128	130311	1878
1879	203	261	12293	14151	6699	137010	1879
1880	452	460	12746	14611	2676	139685	1880
1881	374	418	17120	15029	0	139685	1881
1882	240	240	13360	15269	2624	142309	1882
1883	39	39	13399	15308	3523	145832	1883
1884	114	133	13513	15441	665	146497	1884
1885	277	287	13789	15728	851	147348	1885
1886	123	133	13912	15861	21	147368	1886
1887	180	204	14092	16065	144	147512	1887
1888	117	133	14209	16197	21	147368	1888
1889	42	53	14251	16250	11	147804	1889
1890	127	127	14378	16377	10	147814	1890
1891	228	234	14607	16611	11	147825	1891
1892	308	317	14915	16927	0	147815	1892
1893	141	141	15055	17068	0	147825	1893
1894	141	151	15197	17219	9	147834	1894
1895	94	94	15291	17313	0	147834	1895
1896	58	58	15349	17372	38	147872	1896
1897	73	73	15422	17445	77	147948	1897
1898	216	228	15638	17673	0	147948	1898
1899	204	208	15842	17881	0	147948	1899
1900	112	112	15954	17993	24	147972	1900
1901	29	29	15983	18022	3	147976	1901
1902	132	132	16115	18154	4	147980	1902
1903	95	95	16210	18249	5	147985	1903
1904	68	74	16278	18323	6	147990	1904
1905	86	93	16365	18415	32	148022	1905
1906	36	36	16401	18451	0	148022	1906
1907	70	70	16471	18521	51	148074	1907
1908	33	33	16504	18554	0	148074	1908
1909	10	10	16514	18564	50	148124	1909
1910	16	16	16530	18580	68	148192	1910
1911	30	30	16560	18610	0	148192	1911
1912	0	0	16560	18610	30	148222	1912
1913	0	0	16560	18610	20	148242	1913
1914	40	40	16600	18650	8	148250	1914

Appendix 3

Statistical Graph: Price of Whale Oil and Baleen,
Number of Whaleships in the Fishery, Catch per Unit of Effort

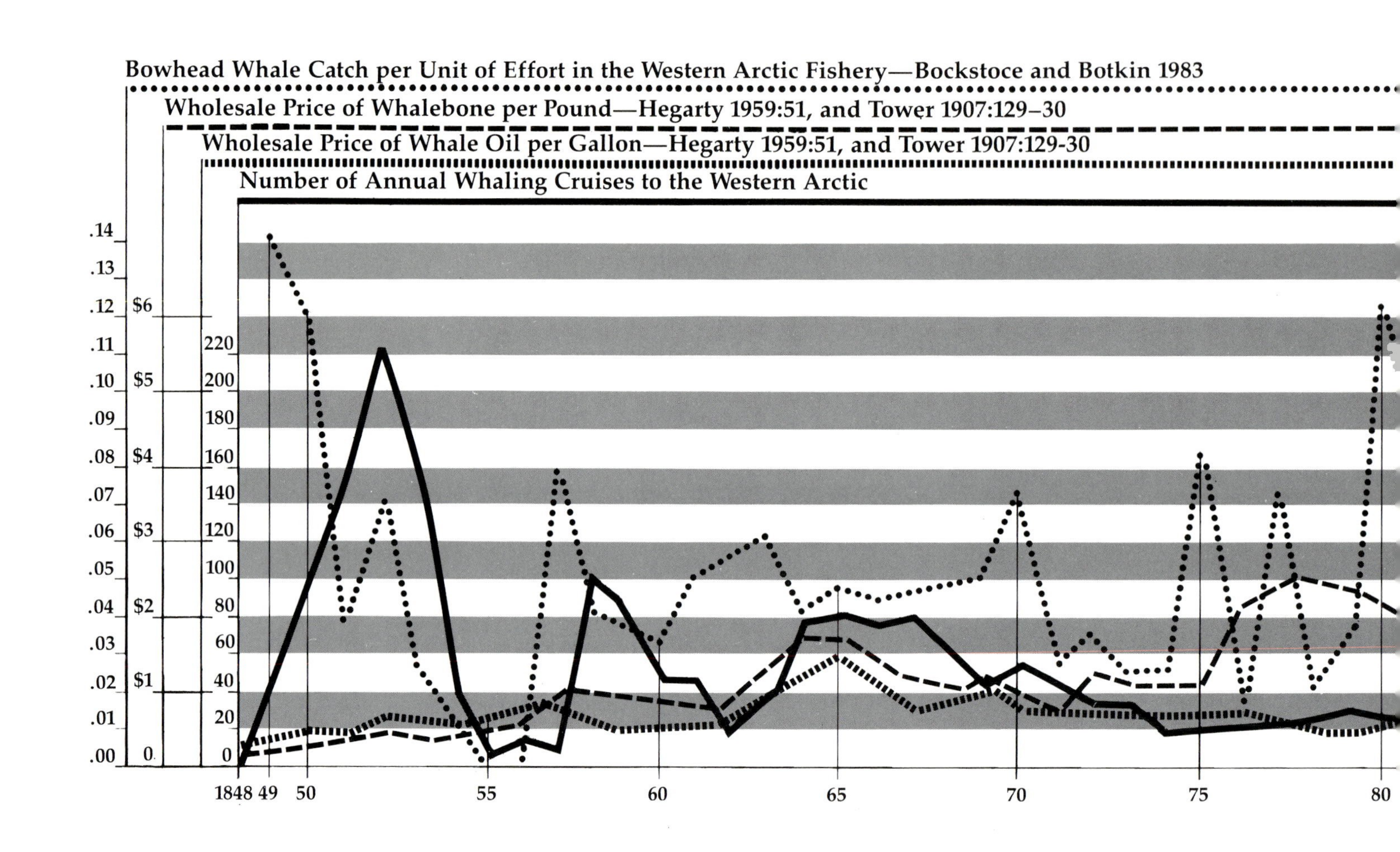

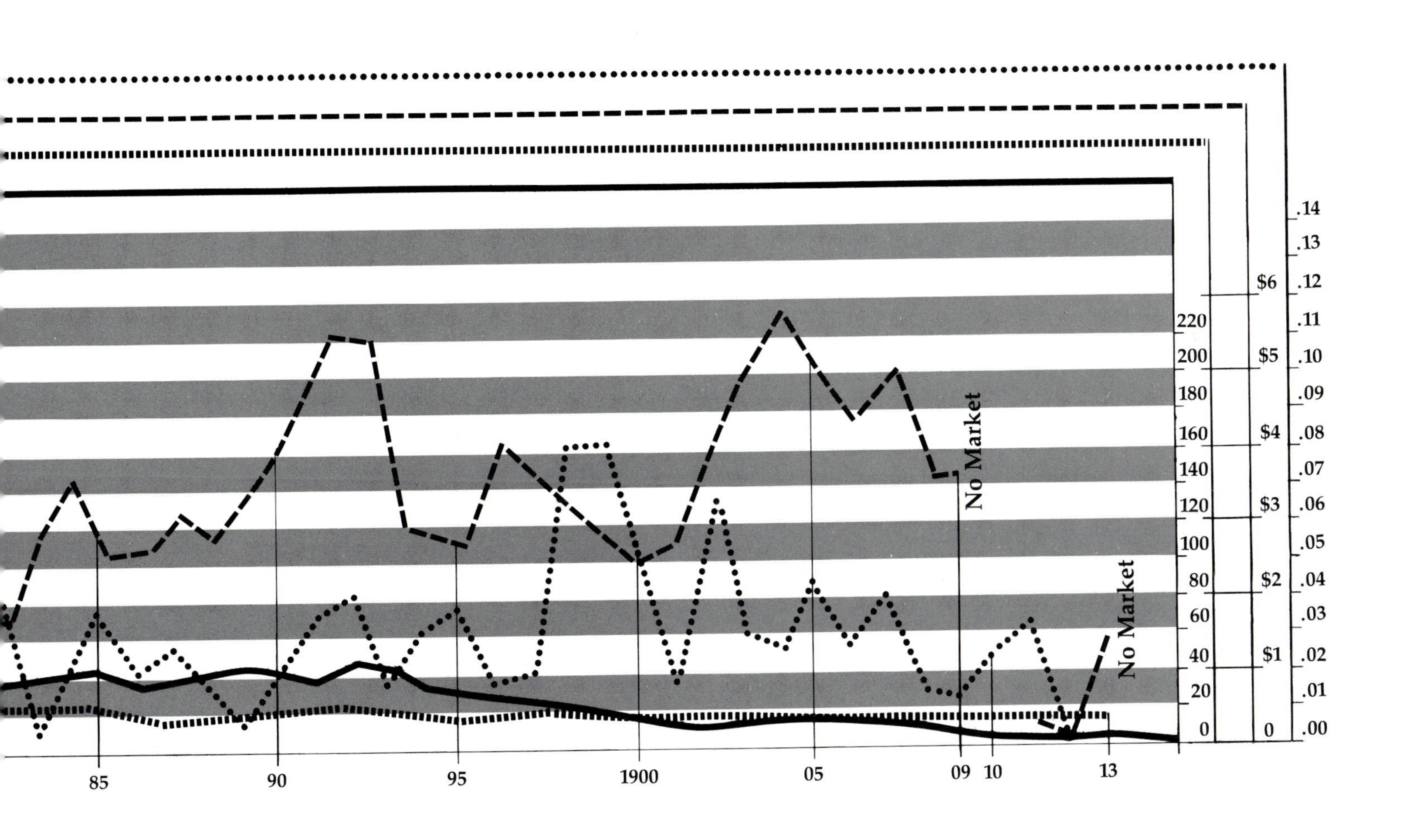
No Market
No Market
220
200
180
160
140
120
100
80
60
40
20
0
$6
$5
$4
$3
$2
$1
0
.14
.13
.12
.11
.10
.09
.08
.07
.06
.05
.04
.03
.02
.01
.00
85
90
95
1900
05
09
10
13

Appendix 4

Gazetteer of Whalers' Place Names in the Western Arctic

by John R. Bockstoce and Charles F. Batchelder

American Side. Generally used to refer to the waters of Chukchi Sea—and less often to the waters of the Bering Sea—near the coast of Alaska

Anadir Sea (also the Gulf, Anadyr Sea, Anadair Sea, and similar variations). Gulf of Anadyr

Armata Bay, Siberia (also Aramata Bay). Proliv Senyavina (65°40′N, 172°30′W), where several natives and one white man were killed aboard the ship *Armata* in a drunken brawl in 1851

The Arctic. The waters north of Bering Strait. *See* Arctic Ocean

Arctic Ocean. To American whalemen this term meant only the part of the Arctic Ocean they reached via Bering Strait: the Chukchi Sea, Beaufort Sea, Amundsen Gulf, and the easternmost waters of the East Siberian Sea—names which came into use after the end of commercial whaling in those waters.

Asiatic Side. Generally used to refer to the waters of the Chukchi Sea—and less often to the waters of the Bering Sea—near the Asian Coast

Autumn Grounds, Chukchi Sea. *See* Western Grounds

Baillie Island, Canada (also Bailey Island). Whalers used this term to refer to the anchorage under the southwest sandspit of the larger of the two Baillie Islands (70°32′N, 128°15′W)

Point Barrow Shoal, Alaska (also Owen's Shoal). The 2 1/2 fathom shoal (71°23′00″N, 156°24′30″W) east of Point Barrow onto which ships were occasionally swept in severe southwest gales

Barter Island, Alaska (also Point Manning). 70°07′N, 143°40′W. Leffingwell (1919:93) reports that this name originally applied to a low island nearby. It is likely that whalemen transferred the term to Point Manning—named by Sir John Franklin in 1826—because its steep bluffs are conspicuous to seaward

Behring's Sea. Before the general acceptance of the term Bering Sea to include the waters from the Aleutian Islands to Bering Strait, whalemen used the term to delineate the waters approximately east of the Pribilof Islands and distinct from the Kamchatka Sea, which were the waters west of there.

Bering Straits (also Beerings Strait, Berings Strait, Behring Strait, Berren Straits, Bherings Straits, and other variations). Modern convention restricts this name to the passage between Cape Prince of Wales and Mys Dezhneva, but to many whalemen the "straits" were the narrows bounded on the south by Saint Lawrence Island and Indian Point and on the north by East Cape, the Diomede Islands, and Prince of Wales Shoal

Bering Strait, mouth of. The strait between Saint Lawrence Island and the Chukchi Peninsula

Big Island, Siberia (also Arakam Island, Kayne Island). Ostrov Arakamchechen (64°45′N, 172°25′W)

Blow Hole, Canada (also Wind Hole). The narrow pass at the junction of Yoyak Creek and Firth River (about 69°18′30″N, 139°34′W). Whalers wintering at Herschel Island often encountered violent winds when traveling up this valley in search of caribou.

Bristol Bay, Alaska. To whalemen the body of water circumscribed by Nunivak Island, Saint Matthew Island, the Pribilofs, and the Alaska Peninsula

Compass Grounds, Chukchi Sea. Whaling grounds in the southwestern Chukchi Sea at about 68°N, 171°W, so named because the ships' compasses behaved erratically there

Corwin Coal Mine, Alaska. The vein of low-grade coal (68°52′05″N, 165°08′15″W) from which American steam whalers often topped-up their bunkers. It was named for the U.S. revenue cutter *Corwin* that visited there in 1880.

Cow Yard, Chukchi Sea. An area near Cape Serdze Kamen, Siberia, where large cow whales were taken in 1848 and 1849

Cross Island, Alaska. So named because of the large wooden cross, apparently erected by whalemen, that distinguished this island (70°30′N, 147°55′W) from the other low, featureless islands off Prudhoe Bay

East Cape, Siberia (also Cape East, Yeast Cape). Mys Dezhneva (66°05′N, 169°40′W). Captain James Cook, R.N., named the easternmost projection of Asia, East Cape, a name that was in standard usage until 1898, when Czar Nicholas II, on the 250th anniversary of the voyage on which Semen Dezhnev became the first foreigner to pass through Bering Strait, named the Cape in his honor.

East Shore, Gulf of Anadyr. The coast between approximately Mys Chaplina and Mys Beringa

East Shore, Chukchi Sea. The coast of Alaska

Eastern Ice, Chukchi Sea. In June and early July the warm waters passing north through Bering Strait create an area of open water in the southern Chukchi Sea. Whalemen entering these waters tacked back and forth between the ice on the eastern and western sides.

Eastern Pack, Chukchi Sea. In late September and October, while whalemen cruised near Herald Shoal, the ice pack, driven by northerly winds, would begin to creep south around them. One body of ice, the eastern pack, moved south at about 168°W; and the other, the western pack, came down along the coast of the Chukchi Peninsula. The whalers, mindful of this danger, hunted on Herald Shoal as long as possible and then worked their way south between the converging bodies of ice.

False East Cape, Siberia (also False Cape). Mys Intsova (66°17′N, 170° 11′W), so-named because whalemen, heading toward Bering Strait from the northwest in the autumn often mistook this headland for East Cape (Mys Dezhneva), the westernmost headland of Bering Strait. Several vessels were wrecked just east of this point.

Cape Hawaii, Wrangel Island. Mys Gavai (71°02′N, 177°50′W). Named in 1867 by Captain Thomas Long of the New London whaling bark *Nile* who was the first to pass close to Wrangel Island.

The Hole, Chukchi Sea. An area of open water amid the pack ice northeast of Herald Island where whales were often found in the autumn (about 72°N, 174°W)

Holy Cross Bay, Siberia (also Gulf of Saint Croix, Bay of the Holy Cross). Zaliv Kresta (about 65°45′N, 179°00′W)

Indian Point, Siberia (also Cape Chaplin and other variations, Injun Point, India Point, Long Point). Mys Chaplina (64°24′N, 172°14′W)

Jabbertown, Alaska. 68°20′30″N, 166°37′30″W. The polyglot community of Eskimos, Europeans, and other foreigners clustered around several shore-whaling stations near Point Hope, Alaska

John Howland Bay, Siberia (also Salmon Bay). Laguna Imtuk (64°23′N, 173°50′W) named apparently for the New Bedford bark *John Howland*

Kayne Island, Siberia. *See* Big Island

Koryak Coast, Siberia. The coast between, roughly, Mys Olyutorski (59°55′N, 170°21′E) and Mys Navarin (62°15′N, 179°07′E)

Kotzebue Sound, Alaska. To the whalemen, all the waters east of a line from Point Hope to Shishmaref Inlet

Manning Point, Alaska (also Point Manning). Originally this term was applied to the present-day Barter Island (70°07′N, 143°43′W), but Leffingwell (1919:97), accepting the common usage of whalemen and others for Barter Island, transferred the name to a point a few miles east of there. *See* Barter Island

Marcus Bay, Siberia (also Big Marcus Bay, Markes Bay, Marquis Bay, Martin's Bay). Zaliv Tkachen (64°25′N, 172°44′W), where the Fairhaven whaleship *Marcus* was condemned on September 11, 1853

Masinka Bay, Siberia. Probably Proliv Chechekuyum (64°35′N, 172°35′W)

Mercury Harbor, Siberia. (also Rudder Bay). The harbor (65°26′N, 176°00′W) where the bark *Mercury* and ship *Corinthian* hid from the Confederate raider *Shenandoah* while it cruised in Bering Strait in 1865

Midway Islands, Alaska. A group of low sand islands (approximately 70°29′N, 148°25′W) about eight miles north of Prudhoe Bay that lie about halfway between Point Barrow and Herschel Island

North Head, Siberia. Mys Nunyamo (65°36′N, 170°34′W), the headland at the northern entrance to Zaliv Lavrentiya (Saint Lawrence Bay)

North Shore, Gulf of Anadyr. The coast between approximately Zaliv Kresta and Mys Beringa

Pacific Shoal, Alaska. The shoal (approximately 70°41′N, 151°55′W) near Cape Halkett in Harrison Bay discovered by the New Bedford bark *Pacific* when she ran aground here on August 18, 1878

Pea-soup Grounds, Beaufort Sea. The whaling grounds east of the Mackenzie River Delta near McKinley Bay where whales were often found in silty, "green water"

Plover Bay, Siberia (also Plubber Bay, Plover Harbor, Port Providence, Providence Bay). Name used by whalemen to indicate the whole of Bukhta Provideniya, not merely the roadstead behind the sandspit (Reyd Plover, 64°22′N, 173°21′W), on the east side of the bay

Post Office Point, Chukchi Sea (also Point of the Tongue of Ice). A nearly stationary point of the ice pack found in the summer between 69°45′N and 70°30′N at 170°00′W to 171°30′W. Ships cruising along the edge of the pack often met here and exchanged mail.

Rudder Bay, Siberia (also Mercury Harbor, Port Rescue). Bukhta Rudera (65°26′N, 176°00′W), where the whaleship *Winslow* of Havre repaired her rudder in June 1866

Saint Lawrence Bay, Siberia. Zaliv Lavrentiya (65°38′N, 171°00′W)

Cape Serge, Siberia (also Cape Kamen, Cape Surds, Cape Serdze and Cape Surge). Mys Serdze Kamen (66°57′N, 171°43′W)

South Head, Siberia. Mys Kryguygun (65°28′N, 171°02′W), the southern entrance to Zaliv Lavrentiya

Southwest Grounds, Chukchi Sea. Autumn whaling grounds in the southwestern part of the Chukchi Sea near 68°N, 174°W

West Shore, Chukchi Sea. The north coast of the Chukchi Peninsula between East Cape and, approximately, Cape North

Western Ice, Chukchi Sea. *See* Eastern Ice

Western Grounds, Chukchi Sea (also Autumn Grounds). The waters near Herald Island where whales were often found in late September and early October

Appendix 5 / Glossary

Advance. Money allowed to whalemen before starting on a voyage, for the purchase of outfits. An advance was charged against subsequent earnings.

Agent. The managing owner of a whaleship.

Baleen. The keratinous plates that hang from the upper jaw of whalebone whales and serve to filter plankton from the water, also called whalebone.

Baleen whale. A whale characterized by the keratinous plates hanging from its upper jaw. With the exception of the sperm whale, all whales hunted commercially in the nineteenth century were baleen whales (principally right and bowhead whales).

Barrel. A barrel existed aboard a whaleship only as a unit of measurement for oil: 31½ U.S. gallons. Oil was carried in casks.

Beachcomber. An unemployed whaleman or other person who eked out a meager existence on shore near one of the foreign harbors where whaleships called. Frequently a beachcomber was a whaleman who had deserted his ship.

Before the mast. Ordinary seamen bunked in the part of the ship forward of the foremast (in the forecastle); hence they were said to sail "before the mast," a term differentiating seamen and officers.

Beluga (Delphinapterus leucas). A small, white, toothed whale that lives in the Arctic and was rarely taken by whalemen.

Beset. A ship surrounded and immobilized by tightly packed ice was said to be "beset."

Blanket piece. The long piece of blubber, cut approximately six feet wide, stripped from the whale's carcass and hoisted aboard by heavy cutting tackle.

Blasted whale. A dead whale swollen from the gasses formed by putrefaction.

Blink. The reflection of ice beyond the horizon seen on the underside of low clouds.

Blow. A whale's spout.

Blubber. A whale's subcutaneous fat.

Blubber chain. A chain rove through two holes in the blanket piece to hoist the blanket piece aboard.

Blubber hook. An iron hook, weighing from 75 to 100 pounds, suspended from the cutting tackle and thrust through a hole in the blanket piece to hoist the blanket piece aboard.

Blubber room. The space in the 'tween decks near the main hatch where the blubber was temporarily stored and cut into smaller pieces called horse pieces (about six inches wide and several feet long).

Blue whale (Balaenoptera musculus). The largest whale, rarely taken by American whalemen.

Boarding knife. A two-edged, long handled, sword-like knife used to sever the first length of the blanket piece from a second to bring the first aboard.

Boatheader. The officer in charge of a whaleboat. The boatheader steered the boat to the whale. After the boatsteerer (harpooner) fastened to the whale, he changed places with the boatheader whose duty it was to kill the whale.

Boatsteerer. The harpooner (a term seldom, if ever, used by whalemen). The boatsteerer manned the first oar and struck the whale. After fastening to the whale, he changed places with the boatheader and steered the craft while the boatheader killed the whale.

Boiling. Trying out the blubber. Pieces of blubber were placed in pots set in a brick oven on deck. A fire was lit beneath the pots and the fat melted from the blubber's connective tissue.

Bomb gun. A shoulder gun. The heavy smooth bore that fired a bomb lance.

Bomb lance. A brass tube about fourteen inches long that was filled with explosive and fitted with a time fuse. When a bomb lance was shot from a darting gun or bomb gun, the thrust ignited the fuse. Bomb lances that were designed for use in bomb guns were fitted with brass or rubber vanes to steady the flight; those fired from darting guns had no vanes.

Bone. Whalebone.

Book. A section of horsepiece sliced into thin, parallel leaves to speed rendering. The white blubber was sliced down to the black skin (which was left intact), resulting in a piece that looked something like white pages bound by a black cover.

Bowhead birds. Phalaropes, often found near whale feed.

Bowhead whale (Balaena mysticetus). The object of the Arctic fisheries. The whale that carried more oil and longer plates of baleen than any other whalebone whale. So named because of the great arc of its rostrum. Also called in the eastern Arctic, the Greenland whale and the Arctic right whale; and in the western Arctic, the steeple top, polar whale, and Roys's whale.

Box. The sunken cuddy-board in the bows of a whaleboat, where the forward end of the whaleline was coiled.

Bug light. The torch lighted to allow the boiling to continue at night, often hung between the trywork's smoke stacks. The bug light was fed with scraps of skin from the trypots.

Bull. A male whale.

Calf. A very young whale.

Carcass. The whalemen's term for a whale from which the blubber and baleen had been removed.

Catch bone. Whalebone taken from whales that a ship caught in the Arctic, a phrase used to differentiate it from "trade bone," which was obtained in trade from the natives.

Clean ship. A whaleship that had taken no oil.

Clumsy cleat. A thick thwartship plank that formed the after edge of the box. The edge was notched with a deep U-shaped depression in which the boatsteerer braced his thigh for balance when striking a whale.

Cooler. A copper tank placed on the starboard side of the tryworks. Hot oil from the pots was ladled into the tank to cool before it was transferred to another cooling tank and then to a cask in the hold.

Copper. The bottom plating on a whaleship used to protect the hull from teredo worms and marine growth. The coppering was usually a cheaper alloy called Muntz metal.

Cow. Female whale.

Cranes. Hinged triangular brackets that swung out from the bearers to support the whaleboats.

Crow's nest. The enclosed shelter at the foretopmast crosstrees where a lookout stood to watch for ice and whales. The crow's nest was usually made of a wooden framework covered by a canvas dodger.

Cruise. The whaleship's journey on a whaling ground.

Cut in, to. The process of removing the blubber and whalebone from a whale.

Cutting gear. Apparatus used for cutting in.

Cutting spade. A spatulate-headed, long-handled implement for cutting blubber off a carcass from the cutting stage.

Cutting tackle. Massive blocks and falls hung under the maintop and led to the windlass; used to hoist the blanket piece and headbone.

Dart, to. The iron was darted, hove, pitched, or tossed—but never thrown.

Darting distance. Near enough to the whale to strike. The limit was about thirty feet, but fifteen feet was considered good darting distance.

Darting gun. A small smooth bore attached to the end of a shaft that bombed a whale the same instant that the iron struck.

Darting iron. A toggle iron with a pointed base that was fixed in gudgeons alongside the darting gun.

Deck pot. A pot with feet and handles that was placed near the tryworks to serve as a cooler or to hold scrap or pieces of untried blubber called "books."

Donkey's breakfast. The corn husk or straw mattress used by ordinary seamen.

Donkey engine. The steam auxiliary engine found on both sailing and steam whaleships, which was used for powering the windlass to assist in hoisting the anchor and raising the blanket piece, among other tasks.

Draw, to. Said of an iron when it pulled out of a whale.

Drift. The set of a current, with or without ice.

Drift ice. Ice moving with the current.

Dry skin. A whale whose blubber yielded very little oil.

Drogue. A square piece of wood that was fastened to a whaleline to slow a running whale.

Duff. A mixture of flour, yeast, raisins, and lard that was served as a pleasant change from salt horse and salt junk (perishable food preserved with salt).

Fall catch. The cargo taken after the tenders had resupplied the ships in the summer.

Fast boat. A boat with a line attached to a whale.

Fast whale. A whale that had been struck but not yet killed.

Fasten, to. To strike a whale successfully.

Feed slick. A section of oily surface indicating the presence of krill—the zooplankton on which baleen whales feed.

Finback (Balaenoptera physalus). A fast swimming whale seldom pursued or caught by nineteenth-century whalers.

Fin chain. A short chain fastened around the fin for starting the blanket piece.

Fins. The whale's lateral fins.

Fishery. Applied collectively to the ships whaling on an important ground or seeking a particular type of whale; hence "Arctic fishery" and "sperm whale fishery."

"Five and forty more!" Shouted by the crew when the last blubber from a whale was hoisted aboard. The phrase developed in the sperm whale fishery and refers to forty-five barrels, the size of an average sperm whale.

Floe. A large sheet of sea ice.

Flue. The barb of a whaling iron.

Flukes. The whale's tail.

Fluke chain. The chain put around the small, the part of the whale just forward of the flukes, to secure the whale alongside a whaleship.

Fluke spade. The spade used for hamstringing a whale by cutting the tendons controlling the flukes.

Flurry. The dying struggle of a whale.

Foul line. A whaleline that has jammed while running out of the boat.

Gally, to. To frighten a whale.

Gam. A visit between whaleships at sea.

Get fast, to. To fasten to a whale.

Giant bottlenose (Bernardius bairdii). The second largest of the toothed whales, rarely taken by whalemen.

Go on the whale, to. To approach very close to the whale to strike.

Gray whale (Eschrichitus robustus). A baleen whale that winters in the lagoons of Baja California. It yielded only 24 to 35 barrels of oil and no commercial bone. A vicious fighter also called devil fish, scrag whale, rip sack, and mussel digger.

Greener gun. A heavy swivel gun mounted in the bow of a whaleboat for shooting bombs and whaling irons. The Greener gun was rarely used in the western Arctic.

Green hand. A sailor on his first whaling voyage.

Grounds. The areas where whales were known to be found.

Ground tier. The bottom layer of casks in the lower hold.

Guamies. Whalemen who were natives of Guam.

Gun warp. The line fixed to the shaft of a darting gun and the whaleboat that allowed the boatsteerer to retrieve the gun after fastening to a whale.

Gurry. The greasy sludge left on deck after cutting and boiling.

Hamstring, to. To stop a running whale by using a fluke spade to cut the fluke tendons at the small.

Hay hooked. A whale that was fastened to only lightly, so that the iron drew.

Head, to. To command a whaleboat.

Headbone. The whale's rostrum, holding the whalebone.

Head chain. The chain that was rove beneath the spout holes for hoisting the headbone.

Heave down, to. To winch a ship over far enough to expose half her bottom. Whaling vessels, being of heavy construction, were hove down long after the practice was given up for other vessels. The last vessel to be hove down in New Bedford was the *Josephine* in 1893.

High hook. The vessel that took the largest number of whales in a season.

Hook on, to. To get fast.

Horse piece. A piece of blubber usually several inches wide and a few feet long, cut from the blanket piece.

Humpback (Megaptera novaengliae). A whale usually disdained by the Arctic whalemen because of its speed, small amount of blubber, and useless baleen. Humpback blubber had no resiliency and hence the blubber hooks tended to tear right through it. Humpbacks also usually sank after death.

Ice anchor. An anchor with only one flue that allowed the ships to tie up to ice floes.

Infit. Clothes and credit given a whalemen by a landshark at the conclusion of a voyage.

Inside catch. The catch made by the inside fleet.

Inside fleet. The ships that were wintering at Herschel Island or east of there. A term used to differentiate the inside ships from those that were merely cruising in the eastern Beaufort Sea in the short summer season, called the outside fleet.

Iron. The name applied to the harpoon. Whalemen, rarely, if ever, referred to the iron as a harpoon. "Iron" meant not only the iron barb and foreshaft but also the entire instrument, including the shaft and iron strap.

Itkillicks. The name used by whalemen and Eskimos to denote Indians.

Kanakas. Used primarily to refer to Hawaiian natives, but also to all Polynesians.

Kicking strap. The line fixed across the top of the clumsy cleat under which the whaleline was led. It prevented the whaleline from jumping and towing the boat sideways or stern first.

Killer whale (Orcinus orca). Whales that kill many species of whales, as well as walruses and seals.

Kogmullicks. Eskimos of the Mackenzie delta.

Knuckle joint. The joint of the whale's fin (humerus) that connects with the shoulder socket. The knuckle bone had to be unjointed from the shoulder to begin hoisting the blanket piece.

Lance. The instrument used for killing whales. About eleven feet long, it consisted of a wood shaft, a palmate blade at the end of a six foot metal shaft, and a lance warp.

Landshark. A sailor's outfitter.

Larboard. The port side of a vessel, a term that became obsolete aboard all vessels except whaleships.

Larboard boat. The boat on the port quarter of a whaleship.

Lay. The whaleman's share of the profits of a whaling voyage.
Lead. A lane of open water in the ice.
Lean, to. To remove any bits of muscle that remained attached to the blubber.
Life. The heart and lungs of a whale, where the boatheader aimed his lance.
Line tub. A tub in which the whaleline was stowed in the whaleboat.
Live irons. The two irons attached to the whaleline.
Lobtail, to. The act of a whale beating its flukes on the surface of the water when head-down in a nearly vertical position.
Loggerhead. The projecting bollard in the stern of a whaleboat around which the whaleline was snubbed.
Loose boat. A whaleboat that was not fast to a whale.
Loose irons. Irons that were not bent on the whaleline. Occasionally a drogue was bent onto one and this loose "drug iron" was hove into the whale to slow its flight.
Loose whale. A whale with irons fast but loose from a boat.
Lower away. The order to lower boats for whales.
Make a passage, to. To pass from one whaling ground to another, often with the boats unrigged for whaling.
Maori. A native of New Zealand.
Masinker. A Siberian Eskimo.
Masinker house. The house on a ship's deck, or ashore when wintering, where the ship's natives berthed.
Masinker boots. Boots made by Siberian Eskimos or Eskimos of the Diomede Islands. They were highly prized as footgear by the Arctic whalemen.
Mason and Cunningham gun. A swivel gun mounted in the clumsy cleat that fired either an iron or a large bomb lance. In the western Arctic it was used primarily in the shore fishery, but it never gained wide acceptance.
Messenger iron. A very large toggle iron that was used for raising whales.
Mince, to. To thinly slice the blubber of a horse piece, leaving the skin intact to hold the piece together.
Mincing horse. A plank for holding the horse pieces while they were being minced.
Mincing knife. A knife with a blade about two feet long and a handle at each end, for drawing longitudinally across the horse pieces to cut them into books.
Mincing tub. The tub into which the minced books fell from the mincing horse.
Monkey rope. A rope fastened around the waist of the boatsteerer whose duty it was to stand on the whale and chop through the skull to sever the headbone.
Nakooruks. The whalemen's term for the coastal Eskimos of northwestern Alaska.
Nantucket sleigh ride. A layman's term for a fast boat being towed by a running whale.
Nib. The tip of the whale's upper jaw.
Noble whale. A very large whale.
Nunatarmiuts. Inland Eskimos from the Brooks Range of Alaska, many of whom migrated to Herschel Island to serve as hunters for the wintering whaling fleet.
Outfit. The equipment for a whaling voyage, either a sailor's or a ship's.
Outside catch. The whales caught by the fleet that was not wintering in the eastern Beaufort Sea. A phrase often further restricted to denote whales caught only in the Bering and Chukchi seas.
Outside fleet. Whaleships that had not wintered in the eastern Beaufort Sea, whether or not they briefly cruised in those waters.
Oyster. The white gum in which the plates of baleen are embedded.
Pack. A large body of concentrated drift ice.
Pitch pole, to. To pitch an iron a long distance by tossing it upward so that it described a considerable arc before striking the whale.

Poke. An inflated sealskin used as a drogue.
Polar whale. A bowhead.
Pole. The shaft of an iron, lance, or spade.
Polynya. Not a whalemen's term. A large body of open water amid pack ice.
Preventer boatsteerer. A substitute boatsteerer. A foremast hand who was the next in line for promotion.
Raise whales, to. To sight whales.
Raft, to. To wrap casks or pieces of blubber together and tow them to the whaleship.
Recruits. New men taken aboard a whaleship at a foreign port to make up for men lost to illness, disability, death, or desertion.
Right whale (Eubalaena glacialis). A whale similar in many respects to the bowhead, it is easily differentiated by the large barnacle-like callosity on its nib. It is found in the temperate waters of the Atlantic and Pacific. A vicious fighter.
Ripsack. A gray whale.
Roys's whale. A bowhead.
Run, to. A whale seeking to escape on the surface from a whaleboat.
Salt horse. Salt beef.
Salt junk. Salt pork.
Save, to. To capture a fast whale.
Scarf, to. To delineate the width of the blanket piece during cutting in.
Scrag whale. A gray whale.
Scrap. The twisted bits of skin left after the blubber has been tried out. Scrap was put in a hopper to allow the remaining oil to drain, then used to fuel the trywork's fire.
Seasoner. A recruit taken aboard a whaleship for only a particular part of the voyage. Seasoners taken aboard at Hawaii usually served only for the Arctic cruise.
Set up, to. To assemble a cask or other object.
Settle, to. A whale sinking without turning flukes or making another observable movement.
Settle a voyage, to. To pay off the owners and crew.
Shipkeeper. In the early years of the Arctic fishery, when the captain headed one of the whaleboats, the shipkeeper was usually the cooper, who acted as sailing master while the boats were down.
Ship's natives. Eskimos who were signed aboard for the cruise, either for the summer or for wintering.
Shooks. Barrel staves. They were sent out numbered and knocked down in casks for ease of stowage.
Short lay. The large fraction of the profits awarded to the captain and senior officers.
Short warp. The line with which the second live iron was bent to the whaleline.
Shoulder gun. The heavy smooth bore that fired a bomb lance through the air.
Shy. Whales that were difficult to approach and easily gallied.
Single-flued iron. A early iron with only one barb.
Size bone. Whalebone that was six feet or longer and thus drew premium prices.
Slop chest. The ship's store, which sold clothes, knives, tobacco, and other items.
Small. The narrowest part of a whale's tail, just before the flukes.
Sound, to. To dive.
Sperm whale (Physeter catodon). The largest of the toothed whales, rarely taken north of the Aleutians by American whalemen.
Spring catch. The whales taken by the Arctic whaleships before they met their tenders in the summer.
Steam digester. Steam tryworks carried aboard some of the San Francisco steam whalers.
Steeple top. An early and rare term for the western Arctic bowhead.
Steerage. The quarters for the boatsteerers.

Stove. A ship, boat, or other object that is badly smashed.
Stowing down. Putting oil into casks below decks.
Strike, to. To get fast.
Suds. The foam on the surface of the water from a whale running on or below the surface.
Swivel gun. A heavy gun mounted on a post in the bow of a whaleboat and used for shooting irons or heavy bomb lances.
Tender. A vessel attending to the needs of a ship or fleet, often bringing mail and fresh provisions to the whaleships and taking accumulated oil and bone back to the home port.
Toggle head. The hinged head of a toggle iron.
Toggle iron. A whaling iron with a hinged head and off-set barb that allowed the head to turn at right angles to the shank when strain was taken on the whaleline.
Towing toggle. A toggle that was attached to the end of a line or chain and put through a hole in a whale's lip to tow the dead whale back to the ship.
Trade. Used to encompass either the manufactured goods used in trading or the furs, whalebone, and ivory obtained in trade from the natives.
Trade bone. Whalebone obtained in trade from the natives.
Try out, to. To render oil from blubber.
Trypot. An iron cauldron set over a fire box for trying out blubber.
Tryworks. A brick oven, usually set aft of the foremast, in which pots were set for trying the oil from blubber.
'tween seasons. In general this term was used for the short winter cruise between visits to the Arctic. In the Arctic it was used to denote the summer season when whales were scarce and whalemen often turned to hunting walruses.
Ullage. The loss of whale oil from the casks.
Undersize bone. Baleen that was less than six feet long and consequently commanded a lower price.
Voyage. The entire whaling trip from home port to home port, comprising passages and cruises.
Wash bone, to. To clean whalebone of bits of gum and dirt.
Water sky. The black reflection of water amid the pack ice on the under side of low clouds.
Whalebone. Baleen.
Whalecraft. The equipment used for fastening to and killing a whale, cutting in, and trying out and stowing down the oil.
Whale feed. The plankton on which baleen whales feed.
Whaleline. The line carried in tubs in a whaleboat and fixed to the whaling iron.
Whale oil. The oil from baleen whales, distinct from the more highly prized sperm oil.
White water. The spray made by a breeching or thrashing whale.
Wood to blackskin. A boat in contact with a whale.
Wooding. Collecting firewood on shore.

Key to Repository Symbols

Albany Institute of the History of Art, Albany, New York.	NAII
Anglican Church Archives, Toronto	CaOTCHAr
Baker Library, Harvard University, Cambridge, Massachusetts	MH-BA
Bancroft Library, University of California, Berkeley	CU-BANC
Beineke Library, Yale University, New Haven, Connecticut	CtY
Bernice P. Bishop Museum, Honolulu	HHB
Brown University, Providence, Rhode Island	RPB
Buffalo and Erie County Public Library, Buffalo, New York	NBuHi
California Academy of Sciences, San Francisco	CSfA
Church Historical Society, Austin, Texas (Episcopal Church)	TxAuCH
Church Missionary Society, London, U.K.	CMSL
Cold Spring Harbor Whaling Museum, Cold Spring Harbor, New York	CSHWM
Connecticut Historical Society, Hartford	CtHi
Connecticut State Library, Hartford	Ct
Dukes County Historical Society, Edgartown, Massachusetts	MEdDHi
East Hampton Free Library, East Hampton, New York	NEh
Essex Institute, Salem, Massachusetts	MSaE
Falmouth Historical Society, Falmouth, Massachusetts	FHS
General Services Administration. National Archives and Record Service. Franklin Roosevelt Library, Hyde Park, New York	NHyF
Glenbow Alberta Institute, Calgary	CaACG
Hawaii State Archives, Honolulu	HH
Henry E. Huntington Library, San Marino, California	CsmH
Houghton Library, Harvard University, Cambridge, Massachusetts	MH-H
Hudson's Bay Company Archives, Winnipeg, Manitoba	HBC
International Marine Archives, Old Dartmouth Historical Society, New Bedford, Massachusetts	IMA-MNBedfHi
Kendall Whaling Museum, Sharon, Massachusetts	MShaK
Mariner's Museum, Newport News, Virginia	ViNeM
Mitchell Library, Sidney, Australia	MLS
Mystic Seaport, Mystic, Connecticut	CtMyMHi
Nantucket Whaling Museum, Nantucket, Massachusetts	MNanW
National Library of Australia, Canberra	Au
National Maritime Museum, Greenwich, U. K.	NMM
National Maritime Museum, San Francisco, California	CSfMM
National Museum of Natural History, Ottawa	CaOONM
New Bedford Free Public Library, New Bedford, Massachusetts	MNBedf
New Haven Colony Historical Society, New Haven, Connecticut	CtNhHi
New London County Historical Society, New London, Connecticut	CtNlHi
North Carolina State Archives (North Carolina State Department of Archives and History), Raleigh	Nc-Ar

Old Dartmouth Historical Society, New Bedford, Massachusetts	MNBedfHi
Peabody Museum of Salem, Salem, Massachusetts	MSaP
Presbyterian Historical Society, Philadelphia, Pennsylvania	PPPrHi
Private Collection	PC
Providence Public Library, Providence, Rhode Island	RP
Public Records Office, London, U. K.	UkLPR
Rhode Island Historical Society, Providence, Rhode Island	RHi
Rogers Memorial Library, Southampton, New York	NSoa
Royal Canadian Mounted Police Library, Ottawa	CaOOR
Sheldon Jackson College, Sitka, Alaska	AkSJ
Smithsonian Institution, Washington, D.C.	DSI
Southhampton Colonial Society, Southhampton, New York	SCS
State Library of Tasmania, Hobart	AuHS
Stefansson Collection, Dartmouth College, Hanover, New Hampshire	NhD
Stonington Historical Society, Stonington, New York	SHS
Suffolk County Whaling Museum of Sag Harbor, Sag Harbor, New York	SCWM
Suzzallo Library, University of Washington Archives, Seattle, Washington	WaU
University of Alaska Archives, Fairbanks	AkU
U.S. National Archives and Record Service. National Archives Library, Washington, D. C.	DNA
U.S. Naval War College, Newport, Rhode Island	RNN
Wethersfield Historical Society, Wethersfield, Connecticut	CtWetHi

Notes and References

Chapter 1

Roys aboard Josephine

The Friend (Honolulu), September 15, 1845; Roys, nd; Roys 1854; Schmitt, de Jong, and Winter 1980:19-20; Howerton 1971; Beechey 1831:I, 379.

Roys aboard Superior

The Friend (Honolulu), November 1, 1848; Roys 1854; Roys, nd; Wood, ms, vol. 2, p. 619; *Whalemen's Shipping List and Merchants' Transcript* (New Bedford), January 29, June 6, August 29, 1848; February 6, May 5, May 14, 1849; Scrapbook P-K 206, vol. 1, pp. 102-3, Charles Scammon Collection, Bancroft Library; Schmitt, de Jong, and Winter 1980:22-24; Bertrand 1971:219-54.

Roys's accounts of his discovery were written several years after the fact and contain a number of minor contradictions. In the absence of finding his logbook or journal I have relied as much as possible on the abstract of his logbook in the Matthew Fontaine Maury Collection of logbook abstracts in the National Archives. Roys apparently kept his logbook in nautical time. Here and elsewhere I have converted entries in nautical time to civil time.

It has often been claimed that the *Ocmulgee* also cruised in Bering Strait in 1848. Her logbook (MNBedfHi) reveals that she was, in fact, on Kamchatka in the summer of 1848. See also Cyrus Manter's interview in *The Sunday Standard* (New Bedford), January 21, 1909.

Chapter 2

1750 to 1840

Jenkins 1921:177-235; Starbuck 1964; Kugler 1971; Kugler 1980; Stackpole 1953; Bertrand 1971:20-158, 219-54; G. Jackson 1978:54-154; Harlow 1964:293-328; A. G. E. Jones 1981.

expansion into the North Pacific

Tower 1907:128-29; Waugh 1968:139; Uzanne 1898:104-13; *Whalemen's Shipping List,* January 23, 1844, March 28, 1843; *Manhattan* (MNBedfHi); *The Friend* (Honolulu), September 15, 1845, October 15, 1846; *Ocmulgee* (MNBedfHi); David A. Henderson, personal communication.

It is difficult, if not impossible, to determine either the first ships to cruise on the North Pacific rim or the first ships to take bowheads there. Starbuck (1964:98) claims that the *Hercules* and *Janus* were the first to take bowheads, in 1843, and that the *Ganges* was the first to cruise on Kodiak, in 1835. Judging from the *Whalemen's Shipping List* and from the records of Daniel Wood (ms.), the *Hercules* was on a sperm whaling cruise during that season and the *Janus* was in the Atlantic Ocean. Similarly, in the summer of 1835 the *Ganges* was between voyages in Nantucket.

Chapter 3

outfitting

Church 1938:24-28; Baker 1977:64; Ronnberg 1974; H. Hall 1884:22-28; Brown 1887:237-40.

crew and financial

Hohman 1928:84-113, 217-43, 272-88; Brown 1887:218-37; 291-93; W. N. Burns 1913:18-19; Aiken and Swift advice book (MH-H).

beginning the voyage

Davoll 1981; Brown 1887:229-31; Hohman 1928:114-42.

whaleboats

Ansel 1978; Brown 1887:240-46, 250-55; Bodfish 1936:126-28, 145; Davis 1926:158-60; *Narwhal* (MH-BA), July 7, 1908.

Azores

Taylor 1929:22; Hohman 1928:152.

Hawaii and other ports of resupply

West et al. 1965:19-20; C. L. Manchester (bark *Coral*) to Gideon Allen, November 28, 1856 (PC); *Cornelius Howland* (RP), April 5, 1870; *Northern Light* (MNBedfHi), May 4, 1876.

food

Hohman 1928:130-33; *Frances* (MNBedfHi), August 17, 1851.

trypots

Courser (RP), June 9, 1866.

windlass

Northern Light (MNBedfHi), May 3, 1876.

stoves in forecastle

John Wells (MNBedfHi), May 11, 1870; *Lucretia* (MEdDHi), April 11, 1883.

cutting tackle

Northern Light (MNBedfHi), May 2, 1876; *Samuel Robertson* (MShaK), May 7, 1851; *John Wells* (MNBedfHi), May 12, 1870; *Helen Mar* (MNBedfHi), May 16, 1873; *John Carver* (MNBedfHi), April 24, 1886.

cutting stage

John Wells (MNBedfHi), May 12, 1870; *Cornelius Howland* (RP), April 11, 1868; *Henry Taber* (MNBedfHi), May 8, 1869, June 4, 1870; *Lagoda* (MNBedfHi), May 3, 1869; *Helen Snow* (MNBedfHi), June 11, 1872; *Eagle* (MNBedfHi), April 21, 1868.

topmasts and crow's nest

Brown 1887:234; W. N. Burns 1913:119; Williams 1964:349; *Fortune* (MNBedfHi), May 26, 1851; *Henry Taber* (MNBedfHi), May 7, 1869; *Niagara* (MNBedfHI), May 15, 1852; *Europa* (MNBedfHi), May 7, 1870; *Helen Snow* (MNBedfHi), June 4, 1872; *Minerva* (RP), May 29, 1866; *Francis Palmer* (MShaK), May 5, 1887; *Marengo* (MShaK), July 15, 1867; *John Wells* (MNBedfHi), April 23, 1871.

anchors

Helen Snow (MNBedfHi), June 26, 1872; *Cleone* (MNBedfHi), July 6, 1859, June 21, 1865; Starbuck 1964:139; Marine Insurance Company (MNBedfHi).

rudder
Williams 1964:349; *Europa* (MNBedfHi), May 18 and June 8, 1870; *Northern Light* (MNBedfHi), May 4 and 13, 1876, May 28, 1878; Bodfish 1936:39; *Lagoda* (MNBedfHi), May 10, 1869; *Henry Taber* (MNBedfHi), May 20, 1868; *Whalemen's Shipping List*. November 29, 1870; A. H. Clark 1887a:77; West et al. 1965:36-37.

fenders
Seneca (MNBedfHi), May 15, 1870.

physical oceanography of Bering and Chukchi seas
Bockstoce and Batchelder 1978; Bockstoce and Batchelder, Appendix 4, this volume; Orth 1971; Bezrukov 1964; Coachman et al. 1975; Hopkins 1967:451-84; West et al. 1965:21; Mountain 1974:27-41; E. Simpson 1890:8-9; U.S. Coast Pilot, Alaska, 1926:261; Dall 1882:328; Coachman and Aagaard 1981; Bourke 1983.

biological oceanography of Bering and Chukchi seas
McRoy and Goering 1974; Webster 1979; McRoy et al. 1976; Apollonio 1971; Alexander 1974; Johnson 1963:177-80; Sambrotto et al. 1984.

bowhead feeding behavior
Tilton 1969:74; Williams 1964:353-54; Bodfish 1936:89-90; Pivorunas 1979:435; Scammon 1874:54-57; Matthews 1978:54-55; Carroll and Smithhisler 1980; Wursig, Clark and others 1982; Bogoslovskaya and others 1982.

feed slicks and whale feed
Belvedere (CaOONM), July 30, August 2, and September 2, 1912; Bean 1902:262; *Whalemen's Shipping List*, May 17, 1853; *Cleone* (MNBedfHi), July 29, 1859; *Nimrod* (MNBedfHi), June 21, 1860; *John P. West* (ViNeM), July 18, 1864; *Mount Wollaston* (MNBedfHi), June 9 and September 13, 1877.

bowhead birds
Eagle (MNBedfHi), August 16, 1867; *Whalemen's Shipping List*, May 17, 1853; Brina Kessel, personal communication; *Belvedere* (CaO0NM), August 27, 1912.

sea ice
Brower et al. 1977, I:14-18, II:13-20; Dunbar 1967; Burns et al. 1980:6-17; Page 1900; E. Simpson 1890; Jarvis 1899:197-98.

ice navigation in Bering Sea
Dall 1882; West et al. 1965:21, 36; E. Simpson 1890.

stove by ice
Eagle (MNBedfHi), May 4, 1868; *Mercury* (MNBedfHi), June 27, 1867; *Lagoda* (MNBedfHi), May 10, 1869; *Saratoga* (MMBedfHi), May 11-12, 1851; *America II* (MNBedfHi), July 14, 1851.

"oak against ice"
Starbuck 1878:136; Aldrich 1889:22-23.

protect weak bows
Orca (IMA-MNBedfHi), April 29, 1886; *Henry Taber* (MNBedfHi), May 25, 1870; *Cornelius Howland* (RP), May 1 and May 25, 1868; *Mary and Helen* [I] (MNBedfHi), May 10, 1880; *Coral* (MNBedfHi), May 10, 1880; *Mount Wollaston* (MNBedfHi), August 29, 1887.

emergency repairs
Cleone (MNBedfHi), September 19, 1861; Aldrich 1889:40-41: *Hibernia* (MNBedfHi), May 13, 1869; *Mercury* (MNBedfHi), July 1866 passim; *Helen Mar* (MNBedfHi), May 28-29, 1877.

Bauldry aboard Arnolda
Aldrich 1889:220-22.

pounding in the ice
Lagoda (MNBedfHi), May 10, 1896; *Sea Breeze* (MNBedfHi), May 9, 1869; *Europa* (MNBedfHi), May 8, 1873; *Henry Taber* (MNBedfHi), May 11-12, 1870.

escape from ice pack
Cossack (MNBedfHi), May 28, 1851; *Seneca* (MNBedfHi), May 15, 1870; *Eugenia* (MNBedf), June 23, 1870; *Kutusoff* (RP), July 1, 1852; Page 1900:6; *John Howland* (MNBedfHi), June 9, 1862; *Europa* (MNBedfHi), July 29, 1870; *California* (MNBedf), May 24, 1868; *Mary and Susan* (RP), May 26, 1987; West, et al. 1965:52; *Gayhead* (MNBedfHi), June 10, 1864.

fog
Minerva (RP), July 4, 1865; *Lydia* (RP), July 25, 1866; *Condor* (MNBedf), July 2, 1851; Muir 1917:216.

depths
Cleone (MNBedfHi), July 9, 1859; *Europa* (MNBedfHi), June 26, 1870; Healy 1887:16.

collisions
W. N. Burns 1913:176.

Chukchi Sea conditions and cruising patterns
Jarvis 1899:197-98; E. Simpson 1890; Bower et al. 1977, III:13; Kovacs and Mellor 1974:139; *Chronicle* (San Francisco), October 28, 1897; Hooper 1881:45; Williams 1964:286-87.

leave Arctic
Florida (MNBedfHi), September 7, 1852; Henry Pease, in Starbuck 1878:138; *Pacific Commercial Advertiser* (Honolulu), December 1, 1866; Aldrich 1889: 218-19; *Congress* (MShaK), September 28, 1866; *Orca* (PC), October 23, 1883; Tilton 1969:249; *Tribune* (Chicago), February 1886, newspaper clipping, scrapbook T-2 (MNBedfHi); *Narwhal* (CaOTCHAr), October 22, 1901; *Governor Troup* (RP), September 28, 1866; *Antilla* (RP), August 20, 1858; *Pacific Commercial Advertiser* (Honolulu), November 7, 1861; *Milo* (MNBedfHi), September 10, 1853; *China* (MNBedfHi), September 9, 1853; *Seneca* (MNBedfHi), October 19, 1870.

"half way around the compass"
West et al. 1965:21-22.

ballast ship with water
Henry Taber (MNBedfHi), August 12,, 1869; *Almira* (MNBedfHi), September 11, 1866; *Lagoda* (MNBedfHi), October 14, 1869.

clean up
Henry Taber (MNBedfHi), September 24, 1870; *Cornelius Howland* (RP), October 21, 1869; *John Howland* (MNBedfHi), October 5, 1862.

tryworks overboard
Cornelius Howland (RP), October 21, 1868; *Shepherdess* (RP), July 6, 1850.

unrig boats
Aurora (NBedf), October 14, 1869; *Belvedere* (MNBedfHi), October 25, 1885; *Helen Mar* (MNBedfHi), September 25, 1875; *Corinthian* (RP), October 21, 1867; *Barnstable* (RP), September 11, 1862; *Courser* (RP), October 6, 1866.

unrig cutting gear
John Wells (MNBedfHi), October 16, 1870; *Helen Mar* (MNBedfHi), October 15, 1872; *William Baylies* (MNBedfHi), October 15, 1906; *Progress* (MNBedfHi), October 14, 1881; *Henry Taber* (MNBedfHi), October 14, 1869.

unrig crow's nest and send up spars
Grampus (MNBedfHi), October 13, 1888; *George Howland* (MNBedfHi), October 22, 1867; *Eagle* (MNBedfHi), October 24, 1867; *Navy* (MNBedfHi), October 4, 1862; *Herman* (CSfA), October 12, 1914; *Condor* (MNBedf), September 28, 1852; *Samuel Robertson* (MShaK), September 26, 1850; *Montreal* (MNBedfHi), September 21, 1852.
rig anchor
George and Mary (HH), September 29, 1851; *Robin Hood* (CtMyMHi), September 21, 1859; *Helen Mar* (MNBedfHi), October, 20, 1874; *Congress* (MShaK), October 4, 1866; *John Howland* (MNBedfHi), October 9, 1862; *Cleone* (MNBedfHi), September 7, 1865; *Saratoga* (MShaK), September 9, 1858.
tow log
Leo (MNBedfHi), October 6, 1879; *Beluga* (MNBedfHi), July 28, 1897.
glad to leave Arctic
Litherland (AuHS), July 30, 1851; *Robin Hood* (CtMyMHi), September 14, 1859; *Constitution* (MNanW), August 25, 1851; *Nassau* (MShaK), August 22, 1851.
throw clothes overboard
Russell 1898:156.
"Rolling down to old Mohee"
Atkins Adams (MNBedfHi), December 25, 1859.

Chapter 4

spot whales from crow's nest
Brown 1887:256; A. J Allen 1978:172; Barker, ms., 28; Williams 1964:276, 370; *Montreal* (MNBedfHi), August 7, 1851; *Mount Wollaston* (MNBedfHi), June 20, 1877; *Courser* (RP), July 9, 1866.
indications of whales' presence
Eagle (MNBedfHi), October 12, 1868; *Alexander* (MNBedfHi), June 14, 1897; *Beluga* (MNBedfHi), May 6, 1897; *Cape Horn Pigeon* (MShaK), June 20, 1864; *Neva* (NEh), August 28, 1859; *Cornelius Howland* (RP), April 23, 1868; *Robin Hood* (CtMyMHi), July 9, 1859.
mistake other things for bowheads
Betsey Williams (MNBedfHi), May 28, 1852; *Mercury* (MNBedfHi), September 6, 1864; *Franklin* (RP), May 7, 1854; *Pioneer* (RPB), June 1, 1852.
percentage of struck whales captured and struck whales lost
Bockstoce and Botkin 1983; Bockstoce 1980, Table 4.
signals from ship to boats
Brown 1887:257-58.
lower whaleboats for chase
Brown 1887:258-59; Ansel 1978:16; Bodfish 1936:27.
too calm
Cleone (MNBedfHi), August 24, 1859; *Sea Breeze* (RP), September 29, 1870; *Betsey Williams* (MShaK), June 29, 1852; *Barnstable* (RP), July 14, 1863; *Mercury*, August 25, 1864; *Mary and Helen* [I] (MNBedfHi), July 12, 1880; *North Star* (MNBedfHi), June 12, 1882; *Helen Mar* (MNBedfHi), July 26, 1880; *Janus* (RP), August 25, 1864; *Mary D. Hume* (MH-BA), August 29, 1893; *Narwhal* (PC), October 5, 1887.
noise gallies whale
New England (CtHi), August 13, 1859; *Grampus* (MNBedfHi), September 30, 1888; *Helen Mar* (MNBedfHi), June 6, 1877; *Ocean* (MNanW), September 4, 1866; *Narwhal* (MH-BA), August 29, 1903; Pedersen 1952:38.
bowhead's eyesight
Lagoda (MNBedfHi), September 6, 1869; *Helen Mar* (MNBedfHi), August 3, 1877; *Roman* (MShaK), August 16, 1854; *Beluga* (MNBedfHi), August 3, 1898; *Belvedere* (CaOONM), September 20, 1912.
too dark
Helen Mar (MNBedfHi), October 8, 1874.
too foggy
Montreal (MNBedfHi), July 21, 1851; *Alfred Gibbs* (MShaK), June 22, 1852; *Daniel Webster* (MNBedfHi), May 30, 1851; *Mercury* (MNBedfHi), August 15, 1864.
too rough
Navy (MNBedfHi), August 20, 1863; *Eagle* (MNBedfHi), September 23, 1867; *Helen Mar* (MNBedfHi), July 20, 1880.
whale escapes in ice
John P. West (ViNeM), June 1, 1864; *Mercury* (MNBedfHi), August 15, 1864; *Coriolanus* (CtMyMHi), May 26, 1852.
whale escapes to windward
Bowhead [II] (CaACG), August 29, 1905; *John P. West* (ViNeM), July 22, 1866; *Adeline* (MNBedfHi), September 2, 1851; *Cicero* (MNBedfHi), August 3, 1864; *Ocean* (MNanW), August 23, 1867; *Roscoe* (MNBedfHi), August 17, 1866; *Eagle* (MNBedfHi), July 3, 1867.
John Peabody
Robin Hood (CtMyMHi), July 18, 1859.
congestion among ships
Roscoe (MNBedfHi), June 23, 1866; *Eagle* (MNBedfHi), June 19, 1867; *John Carver* (MNBedfHi), September 9, 1885; *Saratoga* (MShaK), June 1, 1858.
competition between ships
Williams 1964:368; *Montreal* (MNBedfHi), June 13, 1852; DeWindt 1899:285.
Lydia*'s ruse*
Montreal (MNBedfHi), June 1, 1852.
wounded whale gallies others
Cicero (MNBedfHi), August 14, 1864; *Cornelius Howland* (RP), September 16, 1868; *Beluga* (MNBedfHi), July 28, 1898; *John P. West* (ViNeM), June 18, 1864; *Belvedere* (CaOONM), August 31, 1912; unidentified newspaper clipping, December 31, 1904, scrapbook 2, pg. 112, (MNBedfHi); Bean 1902:264; Aldrich 1889:34; Anderson 1924:498.
gallied whales
Omega (RP), June 26, 1849; *Alfred Gibbs* (MShaK), July 19, 1852; *Grampus* (MNBedfHi), July 26 and October 2, 1888; *Cicero* (MNBedfHi), August 18, 1864; *Samuel Robertson* (MShaK), June 26, 1850.
observing the sabbath
Tiger (CtMyMHi), August 19, 1849;
Julian (CtY), September 18, 1859.
whaling irons
Lytle 1983; *Betsey Williams* (MShaK), July 9, 1852; *Montreal* (MNBedfHi), July 23, 1851; *Whalemen's Shipping List* (New Bedford), May 31, 1852, July 19, 1853.

bomb lance and shoulder gun
Lytle 1984; West et al. 1965:23; Bodfish 1936:127.
accidents with bomb lance
St. George (RP), September 28, 1867; *Emily Morgan* (MNBedf), September 7 and 8, 1871; *Eliza* (MNBedfHi), December 30, 1856.
shoulder guns explode or fail to discharge
Bodfish 1936:112; *John P. West* (ViNeM), July 29, 1866; *Franklin* (RP), September 7, 1854; *Narwhal* (MH-BA), July 6, 1907.
details of use of shoulder gun and bomb lance
Brown 1883:9-10, 13.
some early dated references to use of bomb lance and shoulder gun in Arctic
Prudent (SHS), May 29, 1851; *Tamerlane*
(IMA-MNBedfHi), September 26, 1851; *Prudent* (MShaK), June 15, 1852; *Rousseau* (MNBedfHi), June 9, 1854; *St. George* (MNBedfHi), July 15, 1854; *Eliza F. Mason* (MNBedfHi), May 23, 1854; *Franklin* (RP), May 10, 1854.
drogue iron
Arab (RP), May 28, 1851; *Columbus* (CtNlHi), July 18, 1853; *South Boston*
(CtWetHi), June 14, 1852.
sealskin poke
Minerva (RP), June 13, 1866.
examples of stimulus for development of darting gun
Samuel Robertson (MShaK), June 17, 1851; *Sophia Thornton* (RP), July 5, 1852, July 21, 1853; *Montreal* (MNBedfHi), June 18, 1852; *Niagara* (MNBedfHi), June 28, 1852; *Arab* (RP), June 7, 1852; *Menkar* (AuHS), May 30, 1852; *Oregon* (MNBedfHi), June 16, 1854.
darting gun history
Lytle 1984; A. J. Allen 1978:91; *Evening Journal* (New Bedford), February 7, 1891; *Pacific Commercial Advertiser* (Honolulu), November 23, 1867.
strike whale
Ansel 1978:20; *Alexander* (MNBedfHi), September 30, 1898; Brown 1887:252; Pedersen 1952:38, 40; West et al. 1965:23; *Rosie H.* (CaOONM), August 23, 1910; Bodfish 1936:91, 93, 127; *Herman* (CaOONM), August 12, 1910.
kill whale outright
Three Brothers (MNBedfHi), September 20, 1866; *Congress* (MShaK), May 11, 1867; *Montreal* (MNBedfHi), May 2, 1851; *Prudent* (MShaK), June 20, 1852; Pedersen 1952:38.
behavior of wounded whale
Bodfish 1936:97; *Whalemen's Shipping List* (New Bedford), May 3, 1853; *Arnolda* (PC), September 20, 1866.
"sound out" whaleline
N. Byron Smith, ms. 89.
"roll out" irons
Montreal (MNBedfHi), July 22 and 23, 1851.
whale escapes into ice
Arab (RP), May 28, 1851; *Tamerlane*
(IMA-MNBedfHi), May 26, 1851; *Corinthian* (MShaK), June 11, 1865; *Navy* (MNBedfHi), July 29 and 30, 1862; *Mercury* (MNBedfHi), August 31, 1865; *Cape Horn Pigeon* (MShaK), August 20, 1864; *Gayhead* (MNBedfHi), June 22, 1864; *Covington* (MNBedfHi), p. 53.
fighting whale
Mary D. Hume (MH-BA), September 3, 1893; *Roscoe* (MNBedfHi), September 28, 1867; West et al. 1965:62; DeWindt 1899:284; *Pioneer* (RPB), June 18, 1852; *Henry Taber* (MNBedfHi), August 16, 1870; *Arab* (RP), August 30, 1851; *Montreal* (MNBedfHi), July 22, 1851.
whale outwaits boats—too dark, too rugged
Montreal (MNBedfHi), July 20, 1852; *Helen Mar* (MNBedfHi), October 3, 1878.
boatsteerer misses whale
Emily Morgan (MNBedf), August 28, 1871; *Marcia* (MNBedfHi), September 26, 1860; *Navy* (MNBedfHi), July 24, 1862; *Helen Mar* (MNBedfHi), October 4, 1877; *Northern Light* (MNBedfHi), June 4, 1878.
poor darting by boatsteerer
Thrasher (PC), June 12, 1884; *Navarch* (MNBedfHi), September 4, 1894; *Trident* (MShaK), September 3, 1870; *Navy* (MNBedfHi), September 4, 1894; *Arab* (RP), July 10, 1851; *Northern Light* (MNBedfHi), September 17, 1875; *Helen Mar* (MNBedfHi), October 5, 1879; *Governor Troup* (RP), August 25, 1864; *Cape Horn Pigeon* (MShaK), August 18, 1864.
iron draws
Arab (RP), August 19, 1851; *Gayhead* (MNBedfHi), July 31, 1864; *Navy* (MNBedfHi), July 26, 1862; *William Baylies* (MNBedfHi), September 25, 1905; *Sophia Thornton* (RP), July 4, 1852; *Bowditch* (RP), July 11, 1850; *Francis A. Barstow* (RP), October 6, 1891.
whaleline parts
Courser (RP), June 1, 1867; *Montreal* (MNBedfHi), July 27, 1852; *Cornelius Howland* (RP), June 1, 1869; *Beluga* (MNBedfHi), May 6, 1897; *Navy* (MNBedfHi), May 23, 1863; *Harrison* (RP), June 28, 1851; *Saratoga* (MNBedfHi), August 21, 1851; *Helen Mar* (MNBedfHi), July 12, 1880; *Cambria* (RP), June 2, 1850.
whaleline draws from iron
Gayhead (MNBedfHi), June 2, 1858.
iron breaks
Pacific (RP), August 9, 1852; *Arnolda* (PC), September 7, 1866; *Marengo* (MSaP), June 26, 1858.
Sophia Thornton *mishap*
Sophia Thornton (RP), September 5, 1853.
guns frozen
Herman (CSfA), October 8, 1914.
shoot bow of boat off
Narwhal (MH-BA), August 27, 1905; *Gayhead* (MNBedfHi), July 9, 1857.
bomb blows out iron
Mary D. Hume (MH-BA), July 14, 1895.
guns wound and kill men
Emily Morgan (MNBedf), September 7, 1871; *St. George* (RP), September 28, 1867.
lost boats
Sophia Thornton (RP), July 24, 1852; *Majestic* (RP), July 7, 1858; *Arab* (RP), July 11, 1852; *Columbus* (RP), September 8, 1852.
whaleline fouls boat
John P. West (ViNeM), September 27, 1865; *Elizabeth Swift* (MNBedf), September 20, 1867; *Arab* (RP), September 2, 1852; *Beluga* (MNBedfHi), September 6, 1898; *John Wells* (MNBedfHi), September 19, 1870; *Mercury* (MNBedfHi), July 14, 1865; *Navy* (MNBedfHi), July 23, 1863.

whaleline fouls man
Active (DSI), August 26, 1868; *Gayhead* (MNBedfHi), August 14, 1867; *Betsey Williams* (MNBedfHi), July 28, 1852; *Beluga* (MNBedfHi), August 26, 1898.
effectiveness of bombs on whale
Herman (CaOONM),August 12, 1912; Bodfish 1936:97.
"Nantucket sleigh ride"
Brown 1887:267-68; Ansel 1978:23; Mason 1900:132; W. N. Burns 1913:202.
spade flukes
Brown 1887:269.
"bowing on"
Ansel 1978:23.
"wood to blackskin"
Ashley 1926:146.
"aft the foreshoulder"
N. Byron Smith, ms., 88.
poke whale's eye
Brown 1887:269.
waif dead whale
West et al. 1965:27; Tilton 1969:60.
tow dead whale
Brown 1883:19; Brown 1887:270.
sunk whale
Rosie H. (CaOONM), August 25, 1910; *Pioneer* (RPB), May 22, 1852; Davidson 1868:477.
messenger iron
Narwhal (MN-BA), August 27, 1905, May 13, 1908; Bodfish 1936:173; *Northern Light* (MNBedfHi), October 21, 1876; *Narwhal* (MEdDHi), August 4, 1898; Cook 1926:114.
haul up sunken whale
Prudent (SHS), July 8, 1851; *Narwhal* (MH-BA), May 13, 1908; *Eagle* (MNBedfHi), July 3, 1867; *Three Brothers* (MNBedf), July 28, 1866; *William Baylies* (MNBedfHi), October 6, 1907. Unidentified clipping, scrapbook 16, p. 190 (MNBedfHi).
wait for whale to "blast"
Arnolda (PC), September 22-26, 1864; Macy, 1871:461-62.
whales killed by killer whales
Marengo (MShaK), June 8, 1852; *Congress* (MShaK), July 6, 1865; Brown 1887:284; *Enterprise* (MEdDHi), August 8, 1851; *Columbus* (RP), June 22, 1852; *Florida* (MNBedf), May 21, 1862; *Mercury* (MNBedfHi), August 26, 1865; *Milo* (MNBedfHi), June 28, 1852; *Navy* (MShaK), July 7, 1863; *South America* (PC), June 20, 1858.
dead whale found killed by natives
Betsey Williams (MNBedfHi), July 8, 1852.
customs regarding ownership of dead whales
Williams 1964:368-69; Brown 1887:251; *Roman* (MShaK), June 19-20, 1854; *South Boston*
(CtWetHi), July 3, 1852; *Two Brothers* (RP), July 9, 1852; *Minerva* (RP), June 13, 1866; *Covington* (MNBedfHi), p. 57; *Horatio* (PC), July 9, 1891; Smith 1978:77.
some examples of picking up dead whales
Benjamin Tucker (MShaK), July 24, 1850; *Samuel Robertson* (MShaK), July 12, 1850; *Montreal* (MNBedfHi), June 30, 1852; *Gayhead* (MNBedfHi), June 23, 1858; *Pioneer* (RPB), July 7 and 8, 1852, June 16, 1853; *Cornelius Howland* (RP), October 8, 1868; *Cambria* (RP), June 8, 1850; *Tiger* (CtMyMHi), August 19, 1849.
whale behavior in close ice
Bodfish 1936:38.
strike whale from ship
Cornelius Howland (RP), June 1, 1865; *George Howland* (MNBedfHi), June 20, 1869; *Helen Mar* (MNBedfHi), June 11, 1877; *Cornelius Howland* (MNBedf), June 2, 1870; *Beluga* (MNBedfHi), May 14, 1903; *Governor Troup* (RP), June 28, 1864; *Governor Troup* (MH-BA), June 28, 1864; *Cape Horn Pigeon* (MShaK), May 24, 1865; *Eagle* (MNBedfHi), May 6, 1868.
send men across ice on foot
Thrasher (MH-BA), April 12, 1896; Williams 1964:349; *Henry Taber* (MNBedfHi), May 18, 1870; *Belvedere* (MNBedfHi), April 19, 1885; West et al. 1965:36; *Trident* (MShaK), May 18, 1870; *William Baylies* (MNBedfHi), May 14, 1890; *Cornelius Howland* (RP), May 10, 1870.
lower boats onto ice
John Carver (MNBedfHi), May 22, 1885; *Cornelius Howland* (RP), April 24, 1868, May 12, 1868; Tilton 1969:249.
whale takes line in ice
Beluga (MNBedfHi), August 11, 1898; C. T. Pedersen 1952:38-39; *John Carver* (MNBedfHi), May 7, 1886.
sealskin floats
Vineyard Gazette (Martha's Vineyard), April 27, 1934; C. T. Pedersen 1952:39; *Aurora* (MNBedf), August 22, 1870; *Orca* (PC), June 6, 1883.
haul whale out of ice with windlass
Helen Mar (MNBedfHi), September 26, 1877; *Vineyard Gazette* (Martha's Vineyard), April 27, 1934.
tow whale out of ice
Montreal (MNBedfHi), June 25, 1852; *Gayhead* (MNBedfHi), June 22, 1864; Jernegan, ms. (MEdDHi); *Mercury* (MNBedfHi), June 20, 1865.
cut in whale in ice
Lagoda (MNBedfHi), August 20, 1872; *Three Brothers* (MNBedf), May 8, 1868; *Cornelius Howland* (RP), May 12, 1868; Jernegan, ms. (MEdDHi).
walrusing
See chapter 6, this volume.
gray whaling
Oregon (MNBedfHi), August 27, 1854; *Ocean* (MNBedfHi), ms. map, 1860s; *George Howland* (MNBedfHi), July 18, 1868; *Thomas Dickason* (RP), June 1871 passim; *Frances* (MNBedfHi), July 3, 1851; *Gayhead* (MNBedfHi), June 22, 1864; Williams 1964:376; Berzin and Rovnin, nd:127; Roys, ms., "Descriptions of whales"; Scammon 1874:21; Clark 1887a:24.
Greener gun
George Howland (MNBedfHi), July 19, 1868; *Cornelius Howland* (MNBedf), May 13, 1868.
humpbacking
Omega (RP), June 9, 1849; *Phillipe de la Noye* (NHyF), July 16, 1851; *Mount Vernon* (MShaK), September 5, 1851; *Nassau* (MShaK), July 15, 1851; *Charles Phelps* (RP), July 6, 1851; *Roman* (MShaK), September 7 and September 25, 1854; *Majestic* (RP), August 15, 1858; *Jireh Swift* (MNBedf), July 30, 1858; *Goethe* (RP), August 8 and 9, 1858; *South America* (MShaK), June 28, 1858; *Polar Star* (MNBedf), June 9, 1858; *Benjamin Tucker* (CtMyMHi), September 8, 1858; *Monte-*

zuma (MNBedfHi), September 19, 1859; *Navy* (MShaK), September 19, 1860; *Awashonks* (MNBedf), October 14, 1867; *Marengo* (MShaK), October 8, 1867; *President* (RP), July 16, 1868; *Roscoe* (MNBedfHi), July 6, 1872; *Mount Wollaston* (MNBedfHi), June 27, 1877; Roys, ms., "Descriptions of Whales"; Scammon 1874:39, 44-46; Berzin and Rovnin, nd:110-14; Stevenson 1904:189.

right whaling
Berzin and Rovnin, nd:127-29; Williams 1964:347, 356; Brown 1887:263; Roys, ms. "Descriptions of Whales."

finback whales
Narwhal (CaOTCHAr), September 26, 1901; *Prudent* (SHS), May 29, 1851; *American* (MShaK), June 13, 1852; *Gayhead* (MNBedfHi), May 22, 1858; Roys, ms., "Descriptions of Whales"; Berzin and Rovnin, nd:115-19; N. Byron Smith, ms., 43; Stevenson 1904:192.

killer whales
Roys, ms., "Descriptions of Whales"; *John Wells* (MNBedfHi), April 24, 1871; *Robin Hood* (CtMyMHi), August 23, 1859; *Samuel Robertson* (MShaK), May 27, 1851; *Marengo* (MShaK), June 8-9, 1852; *Cornelius Howland* (RP), July 21, 1870; *Progress* (MNBedfHi), September 14, 1881.

porpoise
Helen Mar (MNBedfHi), June 4, 1872; *John Wells* (MNBedfHi), May 10-11, 1870.

giant bottlenose dolphin
Majestic (RP), May 25, 1858.

"oleaginous blanket"
Brown 1887:227.

songs at the windlass
Brown 1887:282; W. N. Burns 1913:206-7.

"Whiskey for the Johnnies"
W. N. Burns 1913:206-7.

cutting in
Brown 1887:277-85; Scammon 1874:231-36; *Cornelius Howland* (RP), April 18, 1868; *Covington* (MNBedfHi), pp. 55-57; Aldrich 1889:93-96; W. N. Burns 1913:140-41; Cook 1926:38-41.

blanket piece kills captain
China (MNBedfHi), September 25, 1853.

time required for cutting in
Frances (MNBedfHi), September 7, 1851; *Samuel Robertson* (MShaK), May 18, 1850; *Montreal* (MNBedfHi), June 20, June 25, August 1, 1852.

build blubber room
Columbus (RP), June 19, 1852; *Helen Mar* (MNBedfHi), May 30, 1872; *Europa* (MNBedfHi), July 1, 1870; *Lagoda* (MNBedfHi), August 7, 1869, August 7, 1871; *Courser* (RP), July 23, 1866; May 8, 1867.

remove headbone
Belvedere (CaOONM), August 11, 1912; Bodfish 1936:95-96; W. N. Burns 1913:140-41; Williams 1964:352-53; Taylor 129:62.

section headbone
Narwhal (CaOTCHAr), September 1, 1901.

scrape gum from whalebone
Roscoe (MNBedfHi), June 25, 1867; *Narwhal* (CaOTCHAr), October 29, 1901; *Horatio* (PC), July 29, 1891; *Seneca* (MNBedfHi), October 1870, passim; W. N. Burns 1913:240; Aldrich 1889:47.

soak whalebone
Cook 1926:238; *Rosario* (MEdDHi), July 7, 1897.

wash whalebone
Narwhal (CaOTCHAr), October 30, 1901; *Rosario* (MEdDHi), July 8, 1897.

dry whalebone
Brower, ms., 437; *Thrasher* (PC), June 19, 1884; *Helen Mar* (MNBedfHi), July 19, 1878; *Examiner* (San Francisco), November 12, 1888.

bundle whalebone
Helen Mar(MNBedfHi),July 19, 1878; *William Baylies* (RP), June 26, 1888; *Mary D. Hume* (MH-BA), spring 1894, passim.

rats eat whalebone
California (MShaK), August 22, 1895.

smoke ship
Cook 1926:35: *Helen Mar* (MNBedfHi), August 1, 1877.

"hurrah for five and forty more"
Aldrich 1889:131; *Narwhal* (CaOTCHAr), September 1, 1901.

problems of cutting in
Bowditch (RP), June 18, 1850; *Saratoga* (MNBedfHi), September 16, 1851; *Montreal* (MNBedfHi), July 19, 1851, June 27, 1852; *Sophia Thornton* (RP), July 7, 1853; *Charles Phelps* (RP), May 19, 1851; *Helen Mar* (MNBedfHi), October 1, 1879; *Thrasher* (PC), September 16, 1886; *Whalemen's Shipping List* (New Bedford), May 3, 1853; *Marengo* (MSaP), July 20, 1858; *Navy* (MNBedfHi), August 29, 1862; *Cambria* (IMA-MNBedfHi), July 1, 1852; *Java II* (RP), September 9 and 10, 1858.

scrap press
Triton (MNBedf), August 6, 1888.

mincing and boiling
Scammon 1874:238-39; Brown 1887:278-79, 285-86; Stevenson 1904:193-95; W. N. Burns 1913:146-47; Aldrich 1889:97-98; Pease et al. 1889:50.

mincing machine
Cornelius Howland (RP), October 2, 1868.

"dry skins"
George and Mary (MShaK), May 16, 1852; *Adeline* (MNBedf), August 1866; *Betsey Williams* (MShaK), June 13, 1852; *Arab* (RP), June 10, 1852; *Arnolda* (PC), September 26, 1864; A. H. Clark 1887a:17; Lawrence 1966:108-9; Scammon 1874:18; Stevenson 1903:189.

oil from young whales
Thrasher (PC), September 16, 1886; Stevenson 1903:189; *Saratoga* (MShaK), August 20, 1858.

cooling tank between decks
Thomas Dickason (RP), October 15, 1870; *Grampus* (MH-BA), May 29, 1899.

coopering and stowing oil
Eagle (MNBedfHi), October 18, 1867; Brown 1887:289; *George Howland* (MNBedfHi), May 20, 1868.

time required for boiling
Cook 1926:41.

"blubber logged"
Scammon 1874:238; *Arab* (RP), September 6, 1851; *Menkar* (AuHS),

August 16, 1852; *Coral* (MNBedfHi), August 17, 1880; Beane 1905:212.
smoke from tryworks
Belvedere (CaOONM), September 7, 1912.
weight of blubber
Stevenson 1904:190.
color of oil
Trident (MShaK), August 28, 1870; *Roman* (MShaK), September 7, 1854; Stevenson 1903:189; Stevenson 1904:202-3.
full ship
Shepherdess (RP), July 6, 1850; *Pioneer* (RP), August 24, 1850; *Java II* (RP), August 11, 1858.

Chapter 5

rush to Bering Strait
Franklin (PC), May 26, 1849.
number of whaleships in North Pacific fishery
Whalemen's Shipping List (New Bedford), January 6, 1852.
Roys's Arctic cruise in 1848
The Friend (Honolulu), November 1, 1848.
Northwest Coast fishery statistics
Whalemen's Shipping List (New Bedford), January 9 and 16, 1849.
Okhotsk and Baja California fisheries
David A. Henderson, personal communication.
high point of American whaling industry
Scammon 1874:213.
whaleships visit Petropavlovsk in 1849
Kellett 1850:9.
activities of fleet in 1849
Starbuck 1878:99; unidentified newspaper clipping, scrapbook T-2 (MNBedfHi), *The Friend* (Honolulu), October 15 and November 1, 1849; *Whalemen's Shipping List* (New Bedford), February 5, 1850; *Ocmulgee* (MNBedfHi), *Majestic* (MSaP); *Marengo* (MNBedf); *Tiger* (CtMyMHi); *Alert* (CtMyMHi), *Awashonks* (MNBedf); *Franklin* (NEh); *Catherine* (MShaK).
"barren as Buzzards Bay"
Omega (RP), June 8, 1849.
fear and "scowling brows"
Tiger (CtMyMHi), June 26 and July 20, 1849.
Joseph Dias, Jr.
Ocmulgee (MNBedfHi), July 7, 1849.
"hurry of whaling"
The Friend (Honolulu), November 1, 1849.
morphological variation among bowheads
Braham, Durham, et al. 1980.
size of large bowheads
The Friend (Honolulu), November 1, 1848; *Majestic* (MSaP), July 26, 1850; unidentified newspaper clipping, scrapbook T-6, p. 147 (MNBedfHi); unidentified newspaper clipping, scrapbook T-1, n.p. (MNBedfHi), anonymous 1868; Scammon 1874:58; Stevenson 1903:192; Bodfish 1936:95; Vermilyea, 1958:322; *Nimrod* (MNBedfHi), July 5, 1860; unidentified newspaper clipping, scrapbook T-4, p. 178 (MNBedfHi); C. T. Pedersen 1952:40-41; diary, John Bertonccini, Liebes Collection (CSfA); Aldrich 1889:99; *Chronicle* (San Francisco), October 28, 1899; Cameron 1910:286.

events of 1850
The Polynesian (Honolulu), October 19, 1850.
events of 1851
Gelett 1917:76.
"scarse and unusually wild"
Montreal (MNBedfHi), June 20, 1851.
"stove most to pieces"
Whalemen's Shipping List (MNBedfHi), November 25, 1851.
departure from grounds in July and August 1851
Montreal (MNBedfHi), August 11, 1851; *Daniel Webster* (MNBedfHi), July 29, 1851.
"fog, ice and broken down hopes"
Nassau (MNBedfHi), August 17, 1851.
events of 1852
Whalemen's Shipping List (New Bedford), December 7, 1852; *Prudent* (MShaK), May 29, 1852.
events of 1853 season
The Friend (Honolulu), December 8, 1854; *Whalemen's Shipping List* (Lahaina), December 1, 1853.
events of 1854 season
Maguire 1854-55:913; *The Friend (Honolulu),* December 8, 1854; *Whalemen's Shipping List* (New Bedford), November 28, 1854, January 9, 1855; Richard Collinson, journal of HMS *Enterprise,* August 12, 1854 (NMM).
"binnacle lights burning"
Dulles 1934:236.
"Joy to the world"
Roman (MShaK), June 4, 1854.
events of 1855
Oregon (MNBedfHi), *Saratoga* (CU-BANC); *The Friend* (Honolulu), September through December 1855 and January 1856, passim.
events of 1856
Pacific Commercial Advertiser (Honolulu), December 18, 1856; *The Friend* (Honolulu), September through December 1856 and January 1857, passim; anonymous 1858:603.
depletion and southern range of bowhead population
Bockstoce and Botkin 1983, Table 2; *Ocmulgee* (MNBedfHi), May 1849, passim.
learning behavior of bowhead in response to threat
Bodfish 1936:90, 191, 203; *Whalemen's Shipping List* (New Bedford), December 2, 1851.
"slow and sluggish beast"
Whalemen's Shipping List (New Bedford), July 12, 1853.
"don't like cold iron"
Bowditch (RP), June 12, 1850.
"deserted village"
Whalemen's Shipping List (New Bedford), December 2, 1851.
hearing of bowhead and evasive maneuvers
Bodfish 1936:94; *William Baylies* (MNBedfHi), September 11 and 27, 1905; Hendee 1930:386.
"whistling whale"
A. J. Allen 1978:100-102.
the Montezuma*'s iron*
Dall 1899:136-7.
events of 1857
The Friend (Honolulu), September 26, 1857; unidentified newspaper

clipping, scrapbook T-1, n.p. (MNBedfHi); unidentified newspaper clipping, scrapbook T-2, p. 65 (MNBedfHi).

events of 1858

Pacific Commercial Advertiser (Honolulu), October 14, 1858; *South America* (PC), July 2, 1858; *Java* (PC), July 1859, passim; *Saratoga* (MShaK), July and August 1858, passim; *South America* (MShaK), July and August 1858, passim; *Majestic* (RP), August 29, 1858; *Gayhead* (MNBedfHi), July 19, 1858; *The Friend* (Honolulu), November 8, 1858.

"Arctic ground is a humbug"

Pacific Commercial Advertiser (Honolulu), October 14, 1858.

events, 1859 through 1864

Unidentified newspaper clipping, scrapbook T-2, p. 190 (MNBedfHi); *Marcia* (MNBedfHi), June through September 1859, passim; *Covington* (RP), July 22, 1859; *The Friend* (Honolulu), October 11, 1859, November 18, 1861, October 1, 1862; *Pacific Commercial Advertiser* (Honolulu), October 25, 1860, October 10, 1861.

Chapter 6

Throughout this chapter I have relied primarily on the logbooks and journals kept aboard CSS *Shenandoah* and several whaleships. Also useful have been Cornelius Hunt's and James Waddell's accounts of events aboard the raider. A large amount of newspaper and other printed material is available on the cruise of the *Shenandoah*, but I have found that the accuracy of the personal accounts of participants in those dramatic months decreased in proportion to the distance of years from the actual events when the narratives were written. Where conflicts with dates have arisen I have relied on the *Shenandoah's* logbook.

events of June 28, 1865

Brunswick (MNBedf); unidentified newspaper clipping, scrapbook T-1, May 26, 1907 (MNBedfHi); *James Maury* (MNBedf), 1867:196-208.

outbreak of war and outfitting of Sea King

Bulloch 1959:II, 122-29, 197-98; Scharf 1977:783; U.S. Secretary of the Navy, *Official Records of the Union and Confederate Navies* (hereafter cited as "Official Records"), 2d series, II:613; Official Records, 1st series, III:10-11, 445-46, 750;

Whittle 1910:6; J. T. Mason 1898:610; Stern 1962:151-52; Waddell, nd:1.

Laurel *and* Shenandoah *at Madeira*

J. T. Mason 1898:603; *Bulloch's orders*—Official Records, 1st series, III:749. Bulloch came to this conclusion after having been given a set of Commander Matthew Fontaine Maury's "Whale Charts," which located the major whaling grounds throughout the world. Maury, who joined the Confederate Navy, had compiled these charts by extracting data from whalers' logbooks while he was hydrographer of the U.S. Navy.

operations before reaching Tristan da Cunha

Waddell, nd:6-9, 20-29; Hunt 1867:29-44 and passim; *Shenandoah* (Nc-Ar); Whittle 1910:10; Official Records, 1st series, III: 403-4, 749-55; Cook 1975:15; U.S. Senate, *Correspondence Concerning claims Against Great Britain Transmitted to the Senate of the United States . . .* (hereafter referred to as "Claims Against Great Britain"), III:316-21; Chew, ms.

voyage to Melbourne

Waddell, nd:30-57.

events in Melbourne

Claims Against Great Britain, V:598-9; Claims Against Great Britain, III:391-420, 477-91; Official Records, 1st series, III:761-70; U.S. Senate *The Case of the United States, to be laid before the Tribunal of Arbitration, to the convened at Geneva . . . ,* 42d Congress, 2d session, Executive document no. 31, 177; Waddell, nd:78; Hunt, 1867:113-15.

Melbourne to Ascension

Morgan 1948:157; *Whalemen's Shipping List* (New Bedford), July 25, 1865; Waddell, nd:82; unidentified newspaper clipping, scrapbook 2, December 8, 1890 (MNBedfHi); Honolulu Consular Dispatches, November 20, 1865, (DNA); Hunt 1867. Great Britain paid the owners of the *Harvest* for her loss.

Okhotsk Sea

Hunt 1867:152-59; Waddell, nd:49; Claims Against Great Britain, III, 482.

events through June 22, 1865

Hunt 1867:171 and passim; Waddell, nd.

June 23, 1865

Hunt 1867:173-81; Waddell, nd:105-8; *Shenandoah* (Nc-Ar); *Mercury* (MNBedfHi), June 15, 1865; the *Standard* (New Bedford), December 12 and 22, 1865. Johnston surrendered the Southern Army to Sherman on April 26; Jefferson Davis was captured in Georgia on May 10; and in Shreveport on May 26, Kirby Smith surrendered the last Confederate army.

June 24 to June 26, 1865

Mercury (MNBedfHi); *Minerva* (RP); *Splendid* (IMA-MNBedfHi); unidentified newspaper clipping, scrapbook T-2, p. 123 (MNBedfHi); Chew, ms.; Hunt 1867:192; unidentified newspaper clipping, scrapbook T-4, p. 161 (MNBedfHi).

June 27 and 28, 1865

Chew, ms.; Starbuck 1964:102-3; Hunt 1867:202; *James Maury* (MNBedf); account of John Heppingstone, Agard Papers, American Antiquarian Society, Worcester, Mass.; and see, for instance, Pasquier 1982:190. The bark *Congress 2nd* should not be confused with the ship *Congress* (MShaK), also of New Bedford and also in Bering Strait that year.

ships scatter

Splendid (IMA-MNBedfHi); *Governor Troup* (RP, MH-BA); *Oliver Crocker* (IMA-MNBedfHi); *Eliza Adams* (MShaK); bark *Martha* (MNBedf); *Cape Horn Pigeon* (MShaK); *Minerva* (RP); *Almira* (MNBedfHi); *Mercury* (MNBedfHi); *Cleone* (MNBedfHi); *John P. West* (ViNeM); *Camilla* (MNBedf); *Corinthian* (MShaK); ship *Congress* (MShaK); *Cornelius Howland* (RP); *Joseph Maxwell* (MSaP); Hooper 1881:47.

Waddell's departure from Arctic

Hunt 1867:208; Waddell, nd:113-15; Chew, ms., July 9, 1865.

Shenandoah in the Pacific

Chew, ms., August 6, 1865.

news of and response to Shenandoah

Official Records, 1st series, III:776-77; unidentified newspaper clipping, scrapbook T-2, p. 203 (MNBedfHi); *The Friend* (Honolulu), September 1, 1865; Gilbert 1965:175-76, 178.

Shenandoah*'s return to England*

Bulloch, 1959, II:152-53; A. Cook 1975:31; *The Times* (London), November 8, 1865, quoted in *The World* (New York), November 21, 1865.

Geneva arbitration and Alabama claims

A. Cook 1975; Hackett 1882:iii; Official Records, 1st series, III:749.

analysis and subsequent events

Bulloch 1959, II:197-98; *California* (MNBedf), June 18, 1867; Purrington 1961:3; Horan 1960:188-89; Kugler 1971:26; Starbuck 1964:692.

Chapter 7

natural history and description

Fay 1957, 1982; *South Boston* (CtWetHi), August 17, 1852; Francis Fay, personal communication; J. J. Burns 1970:446-47; J. J. Burns 1965; John Burns, personal communication; Sam Stoker, personal communication; J. A. Allen 1880; Bee and Hall 1956; author's field notes, 1969-77.

development of walrus hunting

Bockstoce and Botkin 1883; *Tiger* (CtMyMHi), July 23, 1849; *Samuel Robertson* (MShaK), July 12, 1850; *Montreal* (MNBedfHi), June 7 and 8, 1852.

elephant sealing

Scammon 1874:118; A. H. Clark 1887c:436-37; Laws 1979; Le Boeuf 1979; King 1964:78-83.

decline of bowhead population

Bockstoce and Botkin 1983.

commencement of intensive walrusing

Cleone (MNBedfHi); *Whalemen's Shipping List* (New Bedford), February 1, 1870; Bockstoce and Botkin 1983; Williams 1964:226; *Trident* (MShaK), July 16, 1870.

gray whales

Henderson 1972:184-85; David Henderson, personal communication; Henderson 1984:159-85.

whale oil prices

Tower 1907:128-29.

ivory prices

Pacific Commercial Advertiser (Honolulu), November 27, 1869; *Commercial Herald and Market Review* (San Francisco), January 31, 1873; A. H. Clark 1887b:317.

walrusing equipment

Helen Mar (MNBedf), August 11, 1868; *Europa* (MNBedfHi), July 11, 1870; *Whalemen's Shipping List* (New Bedford), November 29, 1870; *Northern Light* (MNBedfHi), May 5, 1876; *Cornelius Howland* (RP), July 1, 1869, and May 2, 1870.

walrus boats

North Star (PC), May 21, 1882; *Helen Mar* (MNBedfHi), August 3, 1872, and May 12, 1874; *Leo* (MNBedfHi), October 6, 1879; *Rainbow* (PC), June 25, 1883; *Sea Breeze* (RP), June 14, 1870; *Northern Light* (MNBedfHi), May 20, 1876; *North Star* (MNBedfHi), May 21, 1882.

hunt

A. H. Clark 1887b:314-15; Williams 1964:226; Hooper 1884:46; Bodfish 1936:20-21.

cutting in

Bodfish 1936:20-21; A. H. Clark 1887b:314-17; Stevenson 1903:214.

mincing, boiling, and stowing

Cornelius Howland (RP), July 3, July 11, and July 25, 1870; *Helen Mar* (MNBedfHi), June 26, 1875, and July 13, 1878; *Northern Light* (MNBedfHi), June 2, 1876; U.S. Commission of Fish and Fisheries 1904:215; A. H. Clark 1887b:317.

conclude walrusing season

Cornelius Howland (MNBedf), August 4, 1869; Williams 1964:226.

ivory market

A. H. Clark 1887b:317-18; Penniman, nd; *Commercial Herald and Market Review* (San Francisco), January 31, 1873; *Pacific Commercial Advertiser* (Honolulu), November 27, 1869.

history of suppression of population

Bockstoce and Botkin 1983; *Whalemen's Shipping List* (New Bedford), August 20, 1878; *Bulletin* (San Francisco), November 8, 1882; Coachman, Aagaard, and Tripp 1975:149; Murdoch 1885:97-98.

outcry against hunt

F. A. Barker journal; Wilkinson, nd; E. S. Allen 1973:200ff; *Whalemen's Shipping List* (New Bedford), April 8, 1872; "Shipmaster," *New Bedford Standard,* reprinted in *The Friend* (Honolulu), March 1872; *The Friend* (Honolulu), November 1, 1871; *Hawaiian Gazette* (Honolulu), November 1, 1871.

winter of 1877-78

Pacific Commercial Advertiser (Honolulu), October 4, 1878.

Saint Lawrence Island starvation

Leo (MNBedfHi),1879; *Pacific Commercial Advertiser* (Honolulu), September 6, 1879; G. W. Bailey 1880;19; Hooper 1881:11; Hooper 1884:8-9, 22-23; Muir 1917:117-23; Nelson 1899:269; Wardman 1884:147-51; unidentified clipping (Ebenezer Nye's letter), p. 12, Bartlett scrapbook (MNBedfHi).

There is no evidence that the Eskimos had stills on the island at the time; nor did any of the observers, including Irving C. Rosse, M. D. (Rosse 1884), attribute the cause of the deaths to epidemic disease.

hardship at other settlements

East Cape, W. H. Dall, Hydrographic Notes, September 12, 1880 (DSI); Nelson 1899:296; Plover Bay, Krause and Krause 1882:130-35; Indian Point, Krause and Krause 1881:263-64, Aldrich 1889:54; Saint Lawrence Bay, *North Star* (PC), May 27, 1882; King Island, DeWindt, 1899:191, Report to U.S. Commissioner for Education for 1890-91:945; M. A. Healy to Secretary of the Treasury, report of U.S.R.C. *Bear,* 1891, Healy Papers, Sheldon Jackson College, Sitka, Alaska; Point Barrow, P. H. Ray 1885:46, 48; Woolfe 1893:146.

calls for end of walrus hunting

Chronicle (San Francisco), January 29, 1893; unidentified newspaper clipping, scrapbook 15, p. 66 (MNBedfHi).

prohibition of commercial hunt

Department of Commerce, Office of the Secretary, Department Circular no. 286, Bureau of Fisheries, Alaska Fisheries Service, April 21, 1921.

end of whalers' hunt and subsequent events

Murdoch 1885:97-98; Fay 1957; T. Jones 1979:85; Fay et al. 1977; Francis Fay, personal communication.

Chapter 8

cruising pattern in 1850s and 1860s

The Friend (Honolulu), October 11, 1859, November 18, 1861; *Pacific Commercial Advertiser* (Honolulu), October 10, 1861.

Thomas Long
Schmitt, de Jong, and Winter 1980:23; Bertrand 1971:229.
Wrangel Island before 1867
John Howland (MNBedfHi), September 10, 1862; *Adeline* (MNBedf), August-September 1863; *Navy* (MNBedfHi), August 11 and 12, 1863; *Thomas Dickason* (RP), August 27, 1863; *Stephania* (MSaP), September 28, 1865; *John P. West* (ViNeM), August 12, 1865; *Helen Mar* (MNBedf), July 27, 1866; *Camilla* (MNBedf), July 27, 1866; *Adeline* (MNBedf), August 10, 1866; *Pacific Commercial Advertiser* (Honolulu), November 4, 1865.

In 1881 a German whaleman, Captain Edouard Dallman, claimed to have been the first to have landed on the island, when he sailed there aboard the Honolulu trading schooner *W. C. Talbot* in 1866. His claim, however, was rejected by most of the world's geographical societies both because his description of his voyage contained some obvious inaccuracies and because it conflicted with the descriptions of those who did land there in 1881 (Royal Geographical Society 1881; Nordenskiod 1882:I,448-49; Spengemann 1952:152-59).
Wrangel Island, 1867
President (RP), June 4, 1867; *Courser* (RP), July 16, 1867; *Martha* (MNBedf), August 24, 1867; *John P. West* (ViNeM), September 2, 1867; *George Howland* (MNBedfHi), July 15, 1867; *James Maury* (MNBedf), August 1, 1867; Lebedev and Grekov 1967:170; *The Friend* (Honolulu), December 2, 1867; *Pacific Commercial Advertiser* (Honolulu), November 9, 1867; Alaska Commercial Company to Nordenskiold, April 3, 1878, Nordenskiold Papers (University Library, Stockholm, Sweden); Wrangell 1840; Seemann 1853:114-17.
Wrangel Island after 1867
I have records of the following ships having sighted the island; of course, it is likely that more did so: *Japan*, 1870 (Barker, ms., 76); *Helen Mar* (MNBedfHi), October 3, 1874; *Belvedere* (a landing), August 1881; *The Chronicle* (San Francisco), August 16, 1882; *Navarch* (MNBedfHi), September 14, 1893; *Grampus* (MH-BA), September 18, 1900; *Alexander* (MNBedfHi), October 4, 1902; *Beluga* (MNBedfHi), October 3-9, 1902; *Beluga* (MNBedfHi), October 7, 1903; *Alexander* (MNBedfHi), September 17, 1903; *Bowhead* [II] (CaACG), October 7, 1906; *Belvedere* (MShaK), September 6, 1907; *Polar Bear*, August 24, 1910, Lane to Anderson, February 14, 1928, R. M. Anderson Papers (CaOONM); *Herman* (MNBedfHi), October 4, 1911.
cruise in northeastern Chukchi Sea
Aurora (MNBedf), 1866-70.
1868
Cleone (MNBedfHi), 1861; *Eagle* (MNBedfHi); *Aurora* (MNBedf); *Cornelius Howland* (RP); *Corinthian* (MNBedf); *Henry Taber* (RP), July 21, 1869; *Active* (DSI), September 22-24, 1868; *The Friend* (Honolulu), November 2, 1868; *Pacific Commercial Advertiser* (Honolulu), October 31, 1868; ms. map, *Ocean*, 1868 (MNBedfHi).
1869 and 1870
Eugenia (MNBedf); *California* (MNBedf); *Cornelius Howland* (RP); *Hibernia* (MNBedfHi); *Aurora* (MNBedf); *Eagle* (MNBedfHi); *Thomas Dickason* (RP); *Seneca* (MNBedfHi); *Onward* (RP); *Trident* (MShaK); *Sea Breeze* (RP); *John Wells* (MNBedfHi); *Fanny* (MNBedf); *Henry Taber* (MNBedfHi); *Henry Taber* (PC); *Navy* (MShaK), August 28, 1870; *The Friend* (Honolulu), November 1, December 1, 1869; *Pacific Commercial Advertiser* (Honolulu), October 30, 1869, November 5, November 12, 1870; Williams 1964:367; E. Simpson 1890:18.
April—June, 1871
Harper's Weekly (New York), December 21, 1871; *Navy* (MShaK); *Seneca* (MNBedfHi); *Thomas Dickason* (RP); Wilkinson, nd:282.
Oriole
Barker, ms.; 122; *Emily Morgan* (MNBedf), July 16, 1871; Williams 1964:352; David A. Henderson, personal communication.
July—September, 1871
Comet (HHB); *Emily Morgan* (MNBedf); *Eugenia* (MNBedf); *Fanny* (MNBedf); *Gay Head* (MNBedfHi), *Henry Taber* (MNBedfHi); *Henry Taber* (PC); *John Wells* (MNBedfHi); *Lagoda* (MNBedfHi); *Seneca* (MNBedfHi); *Thomas Dickason* (RP); *Thomas Dickason* (PC); *Harper's Weekly* (New York), December 2, 1871.
ice and weather, 1871
Winchester and Bates 1957; Swithinbank 1960; Barnett 1976; Coachman, Aagaard, and Tripp 1975; Zubov 1947:336-57.
abandonment, 1871
E. W. Hall 1981; Jernegan, ms. (MEdDHi); *Emily Morgan* (MNBedf); Ellis 1892:427; *Lagoda* (MNBedfHi); Williams 1964:221-40; *Fanny*, typescript, 109 (MNBedfHi); James Dowden, unidentified newspaper clipping, scrapbook 2, p. 77 (MNBedfHi); *The Friend* (Honolulu), November 1, 1871.

From north to south in approximate order the abandoned ships were: *Roman, Comet, Concordia, Gay Head, George, John Wells, Massachusetts, J. D. Thompson, Contest, Elizabeth Swift, Henry Taber, Emily Morgan,* "big" *Florida, Champion, Oliver Crocker, Navy, Seneca, Fanny, Reindeer, Monticello, George Howland, Paiea, Carlotta, Kohola, Eugenia, Awashonks, Julian, Thomas Dickason, Minerva, William Rotch, Victoria,* and *Mary*.
cost of 1871 disaster and payments
"Old Dartmouth Historical Sketches," no. 45, 1916:42-47; Hohman 1928:293-94; Ellis 1892:431; *Hawaiian Gazette* (Honolulu), October 2, 1872.
wisdom of abandonment
The Friend (Honolulu), November 9, 1871.
salvage wrecks, 1872
Alta California (San Francisco), October 15, 1872; *Pacific Commercial Advertiser* (Honolulu), November 23, 1872; *The Chronicle* (San Francisco), September 25, October 16, 1872; Snow 1910:53-55; *The Friend* (Honolulu), September 7, 1872; *Arnolda* (MNBedfHi-IMA); *Louisa* (MNBedfHi); *Roscoe* (MNBedfHi); *Helen Snow* (MNBedfHi); *Helen Mar* (MNBedfHi); *Triton* (MNBedf); *Marengo* (MSaE); *Lagoda* (MNBedfHi); *Northern Light* (MNBedfHi); *Tamerlane* (MNBedfHi-IMA); *The Commercial Herald and Market Review* (San Francisco), January 31, 1873; *Whalemen's Shipping List* (New Bedford), October 1, November 26, 1872; Hare 1960:46; Starbuck 1964:108-9; unidentified newspaper clipping, scrapbook T-2, p. 123 (MNBedfHi); C. D. Brower ms., nd:231.
wreck of Japan
Wilkinson, nd; *The Friend* (Honolulu), November 9, 1871.
whale oil substitutes
R. J. Forbes 1958:102-3; Williamson and Daum, 1959:33-81, 725; *Whalemen's Shipping List* (New Bedford), June 22, June 29, September 28, December 21, 1852, February 8, 1853, July 2, 1861; "The supply of oil," 1856, unidentified newspaper clipping, scrapbook T-1 (MNBedfHi); Ellis 1892:422-23; Hohman 1928:295; A. H. Clark

1887a:148-53, 163; *The California Farmer* (San Francisco), February 4, 1859.

1872

Camilla (RP); *Roscoe* (MNBedfHi); *Helen Snow* (MNBedfHi); *Northern Light* (MNBedfHi); A. H. Clark 1887a:153-54; *Pacific Commercial Advertiser* (Honolulu), November 2, 1872; *The Friend* (Honolulu), November 2, December 2, 1872; *The Chronicle* (San Francisco), October 12, November 5, 1872.

1873

Helen Mar (MNBedfHi); *Triton* (MNBedf); *The Friend* (Honolulu), November 1, December 1, 1873; *The Commercial Herald and Market Review* (San Francisco), October 25, 1873; *Pacific Commercial Advertiser* (Honolulu), October 18, October 31, 1873.

1874

Helen Mar (MNBedfHi); *Triton* (MNBedf); *The Commercial Herald and Market Review* (San Francisco), October 22, 1874, January 14, 1875; Scammon 1874:59, 65.

1875

Northern Light (MNBedfHi); *Triton* (MNBedf); *Europa* (MNBedfHi); Collinson 1889; *Pacific Commercial Advertiser* (Honolulu), October 30, November 6, 1875; Herendeen to Dall, November 10, 1875, box 11, folder 34, William Healey Dall Papers (DSI).

1876

Northern Light (MNBedfHi); *Harper's Weekly* (New York), November 18, 1876; Ellis 1892:431-33; A. H. Clark 1887a:83-84; Aldrich 1889:209-18; *Alta California* (San Francisco), October 22, 1876; *Pacific Commercial Advertiser* (Honolulu), October 21, November 18, 1876; *Whalemen's Shipping List* (New Bedford), August 1, 1876.

The ships that were abandoned off Smith Bay were: *Acors Barns, Camilla, Cornelius Howland, Desmond, James Allen* [I], *Java 2nd, Josephine, Marengo, Onward,* and *St. George.*

1877

Rainbow (DNA), *Cleone* (MShaK), *Mount Wollaston* (MNBedfHi); *Northern Light* (MNBedfHi); *Helen Mar* (MNBedfHi); *The Chronicle* (San Francisco), November 6, 1877; *Whalemen's Shipping List* (New Bedford), October 9, 1877; John Brown to Swift and Allen, October 1, 1877, Swift and Allen Papers (MNBedfHi).

Honolulu and San Francisco

Daws 1968:169-72; *The Friend* (Honolulu), November 1, 1844; *Pacific Commercial Advertiser* (Honolulu), October 9, 1856, November 1, 1860, December 18, 1869, November 4, 1870, November 23, 1872; Starbuck 1964:112; Great Britain, Naval Intelligence Division 1943:319; *San Francisco Prices Current and Shipping List,* November 30, 1854; Chatfield, nd:18-19; unidentified newspaper clippings, scrapbook T-1 (MNBedfHi); *The Polynesian* (Honolulu), August 27, 1857; Twain 1966:91-95; A. H. Clark 1887a:148, 150, 153, 155; *Martha* (MNBedf); *Hawaiian Gazette* (Honolulu), October 2, 1872; G. P. Moore 1934; *Commercial Herald and Market Review* (San Francisco), November 1, 1867, January 14, 1869, January 14, 1870, January 14, 1875, passim 1876-78; *Alta California* (San Francisco), October 26, 1877; Crawford 1981:44.

"primitive insignificance"

The Polynesian (Honolulu), July 6, 1844.

tenders

The Chronicle (San Francisco), September 18, 1884; *Whalemen's Shipping List* (New Bedford), September 25, 1877; *Rainbow* (DNA), July 13, 1877, August 28, 1878; *Northern Light* (MNBedfHi), July 5, 1878; *Syren* (MNBedfHi), 1878; *Helen Mar* (MNBedfHi), July 24, 1878, August 12, 1879.

1878 and 1879

Rainbow (MNBedfHi); *The Weekly Bulletin* (San Francisco), January 30, 1879; *The Commercial Herald and Market Review* (San Francisco), November 11, 1878; Hegemann 1890:431, 434; *Leo* (MNBedfHi); Hooper 1884:27-28, 58-59; A. H. Clark 1887a:21; Thrum 1913:62; Dall 1882:304; *Pacific Commercial Advertiser* (Honolulu), October 4, 1878; *The Chronicle* (San Francisco), September 18, 1884; *Helen Mar* (MNBedfHi); *Helen Mar* protest (MNBedfHi); Herendeen to Dall, September 29, 1878, February 2, 1879, box 11, folder 34, William Healey Dall Papers (DSI); *The Morning Call* (San Fransico), November 11, 1878, November 27, 1879, December 6, 1879, July 14, 1881; Healy 1889:24; Aldrich 1889:45, 222-29; E. W. Nelson 1887:292; unidentified newspaper clippings, scrapbook 17, pp. 7, 96 (MNBedfHi).

Chapter 9

Roys in Bering Strait

Whalemen's Shipping List (New Bedford), February 6, 1849.

early history of foreign trade in Bering Strait

Burch 1976; Stefansson 1914; D. J. Ray 1975a:88-89, 97-101, 179; D. J. Ray 1975b; Bockstoce 1977d; Lydia Black, personal communication; Howay 1973:84, 96-97, 192, 196, 203-6; Tikhmenev 1978:182-84, 208; Foote 1965:71, 112; Zagoskin 1967:286; Blomkvist 1972:131-38: Ernest S. Burch, Jr., personal communication.

"cloud of arrows"

D. J. Ray 1975a:69.

Mary Brewster's account

Tiger (CtMyMHi), June 26-July 3 and August 3, 1849.

native acquaintance with foreign ships and trade

Roys, ms., 14; Montpelier (RP), July 26, 1849; *Catherine* (Ct), July 17, 1849; *Marengo* (MNBedf), July 4 and 26, 1849.

loss of L. C. Richmond

The Friend (Honolulu), October 1, 1849; *Whalemen's Shipping List* (New Bedford), January 1 and 22, 1850.

wreck of the Citizen

Holmes 1857.

early rum trade

D. J. Ray 1875a:179; Zagoskin 1967:286; Foote 1965:187; John Mathews, ms. (RGS); *Harvest* (MNBedf), October 3, 1852; Bogoras 1909:711-12.

Swallow *wintering at Bering Strait*

Tamerlane (IMA-MNBedfHi), June 3, 1851; Osbon 1902:365-68.

trading vessels at Bering Strait in 1851

Whalemen's Shipping List (New Bedford), December 9, 1851, July 4, 1853; *Cossack* (MNBedfHi), August 1, 1851; *Phillipe de la Noye* (NHyF), July 23, 1851; Collinson 1889:132; *Mount Vernon* (MShaK), May 31, 1851; John Mathews, ms. (RGS).

"plenty of rum on board"

Whalemen's Shipping List (New Bedford), July 4, 1853.

requests for rum
South Boston (CtWetHi), July 29, 1852; Munger 1852:31; W. H. Allen, nd:30; *Samuel Robertson* (MShaK), July 25, 1851; *Constitution* (MNanW), July 16, 1851; *Whalemen's Shipping List* (New Bedford), July 4, 1853; *Roman* (MShaK), August 12, 1854; *The Friend* (Honolulu), December 6, 1853; HMS *Enterprise* (NMM), August 13, 1854.
some early examples of trade and cooperation with the natives of the Chukchi Peninsula
Saratoga (MNBedfHi), July 12, 1850; *Cambria* (RP), June 22, 1850; *Charles Drew* (RP), July 31, 1850; *Nimrod* (RP), June 24, 1850; *Samuel Robertson* (MShaK), June 15, 1850, July 25, 1851; *Nauticon* (IMA-MNBedfHi), June 24 1851; *Arab* (RP), July 23, 1851; *Prudent* (SHS), July 30 and August 3, 1851; *Tamerlane* (IMA-MNBedfHi), July 5, 1851; *Betsey Williams* (MShaK), August 19, 1852; *America* (MShaK), July 25, 1852; *Prudent* (MShaK), August 1, 1853;
Roman (MShaK), July 18, August 6, August 12, September 2-3, and September 26, 1854; *Franklin* (RP), July 2, 1854.
Armata *incident*
Tamerlane (IMA-MNBedfHi), July 14, 1851; *Wolga* (RP), July 25, 1851; *Emu* (MLS), July 10, 1851; *Montreal* (MNBedfHi), August 4, 1851; *Cossack* (MNBedfHi), July 21, 1851; *Whalemen's Shipping List* (New Bedford), July 4, 1853; Barker, ms., 65 (MShaK).
Julian *incident*
Julian (CtY), July 31, 1859; *Hunter* (MNBedfHi), June 17, 1883.
trading vessels' activities in the 1850s and 1860s
Dall 1870:502-3; unidentified newspaper clipping, scrapbook T-1 (MNBedfHi); *South America* (PC), July 1, 1858; Hegarty 1959:48-50; *South Boston*
(CtWetHi), July 29, 1852; *The Friend* (Honolulu), November 8, 1858; *Antilla* (RP), June 5, 1858; Rosse 1884:186.
American Eskimos' reputation for truculence
Dymytryshyn and Crownhart-Vaughan 1979:146-50; *Eliza F. Mason* (MNBedfHi); July 21, 1854; *Julian* (CtY), July 31, 1859; Ernest S. Burch, Jr., personal communication; *Fanny*, ms., 15; J. Simpson 1875:250; Kelly 1890:10; E. W. Nelson 1899:299-303; Hooper 1884:41.
"very rude and had to be watched"
Ebenezer Nye to *The Standard* (New Bedford), August 2, 1879.
trade at Port Clarence, 1859
Whalemen's Shipping List (New Bedford), July 4, 1853; *Sharon* (RP), July 19, 1859; Whymper 1869:169. By contrast the *Antilla* (RP) did no trading on the American shore in 1858.
"guns, rum, powder . . ."
John Howland (MNBedfHi), September 20, 1859.
early maritime trade north of Kotzebue Sound
George and Susan (MNBedf), August 4, 1859; *Julian* (CtY), August 23, 1859.
firearms, 1850s
Thornton 1931:139; Wilkinson, nd.:186-87; Driggs 1905; *John Howland* (MNBedfHi), July 31, 1860, August 22, 1861.
wintering voyages—concept and execution
The Evening Standard (New Bedford), July 11, 1903; *Nimrod* (MNBedfHi), June 23 and 27, 1860; *Betsey Williams* (MShaK), June 30, 1852; Newhall 1981:30-32; Bockstoce and Batchelder 1978b; Osbon 1902:365-68; *Tamerlane* (IMA-MNBedfHi), June 3, 1851; *Sharon* (RP), August 9, 1859; *Saratoga* (MShaK), August 13, 1859.
winter of 1859-60
Cleone (MNBedfHi), 1859-60; *Nimrod* (MNBedfHi), June 23 and 27, 1860; *Jireh Swift* (MNBedf), June 19, 1860; *John Howland* (MNBedfHi), June 19-June 23, 1860.
winter of 1861-62
John Howland (MNBedfHi), June 23, 1862; *Thomas Dickason* (RP), September 22, 1861; *Pacific Commercial Advertiser* (Honolulu), October 9, 1862; *Ocean* (CtNhHi), June 22, 1862; *The Friend* (Honolulu), November 2, 1863.
Captain Brummerhoff's murder
Aldrich 1889:229-34; *The Friend* (Honolulu), November 2, 1863. There is some confusion over the spelling of the name of Brummerhoff's brig *Kohola*. It is also listed as *Kohala*, presumably after the northwest coast of the island of Hawaii. Most often, however, it is spelled *Kohola*, which is the Hawaiian word for "whale."
chronological list of commercial wintering voyages
Bockstoce and Batchelder 1978b; Barker, ms., 74, 80; *William H. Meyers* (MNBedfHi), October 2, 1878.
Port Clarence as trade center
The Esquimaux (Port Clarence, Alaska and Plover Bay, Siberia), July 14, 1867.
competition for trade
John P. West (ViNeM), June 21, 1866; *Eliza F. Mason* (MNBedfHi), July 2, 1854.
buy boots, jackets and mittens
Roscoe (MNBedfHi), July 7, 1862; *Helen Mar* (MNBedfHi), June 4, 1874, July 19, 1879.
stake native traders
Leo (MNBedfHi), August 16, 1878.
traders living at Port Clarence and Plover Bay
Barker, ms. 119 (MShaK); *William H. Meyers* (MNBedfHi), October 2, 1878; *The Alaska Herald* (Sitka), November 20, 1869; Krause and Krause 1882:130.
middlemen, 1860-80
Murdoch 1892:193-96; Whymper 1869:169; Hooper 1885:106; Nordenskiold 1882:115, 231-33; Barker, ms., 25, 70 (MShaK); Wilkinson, nd:187-88; G. W. Bailey 1880a:17; Petitot 1876:8-9, 15, 18; Bompas, ms., C.C.1./0. 11 & 12 (CMSL).
difficulties with drunken Eskimos
Ocean (PC), July 20, 1861.
prelude to Cape Prince of Wales massacre
Woldt 1884:268; Kelly 1890:11; Elliott 1886:431; Hooper 1881:20; G. W. Bailey 1880;15; E. W. Nelson 1899:299; Aldrich 1889:30; *Whalemen's Shipping List* (New Bedford), September 25, 1877.
Cape Prince of Wales massacre
unidentified newspaper clipping (Atlantic Companies), September 20, 1877; *The Chronicle* (San Francisco), November 6, 1877; Brower, ms., 226; Kelly 1890:12-13; Thornton 1931:38; Aldrich 1889:30, 142-46; Hooper 1881:20; *Whalemen's Shipping List* (New Bedford), September 25, 1877; *Pacific Commercial Advertiser* (Honolulu), October 27, 1877.

A number of accounts of this incident have been published and many contain half-truths and errors; generally those that were written farther in time from the event contain the most hyperbole. I have therefore relied primarily on the newspaper accounts that appeared in the autumn of 1877. George Gilley gave his own account of the event to Herbert Aldrich in 1877. It closely agrees with the newspaper reports of 1877. It is regrettable that no credible Eskimo account of the event has survived.

aftermath of the Cape Prince of Wales massacre

G. W. Bailey 1880:15; *The Chronicle* (San Francisco), April 13, 1898; Brower, ms., 315; *The Eskimo Bulletin* (Cape Prince of Wales), May 1892; Hooper 1884:16; *The American Missionary,* October 1891, p. 360.

contraband laws in Alaska

D. J. Ray 1975a:185-94.

trade and the Revenue Marine

Otis 1880:43-44; Wardman 1884:148-50; G. W. Bailey 1880; Dall 1881a:104; Healy 1887:15; Hooper 1884:32-33, 38; Hooper 1881:11-12, 21; E. W. Nelson 1899:231; Krause and Krause 1881:277-78; Igor Krupnik, personal communication; Woolfe 1893:145-46.

Eskimo middlemen bring alcohol into Alaska

The Eskimo Bulletin (Cape Prince of Wales), 1892, 1897; Thornton 1931:xxiii, 29, 120.

trade procedures

Woolfe 1893:137; *Montreal* (MNBedfHi), September 15, 1852; *Henry Taber* (RP), July 7, 1869; E. W. Nelson 1899:263; Brower, ms., 212; Rosse 1884:186; Aldrich 1889:49-50; J. A. Cook 1926:229.

trade goods dispensed from ship

Mermaid (MShaK), July 21, 1896.

trade jargon

Rosse 1884:186-88; Murdoch 1892:55; Wells and Kelly 1890; Hammerich 1958:636; Dall 1881c:867; Sheldon Jackson 1902:89.

sexual intercourse with foreigners

Kelly 1890:19, 21; USS *Bear,* James T. White journal (WaU), July 16, 1894; Woolfe 1893:137, 143; Edson to Langford (TxAuCH), July 3, 1895; *Governor Troup* (RP), June 20, 1864; Murdoch 1892:54; G. B. Borden, ms., 18-19 (CtY).

disease and population reduction

Shklovsky 1916:5; G. B. Borden, ms., 18-19 (CtY); Woolfe 1893:143; Burch 1980; Woldt 1884:303; Burch 1981:15-19; Ernest S. Burch, Jr., personal communication; Kelly 1890:21.

"trading chiefs"

A. J. Allen 1978:158.

Goharren

Krause and Krause 1882:140; Bogoras 1909:63-64; unidentified newspaper clipping, M. A. Healy Collection (CsmH), scrapbook 1, p. 51; *The Weekly Bulletin* (San Francisco), October 13, 1886; Aldrich 1889:50-51; USS *Bear,* James T. White journal (WaU), August 18, 1886, June 30, 1894; Herbert Appossingok, personal communication; Serghei Aroutiounov, personal communication; Kelly 1890:9; *The Chronicle* (San Francisco), October 16, 1890; Report of U.S. Commissioner for Education 1894:950; Sheldon Jackson, diary (PPPrHi), 70; Benjamin F. Sharp diary (PC), June 30, 1895; De Windt 1898:197-98, 204-5, 233.

trade chicanery

Frank Russell, diary (DSI), August 14, 1894; Barker, ms., 90 (MShaK); Madsen and Douglas 1957:136.

Attungoruk

Kelly 1890:11-12; Hooper 1884:41, 107; Nelson 1899:303, 305; Brower 1942:34-43, 114-15, 149-50; Brower, ms., 153-54, 309-10, 396; Aldrich 1889:161; Tom Lowenstein, personal communication; Burch 1981:16-19; Ernest S. Burch, Jr., personal communication.

decline in whaleships trading alcohol

Hooper 1884:33; Bogoras 1909:61-62; Aldrich 1889:114-15; *The Chronicle* (San Francisco), November 26, 1889.

native stills

Burns 1913:163-64; Jones 1927;68; Sheldon Jackson 1902:94.

some examples of native employment aboard whaleships

Barker, ms. (MShaK), 81; *Pacific Commercial Advertiser* (Honolulu), November 1, 1860; *Mercury* (MNBedfHi), October 11, 1866; *Arctic* (RP), October 4, 1873; *Northern Light* (MNBedfHi), May 23, 1876; *Rainbow* (DNA), June 12, 1877; *Loleta* (DNA), August 26, 1878; *Leo* (MNBedfHi), July 10, 1878; *Mary and Susan* (RP), June 10, 1887; *William Baylies* (RP), June 12, 1887; A. J. Allen 1978:23; *Beluga* (MNBedfHi), October 18, 1896; J. A. Cook 1926:322.

native pay and advances

Bodfish trade book, 1911 (MNBedfHi); *Narwhal* (MH-BA), June 19, 1905; *William Baylies* (MNBedfHi), June 13, 1906.

natives are paid with whaleboats

Progress (MNBedfHi), October 15, 1881; Aldrich 1889:56; *Grampus* (MNBedfHi), October 17, 1889; *William Baylies* (RP), October 21, 1892; West et al. 1965:63; J. A. Cook 1926:147.

trade for whaleboats

Mary and Susan (PC), May 31, 1885; *Mary and Helen II* (IMA-MNBedfHi), June 1, 1886.

give flukes to natives

Aldrich 1889:133.

some examples of natives traveling to South aboard whaleships

D. J. Ray 1975:159; Wilkinson, nd:273; *Leo* (MNBedfHi), October 15, 1879; Dall 1881c:865; Nelson and True 1887:293; *Thrasher* (PC), June 9, 1884; *Mary and Susan* (PC), June 19, 1887; *Andrew Hicks* (MNBedf), July 8, 1891; *William Baylies* (RP), October 21, 1892; *William Baylies,* June 14, 1893; USS *Bear,* James T. White journal (WaU), July 16, 1894; Kelly 1897:83; *Bowhead* [II] (CaACG), October 16, 1906; *Narwhal* (MH-BA), September; 18, 1907.

later competition for trade

Madsen and Douglas 1957:96-132; John Rosene Papers (WaU); Jenness 1964:13-14; Usher 1971:174-75; Doty 1900b:190-91; Bodfish 1936:195-96.

Eskimo schooners

Bodfish 1936:178-79, 268; unidentified newspaper clipping, scrapbook T-4,
p. 170 (MNBedfHi); Faber 1916:163-64; Nuligak 1971:31; Ted Pedersen, personal communication.

sleazy dealing in Siberia and subsequent events

Madsen and Douglas 1957:96-132; Swenson 1944; Masik and Hutchinson 1935.

influence of maritime trade in western Arctic

Aldrich 1889:134; Bogoras 1909:63, 730-32; J. A. Cook 1926:31; Tolmachoff 1949: 197-98; Krause and Krause 1882:137; Starokadomskiy 1976:98.

Chapter 10

3 percent insurance surcharge
A. H. Clark 1887a:152.
prices of oil and baleen, 1875
Tower 1907:129.
value of an average bowhead whale in 1865 and 1875
Computed as yielding 100 barrels (31½ U.S. gallons) of oil and 1,500 pounds of baleen; prices from Tower 1907:128.
San Fransico captures fleet's business
The Weekly Bulletin (San Francisco), April 27, 1887.
railroad carries product east from San Francisco
The Commercial Herald and Market Review (San Francisco), January 1878, passim; Habershaw 1883:215; A. H. Clark 1887a:163-64.
California whaling rocket
Lytle 1984; Schmitt et al. 1980:181, 202-3; *Helen Mar* (MNBedfHi), September 9 and 16, 1879; *Leo* (MNBedfHi), August 30 and September 15, 1879.
Leander Owen on steam power
Pacific Commercial Advertiser (Honolulu), October 18, 1873.
"inventive genius" etc.
Pease and Hough 1889:32-35.
Leander Owen's losses
Owen letter, testimony for M. A. Healy, February 5, 1889, M. A. Healy Papers (CsmH).
William Lewis sends Cyrus Manter to Newfoundland
Unidentified newspaper clipping, scrapbook 17, p. 10 (MNBedfHi).
corsets are "longer, tighter and more generally constricting"
Ewing 1981:79.
price of whalebone in 1878
A. H. Clark 1887a:160. An average-sized bowhead is estimated to have yielded 1,500 pounds of baleen.
William Lewis and Frank Reynolds
Reynolds family papers (CtMyMHi).
ownership of Mary and Helen
Works Projects Administration 1940 (3):114-15.
details of Mary and Helen*'s design and construction*
Baker 1977; unidentified newspaper clipping, scrapbook 17, p. 10 (MNBedfHi).
1880 season
Dall 1881b:49; Bean 1902:262.
Mary and Helen*'s 1880 cruise*
Reynolds family papers (CtMyMHi).
biographies of the steam whalers
Bockstoce 1977a:75-109.
1883 season
The Examiner (San Francisco), November 5, 1883; *The Call* (San Francisco), July 19, 1883; *The Bulletin* (San Francisco), August 3, 1883; *The Daily Times* (Los Angeles), October 7, 1883; P. H. Ray 1885:101; *Orca* (PC), June-August 1883; *Rainbow* (PC), October 21, 1883.
building Balaena
Alta California (San Francisco), April 28, 1883.
incorporation of Pacific Steam Whaling Company
The Daily Bulletin (San Francisco), October 31, 1883.
Josiah Knowles
Mjelde 1970:32-201 passim.
Arctic Oil Works
The Weekly Bulletin (San Francisco), November 7, 1883, October 29, 1884.
Refining whale oil
Stevenson 1904:199, 202-3; A. H. Clark 1887a:4-5; Holmes 1857:292-96.
refining whalebone
Starbuck 1964:155-56; unidentified newspaper clipping, scrapbook 17, p. 9 (MNBedfHi); *The Standard* (New Bedford), August 29, 1915.
Pacific Steam Whaling Company ships
Alta California (San Francisco), April 18, 1882; *The Weekly Bulletin* (San Francisco), March 1 and November 22, 1882; *The Daily Bulletin* (San Francisco), November 15, 1882; unidentified newspaper clipping, scrapbook 17, p. 57 (MNBedfHi); Tressler 1923:635-36.
reduction of bowhead whales by 1880
Bockstoce and Botkin 1983:Table 2.
examples of whaleships right whaling in spring season, 1870s
Navy (MShaK), 1871; *Northern Light* (MNBedfHi), 1872; *Helen Snow* (MNBedfHi) 1872; *Roscoe* (MNBedfHi), 1872; *Northern Light* (MNBedfHi), 1875; *Mount Wollaston*, 1876, in *Northern Light* (MNBedfHi), August 23, 1876; *Cleone* (MShaK), August 5, 1877.
routine and route of steam whaling cruises in 1880s
The Weekly Bulletin (San Francisco), October 8, 1880, November 8, 1882, August 6, 1890; *The Chronicle* (San Francisco), January 23, 1887; Simpson 1890; P. H. Ray 1885:101; Cook 1926:3 and 120; *Mary and Susan* (PC), August 5, 1886; *Mary and Helen II* (MNanW), July 13, 1886; *Jacob A. Howland* (RP), 1884 passim; *Thrasher* (PC), August 10, 1886; *William Baylies* (RP), May 29, 1887.
examples of mid-season refit—Plover Bay
Antilla (RP), July 1858; *John Howland* (MNBedfHi), June 1860; *Florida* (MNBedf), June-July 1860, 1861, 1862; *Barnstable* (RP), June-July 1861, 1862, 1863; *Navy* (MNBedfHi), June-July 1862; *Thomas Dickason* (RP), July 1861, 1862, 1863; *Adeline* (MNBedf), July 1863; *Aurora* (MNBedf), July 1866, 1867, 1868; *Henry Taber* (RP), July 1869; *Rainbow* (DNA), July 1877.
examples of mid-season refit—Saint Lawrence Bay
Samuel Robertson (MShaK), July 13, 1851; *Thomas Dickason* (RP), July 1870; *Helen Mar* (MNBedfHi), July 1878.
examples of mid-season refit at Port Clarence prior to 1880
The Friend (Honolulu), November 16, 1853; Whymper 1869:169; *Harvest* (MNBedf), July 12-16, 1852; *Antilla* (RP), June 1858; *Montreal* (MNBedf), July 19, 1859; *Rambler* (MShaK), July 29, 1859; *Sharon* (RP), July 19, 1859; *John Howland* (MNBedfHi), September 1859, July 1860; *Onward* (RP), July 5, 1869; *Seneca* (MNBedfHi), July 22, 1871; *Roscoe* (MNBedfHi), July 8, 1872.
tenders
The Chronicle (San Francisco), September 18, 1884; Hooper 1884:72; Brower, ms., 119; Butler 1973:112; *Whalemen's Shipping List* (New Bedford), September 25, 1877; *Rainbow* (DNA), July 13, 1877, August 20, 1878; *Syren* (MNBedfHi), July 1878; *Helen Mar* (MNBedfHi), July 24, 1878, August 12, 1879; *Mary and Helen* (MNBedfHi), July 23, 1880; *Mary and Helen* (PC), July 24, 1880; *Coral* (MNBedfHi), July 29, 1880; *Progress* (MNBedfHi), August 20, 1881; *Thomas Pope* (MNBedfHi), 1882-86; *William Baylies* (RP), 1887-90; *Orca* (PC), August 15, 1883; *Thrasher* (PC), July 23, 1884.

Corwin Coal Mine
West et al. 1965:22.
coaling station at Port Clarence
Healy 1887:6-7, 13; Healy 1889:10-13; D. J. Ray 1975:200-201.
use of Port Clarence for refit after 1880
Woolfe 1893:143; Aldrich 1889:76, 81; Bodfish 1936:43, 205, 243; *Balaena* (IMA-MNBedfHi), June-July 1884; *Andrew Hicks* (MNBedf), July 1890 and 1891; *Thrasher* (IMA-MNBedfHi), July 1, 1885; USRC *Bear,* letterbook (CsmH), July 14, 1886, M. A. Healy Papers.
smoke ship
Aldrich 1889:19-20.
propeller repair
Page 1900:6; *Herman* (MNBedfHi), June 30, 1911; *Beluga* (RP), May 26-June 24, 1893; *Balaena* outfit book (CsfMM), 1888.
decline in quality of crews and procedure for shipping them, etc.
Starbuck 1878:112; Williams 1964:253-54, 272; *The Chronicle* (San Francisco), January 23, 1887; U.S. Commissioner of Fish and Fisheries 1892:74-79; *The Examiner* (San Francisco), November 12, 1888.
Arctic desertions prior to 1880
Eliza F. Mason, Mechigmen Bay, in *Roman* (MShaK), August 15, 1854; *Robin Hood,* Plover Bay, in *Cleone* (MNBedfHi), September 26, 1859; *LaManche* Plover Bay, in *Cleone* (MNBedfHi) October 7, 1859; *Thomas Dickason* (RP), Plover Bay, June 28, 1862.
"bound to heaven . . . stop in Hades"
Jernegan ms. (MEdDHi).
desertions after 1880
Healy 1887:10; *Rosario* (MShaK), July 21, 1888; S. J. Call to M. A. Healy, July 30, 1891, Alaska file, U.S. Revenue Cutter Service (DNA); *Andrew Hicks* (MNBedf), July 11, 1890; Thornton 1931:81-82; *Rosario* (MNBedfHi), August 7, 1891; Report on Cruise of USRC *Bear* of 1891, M. A. Healy Papers, Sheldon Jackson College, Sitka, Alaska; Kelly 1897:87; *Beluga* (MNBedfHi), July 17, 1900; De Windt 1904;168-70; *Alexander* (MNBedfHi), July 13, 1903; *Alexander* (MEdDHi), May 20, 1904; *Belvedere* (RP), June 7, 1904; *William Baylies* (MNBedfHi), July 5, 8, and 24, 1905; *Narwhal* (MH-BA), July 8, 1905.
discipline
J. T. Brown 1887:220-21; Hohman 1928:74-75; Williams 1964:253; Starbuck 1964:114.
punishments prior to 1880 in Arctic
Eliza F. Mason (MNBedfHi), August 13-15, 1854; *Catherine* (Ct), July 16, 1849; *Marcia* (MNBedfHi), July 6, 1860; *Corinthian* (MNBedfHi), September 13, 1853; *Fortune* (MNBedfHi), August 26, 1851; *Rousseau* (MNBedfHi), July 24, 1854; *St. George* (MNBedfHi), August 8, 1854; *Eagle* (MNBedfHi), July 31, 1867; *Northern Light* (MNBedfHi), September 23, 1873, July 4, 1878; *Emily Morgan* (MNBedf), September 23, 1873.
steamer routine and route, summer and autumn
E. Simpson 1890; Healy 1889:22; P. H. Ray 1885:101; J. A. Cook 1926:120; *The Weekly Bulletin* (San Francisco), October 17, 1883; Murdoch 1892:26; *Jacob A. Howland* (RP), August 6, 1882; *Hunter* (MNBedfHi), August 17, 1883; *Thrasher* (IMA-MNBedfHi), August 25, 1885; *Mary and Helen II* (MNanW), August 15, 1886; *Francis A. Barstow* (RP), 1891.
expenses of steam versus sailing whaleships
Hegarty 1959:8-25; Lindeman 1899:71; unidentified newspaper clipping, scrapbook 16, p. 218 (MNBedfHi).

Chapter 11

plan for Vigilant *to winter at Point Barrow*
Hooper 1885:59
Ned Herendeen
Brower 1942:31; Hare 1960:46; Stefansson 1934:110-111; Baker 1893:77-78; Refuge Station Correspondence, Record Group 26 (DNA); Baker et al. 1906:303; Bean 1902:246; Herendeen 1893:78.
After he was fired by the Pacific Steam Whaling Company, Herendeen apparently spent the winter of 1886-87 on his own at Point Barrow. He then returned to San Francisco and, with his brother Louis, invested in the little schooner *Nicoline,* planning to winter near the Mackenzie delta. The voyage was a failure (see chapter 12) and Herendeen returned to San Francisco in 1891 without funds. William Healey Dall, his patron and friend, eventually secured for him the position of doorkeeper at the National Museum in Washington, D.C.
California shore whaling
David Henderson, personal communication.
native whaling at Point Barrow
P. H. Ray 1885:48, 101; Aldrich 1889:176-77; Murdoch 1891; Murdoch 1892:240; Mason 1902:269-70; Brower, ms., 277-78; Hadley 1915:916.
Point Barrow station, 1884-86
Brower, ms., 116, 133, 145; Healy 1889:10.
Point Barrow station, 1886-88
Brower 1942:82, 123-234; Brower, ms., 235-36, 281, 290, 318; *Mary and Helen II* (MNanW), August 22, 1886; *Narwhal* (PC), August 7, 1887; Aldrich 1889:110.
shore whaling catch statistics
Marquette and Bockstoce 1980.
Brower's use of Eskimo whaling techniques
Brower 1942:123-24; Brower, ms., 318.
"squirt of the grapefruit"
Brower 1942:149.
company fires all but Brower and Leavitt
Brower 1942:125-34; Brower, ms., 331, 343, 375; Borden, ms., 7 (CtY).
season of 1890 at Point Barrow
Brower, ms., 378; Brower 1942:145; Borden, ms., 20 (CtY); *The Chronicle* (San Francisco), September 19, 1890; Poole 1971:6-7; Report of U.S. Commissioner for Education 1891:924.
season of 1891 at Point Barrow
Brower, ms., 304, 395, 402-3: Brower 1942:112, 152; Kelly 1890:66; *The Sunday Chronicle* (San Francisco), September 15, 1889; Simpson 1890:4; Borden to Secretary of the Treasury, February 22, 1892, Refuge Station correspondence, RG 26 (DNA); Driggs 1892:99; Mary Healy diary, August 13, 1891, M. A. Healy Papers (CsmH); Hadley 1915:917.
change in Eskimo whaling
Brower 1942:152; Brower ms., 279, 456; Stefansson, diary (NhD), December 8, 1908; Sonnenfeld 1957:283.

season of 1892 at Point Barrow and Point Belcher
Brower, ms., 411, 418, 424-25; *Orca*
(IMA-MNBedfHi), August 14-16, 1891; M. A. Healy to Secretary of the Treasury, report of 1891 cruise of USS *Bear*, Healy Papers (AkSJ); James T. White journal (WaU), August 1, 1894; Point Barrow Refuge Station logbook, April 22, 1892, RG 25 (DNA).

seasons of 1893 to 1896 at Point Barrow
Brower, ms., 443-47, 452-56; Brower 1942:159; *Mary D. Hume* (MH-BA), July 18, 1893; Russell 1898:152; *The Chronicle* (San Francisco), September 7, 1898.

Point Hope Shore Station, 1887-89
Brower, ms., 304, 308-10; Brower 1942:112-15, 137; *The Chronicle* (San Francisco), November 25, 1889; Burch 1981:16-19; Tom Lowenstein, personal communication; Kelly 1890:11; L. M. Stevenson to Sheldon Jackson, June 30, 1891
(PPPrHi); Henry Koenig to Governor of Alaska, December 4, 1888, M. A. Healy Papers (AkSJ); unidentified newspaper clipping, M A. Healy scrapbook, p. 61, M. A. Healy Papers (CsmH); Edson to Langford, July 3, 1895 (TxAuCH); *Mary and Susan* (PC), July 6, 1888; *Rosario* (MShaK), July 10, 1888; Poole 1971:6-7; Woolfe 1893:145.

Peter Bayne
Bayne was a Nova Scotian who spent much of his adult life in the Arctic. In 1867 he wintered aboard the *Ansel Gibbs* in northern Hudson Bay and was hired by the explorer Charles F. Hall to help him search for the remains of the lost expedition of Sir John Franklin. Bayne was apparently the origin of a rumor that Franklin's body had been interred in a cement vault on the Arctic coast (Burwash 1931:112-16).

Bayne was a member of Herendeen's experimental wintering party at Point Barrow in 1884-85. In 1886 he was a boatheader aboard the *Orca* and in 1887 he set up the shore station at Point Hope. Bayne left Point Hope in 1889 and returned to the Arctic the following summer as master of the *Silver Wave*, which he fraudulently condemned (see chapter 13, and Bockstoce 1979).

He was arrested for this in 1892 but apparently was not convicted because he returned to Point Hope in 1893 as the master of the little schooner *Emily Schroeder*, which was wrecked in the great storm of October 1893. Bayne remained at Point Hope running a shore station there until at least 1898. After the discovery of gold at Nome in 1898, he apparently drifted there and continued his shoddy dealings for a number of years.

Point Hope shore station, 1889-92
The Chronicle (San Francisco), November 25, 1889; whaling agreements,
Henry Koenig papers (AkU); John Driggs to Joshua Kimber, September 25, 1890, (TxAuCh), Driggs 1892:99; Walter Howland to wife, July 14, 1890 (MNBedfHi); James T. White journal (WaU), August 1, 1894; S. J. Call to M. A. Healy, Alaska file, Revenue Cutter Service, RG 26 (DNA); Woolfe 1893:145.

Point Hope whaling, 1892-98
MacInnes 1932:70; *The Alaskan* (Sitka), March 10, 1894; USS *Bear* (DNA), August 1, 1894; John Driggs to Joshua Kimber, June 28, 1900,
(TxAuCH); I.O. Stringer diary (CaOTCHAr), September 10, 1895; *The Chronicle* (San Francisco), October 14, 1894; *Mary D. Hume* (MH-BA), August 14, 1893; Henry Koenig diary (AkU), June 12, 1896; Bertholf 1899:25; *The Eskimo Bulletin* (Cape Prince of Wales, Alaska), July 1898.

Saint Lawrence Island shore whaling
The Chronicle (San Francisco), August 31, 1893; USS *Bear* (DNA), June 30, August 25, August 26, 1894; Doty 1900:227-28; *Narwhal* (MN-BA), notes at end of 1902 journal; Faber 1916:63; J. A. Cook 1926:248.

natives involved with the whaling industry
Brower, ms., 448, 482; A. J. Allen 1978:140; Green 1959:2-3; Ernest S. Burch, Jr., personal communication; Burch 1981:16-19; Tom Lowenstein, personal communication; John Driggs to Joshua Kimber, June 28, 1900
(TxAuCH).

"ship" a crew
Hadley 1915:917-18; Henry Koenig Papers (AkU), various whaling agreements; Sonnenfeld 1957:520; Mikkelson 1909:349; Stefansson 1912:60-62.

whaling equipment and procedures
Woolfe 1893:146-48; Bailey and Hendee 1926:23-27; Henry Koenig diary (AkU), various entries; Lytle 1984; A. J. Allen 1978:144-51; J. R. Bockstoce, field notes; Brower, ms., 556; Hadley 1915; Rainey 1940.

decline of the shore fishery
Henry Koenig Papers (AkU), various documents.

subsequent developments in Eskimo whaling
Bockstoce 1977; Bockstoce 1980; Bockstoce and Botkin 1983.

retrieval of wounded whales
Bockstoce 1980A, Table 2, cols. C & F; Marquette and Bockstoce 1980; Anonymous [1906]:163.

Chapter 12

open passage between Davis Strait and Beaufort Sea
Friis 1971: Maury to De Haven, 1851; in Maury (3d ed.) 1851:178; Ross and MacIver 1982.

Richardson sees whales at Cape Bathurst
Captain Pierce, *The Friend* (Honolulu), December 8, 1854.

presumption of whales at mouth of Mackenzie River
Scammon 1874:65; *The Weekly Bulletin* (San Francisco), January 30, 1879; *The Herald* (New York), October 12, 1882.

whaleships reach Barter Island
Collins 1891:79; Aldrich 1889:111-12; *Grampus* (MNBedfHi), August 16, 1888; E. Simpson 1890:17; *The Chronicle* (San Francisco), November 16, 1888.

"a beachcomber . . ."
Hadley 1915:910

Tuckfield's voyage
Grampus (MNBedfHi), August 14, 1888, August 2, 1889; Brower, ms., 328-29, 377; *The Chronicle* (San Francisco), September 30, 1889; Poole 1971:6-7; James T. White diary, USRC *Bear* (WaU), August 6, 1889; G. B. Borden to James McKenna, letterpress copy book (RP), August 17, 1889.

USS Thetis
Stockton 1890; Butler 1891.

"commercial and whaling interests"
Stockton 1890:172.
first whaleships to Herschel Island
Stockton 1890; Butler 1891; Bodfish 1936:44-46; *Grampus* (MNBedfHi), August 1889; USS *Thetis*, August 14-17, 1890 (RNN).
"forbidden sea"
Bodfish 1936:44.
Spy *in Beaufort Sea*
G. B. Borden to James McKenna, letterpress copy book (RP), August 17, 1889.
George Leavitt
Grampus (MNBedfHi), August 26, 1889; James T. White diary, USRC *Bear* (WaU), August 15, 1889.
Orca *and* Thrasher
Gilbert Borden journal (CtY), September 8, 1889.
Nicoline, *1889*
The Alaskan (Sitka), August 31, 1889; *The Chronicle* (San Francisco), July 3, 1889 and October 3, 1893.
Grampus *and* Mary D. Hume, *1890*
The Weekly Bulletin (San Francisco), March 16, 1892; Bodfish 1936:47-49; *Mary D. Hume* (MNBedfHi), 1890-91.
headhunting
Tower 1907:128; Bodfish 1936:47; *Coral* (MNBedfHi), August 12, 1880; *The Bulletin* (San Francisco), September 28, 1880.
Bodfish's amputation
Bodfish 1936:49.
Herschel Island, 1890-92
Bodfish 1936:50-85; *Mary D. Hume* (MNBedfHi), 1890-92.
size of Herschel Island
West et al. 1965:56.
biological productivity of eastern Beaufort Sea
Wadhams 1976:240-41; Grainger 1974:590, 601-3.
climate of Herschel Island
Pilot of Arctic Canada, vol. 3 (2d ed.), 1968:368.
scurvy
I. O. Stringer diary (CaOTCHAr), May 5, 1893; Bodfish 1936:75; *William Baylies* (RP), 1894-95; *Narwhal* (MH-BA), February 10, 1903.
winter routine, 1890-91
Bodfish 1936:53-69; Bodfish letter (MNBedfHi), 1891.
bowhead migration in Beaufort Sea
Braham, Fraker, and Krogman 1980; Fraker and Bockstoce 1980.
season of 1891
The Weekly Bulletin (San Francisco), November 25, 1891; *Mary D. Hume* (MNBedfHi), July-September 1891; Bodfish 1936:69-72.
Mary D. Hume, *winter of 1891-92 and return to San Francisco*
Bodfish 1936:72-85; unidentified clipping, scrapbook 16 (MNBedfHi), p. 168: *The Chronicle* (San Francisco), September 29 and October 1, 1892.
Pacific Steam Whaling Company operations
I. O. Stringer diary (CaOTCHAr), May 5, 1893, September 8, 1894; J. N. Knowles to M. A. Healy, M. A. Healy Papers, (AkSJ); *Narwhal* (PC), September 23, 1892, July 21, 1893; *Newport* (MH-BA), October 9, 1896, January 7, 1897; *Mary D. Hume* (MH-BA), December 11, 1893.
"Cape Bathurst or bust the boiler"
Orca (IMA-MNBedfHi), August 5, 1892.
Cape Bathurst, 1892
Narwhal (MNBedfHi), August 19, 1892.
season of 1893
I. O. Stringer correspondence (CaOTCHAr), January 20, 1894; *Narwhal* (MNBedfHi), July-September 1893; *Mary D. Hume* (MH-BA), September 30, 1893.
season of 1894
Russell 1898:151-52; *Newport* (MNBedfHi) July-September 1894.
set out for wintering voyage
outfit book, *Narwhal* (MNBedfHi); West et al. 1965:26, 54, 57; J. A. Cook 1926:32; *Belvedere* (MNBedfHi), April 8, 1897; *Grampus* (MH-BA), May 16, 1899; *Mary D. Hume* (MH-BA), August 10, 1893; I. O. Stringer diary (CaOTCHAr), May 11, 1893, August 23, 1900.
preparations for winter before freeze-up
Frank Russell diary (DSI), August 19, 1894; West et al. 1965:55; J. A. Cook 1926:55-57; *William Baylies* (MNBedfHi), October 2, 1894; *Mary D. Hume* (MH-BA), September-November 1893 and 1895; *Triton* (MNBedf), September 19, 1894; I. O. Stringer diary (CaOTCHAr), May 4, 1893; *Rosario* (MShaK), October 13, 1894.
preparations for winter after freeze-up
West et al. 1965:55-56; A. J. Allen 1978:21-22; *Newport* (MH-BA), October-November 1896; *Narwhal* (MNBedfHi), October-November, 1892 and 1893; *Mary D. Hume* (MH-BA), October-November 1893-95; *Narwhal* (MH-BA), October-November 1902; *Newport* (MNBedfHi), October-November 1893; I. O. Stringer diary (CaOTCHAr), May 16, 1893; J. A. Cook 1926:55-88; *Rosario* (MShaK), November 1894; Baragwanaih 1908.
vessels wintering at Herschel
Bockstoce and Batchelder 1978b.
trade with indians
Mary D. Hume (MNBedfHi), October 24, 1890; I. O. Stringer to Sadie Stringer (CaOTCHAr), July 2, 1893; I. O. Stringer diary (CaOTCHAr), October 27, 1897; Krech 1979b:110-11; *Narwhal* (MNBedfHi), May 1, 1893; *William Baylies* (RP), November 24, 1894; *Mary D. Hume* (MNBedfHi), December 31, 1893, November 5, 1894, December 1, 1894; J. A. Cook 1926:59, 75-76.
Kogmullicks
J. A. Cook 1926:263; *Narwhal* (MH-BA), May 9, 1903.
meat procurement
West et al. 1965:58; Brower, ms., 427, 482; Edson to Board of Missions (TxAuCH), July 1, 1896; Nuligak 1971:31, 34; *Newport* (MH-BA), February 17, 1897; *Williams Baylies* (RP), November 8, 1894, January 30, 1895; Woolfe 1893:138; Harrison 1908:81-82, 103; Stefansson diary (NhD), 1906, p. 139; Amsden 1979:398-402; J. A. Cook 1926:92, 259, 305; Russell 1898:151; *Rosario* (MShaK), February 23, 1895; Stone 1900:51; *Mary D. Hume* (MH-BA), 1894-96; *Triton* (MNBedf), February 4, 1895; Graham 1935:96; Burch 1980:291, 294.
depletion of caribou herd
Bockstoce 1980b.
"mixture of wooden and canvas buildings . . ."
William Baylies (RP), 1894.
trade jargon
Stefansson 1909; Russell 1898:141.

"sombrero"
Russell 1898:146.
Hudson's Bay Company
Frank Russell diary (DSI), July 13, 1894; Constantine 1903; Inspection Report, Peel River post, March 1, 1893 (HBC); John Firth to Commissioner, May 30, 1897 (HBC); Steward 1955:262; Inspection Report, Mackenzie River District, October 8, 1910 (HBC); St. Matthew's Mission Journal (CAOTCHAr), July 1896; Commissioner to Deputy Minister of Marine and Fisheries, October 20, 1893 (HBC).
"converted Herschel Island . . . into a hive of debauchery"
Jenness 1964:14.
"Down the gangplanks . . ."
Godsell 1941.
D. M. Howard's report
G. B. Ravndal to Secretary of State, June 22, 1906, quoted in Leet 1974:46.
alcohol and prostitution
John Firth to Commissioner, May 30, 1897 (HBC); West 1965:64; Stefansson diary (NhD), March 2, 1907; St. Matthew's Mission Journal (CaOTCHAr), pp. 21, 24; Frank Russell diary (DSI), August 8, 1894; Bodfish 1967; W. S. Mason 1900:72; Report to Commanding Officer, RCMP, August 21, 1903 (CaOOR).
Captain E. W. Newth
Stefansson diary (NhD), August 11, 1906; John Driggs to Bishop Rowe (TxAuCH), January 19, 1904; Brower, ms., 398; Whittaker, nd:240, 248; death certificates, St. Thomas's Mission, Point Hope, Alaska, September 9, 1901; H. R. Thornton to Sheldon Jackson (PP-PrHi), June 28, 1893.
alcohol consumption
I. O. Stringer diary (CaOTCHAr), September 8 and 10, 1894, September 3, 1895, October 30, 1897, August 26, 1899; *Mary D. Hume* (MH-BA), August 25, 1894, September 2, 1896; W. S. Mason 1900:81; St. Matthew's Mission Journal (CaOTCHAr), February 8, 1897; *Newport* (MNBedfHi), September 1, 1896, April 23 and May 24, 1897; Seale 1946:108.
Anglican mission
Peake 1966; *Narwhal* (MNBedfHi), May 1, 1893; St. Matthew's Mission Journal (CaOTCHAr), 1893, p. 15; I. O. Stringer diary (CaOTCHAr), 1893; I. O. Stringer to Sadie Stringer (CaOTCHAr), June 3, 1893.
Pysha
Brower, ms., 547; Brower 1942:221-24; A. J. Allen 1978:36-37, I. O. Stringer diary (CaOTCHAr), August 25, 1901; *Mary D. Hume* (MH-BA), December 3, 1895; *Newport* (MNBedfHi), December 3, 1985; *Beluga* (MNBedfHi), December 3, 1895.
police
The Daily Colonist (Victoria, British Columbia), March 17, 1896; Zaslow 1971:252-69; Alexander (MEdDHi), August 16, 1903; J. A. Cook 1926:259; *Narwhal* (MN-BA), November 18, 1903 and 1906-7.
winter routine
Bodfish 1936:104-5; J. A. Cook 1926: 259; West et al. 1965:55, 59-60.
desertions
Wright 1976:165 ff.; *Narwhal* (MNBedfHi), May 29, 1893, October 24, 1893; *Mary D. Hume* (MH-BA), October 23, 1893, March 25, 1896; *Navarch* (MNBedfHi), January 21, 1896; I. O. Stringer diary (CaOTCHAr), September 14, 1896, April 18, 1898; *Newport* (MNBedfHi), January-March 1896, June 18, 1897; A. J. Allen 1978:24-27; John Bertonccini, ms. (untitled, undated) (MNBedfHi).
games at Herschel
Narwhal (MNBedfHi), 1892-94; *William Baylies* (MNBedfHi), 1894-95; *William Baylies* (RP), 1894-95.
baseball
Bodfish 1936:108; *Newport* (MNBedfHi), February-May 1894; A. J. Allen 1978:38-40; *Mary D. Hume* (MH-BA), April 9, 1894; *Newport* (MH-BA), March 7-8, 1897.
wives and families
J. A. Cook 1926:57-58; *The Bulletin* (San Francisco), June 21, 1894; *Mary D. Hume* (MH-BA), 1894-95; George W. Smith, nd; *William Baylies* (RP), 1894-95; *Newport* (MNBedfHi), 1895-96; *Beluga* (MNBedfHi), 1895-96; Bodfish 1936:114-16; *The Chronicle* (San Francisco), November 22, 1896; Sophie Porter journal (*Jesse H. Freeman*) (MNBedfHi) *Thrasher* (MShaK).
trousers "very full aft"
William Baylies (RP), November 9, 1894.
"Miss Dorothy Porter, aged 5 years"
William Baylies (RP), December 10, 1894.
Captain Whiteside's pig
J. A. Cook 1926:76-81; *Newport* (MNBedfHi), April 8, 1896; West et al. 1965:61-62; Bodfish 1936:137.
spring preparations for cruise
Narwhal (MNBedfHi), April-July 1893; *Triton* (MNBedf), April 11, 1895; J. A. Cook 1926:229; I. O. Stringer diary (CaOTCHAr), May 3-4, 1893; *Mary D. Hume* (MH-BA), May 10, 1894, June-July 1896; *Beluga* (MNBedfHi), May 30, 1896; *Newport* (MNBedfHi), May 1895; Bodfish 1936:123.
Fourth of July
Mary D. Hume (MH-BA), July 4, 1894, July 4, 1895; *Newport* (MNBedfHi), July 4, 1894; *Rosario* (MShaK), July 4, 1895; *Narwhal* (MH-BA), July 4, 1904. *Narwhal* (MNBedfHi), July 4, 1894.

Chapter 13

enmity between Mrs. Whiteside and Mrs. McGregor
Brower, ms., 497; West 1965:54.
Brower's description of Navarch*'s imprisonment*
Brower, ms., 497.
events leading to and including Navarch*'s first abandonment*
Navarch (MNBedfHi), July and August 1897; Brower, ms., 496-97; *The Chronicle* (San Francisco), September 17, 1898. Several accounts disagree in the dates of the two abandonments of the *Navarch*. I have relied on the logbook.
return to Navarch *and second abandonment*
Brower, ms., 500; Buxton, ms. (PC), 25.
Brower's ice drift
Brower, ms., 500-509; *Rosario* (MEdDHi), August 17, 1897.
rescue of Whiteside's party
USRC *Bear* (DNA), August 17, 1897
other events of late August 1897.
Belvedere (MNBedfHi, no. 222), August 1897.

ice movements near Point Barrow
Hegemann 1890:425, 433.
events of early September 1897
Belvedere (MNBedfHi, no. 222), September 1897 passim; *Belvedere* (MNBedfHi, no. 545), September 1897; *Alexander* (MNBedfHi), September 1897; Brower, ms., 510; McIlhenny, ms.(PC), 40; A. J. Allen 1978:46-52.
the Alexander's *escape*
The Call (San Francisco), November 4, 1897.
"get out without bucking"
McIlhenny, ms.(PC), 39-40.
Rosarios's *fate*
Rosario (MEdDHi), 1897-98.
fate of Belvedere, Orca *and* Jesse H. Freeman
Belvedere (MNBedfHi, no. 222), September 1897; Brower, ms., 510 and 532; McIlhenny, ms.(PC), 47 and 55-57; *The Call* (San Francisco), April 9, 1898; Buxton, ms.(PC), 50-51.
burning of Jesse H. Freeman
A. J. Allen 1978:52
Roys's report of his discovery
The Friend (Honolulu), November 1, 1848.
Maury and E. and G. W. Blunt
Friis 1971:9; Burstyn 1957.
Blunts's chart of Bering Strait
"Polar, Behring Sea and Strait, from english & russian surveys," New York, E. & G. W. Blunt, 1849.
Senate resolution and Maury's reply
Hunt's Merchants' Magazine, August 1852, pp. 227-28; Seward 1852.
Vincennes *expedition*
USS Vincennes (DNA), July-September 1855; Nourse 1884:108-31; Friis 1971:14-16; Cole 1947; Cole 1968:136-56.
primary Vincennes *chart*
"Behring's sea and Arctic ocean from surveys in the U.S. North Pacific surveying expedition in 1855, and from russian & english authorities. Compiled by E. R. Knorr." U.S. Navy Department, Bureau of Navigation, Hydrographic Office, 1855.
Alaska purchase
Kushner 1972:90-95; Henry W. Clark 1972:62.
government of Alaska
D. J. Ray 1975a:186.
history of Revenue Marine's northern cruises
Bailey 1880a:17; Murphy 1968:3-6, D. J. Ray 1975a:190; Ransom and Engel 1964:111.
whalemen's attitude to revenue cutter's surveillance
Healy 1889:19.
"whiskey traffic . . . ceased "
Healy 1889:17.
Michael A. Healy
Murphy 1968:12-14; M. A. Healy Papers (CsmH); M. A. Healy Papers, Sheldon Jackson College, Sitka, Alaska.
Mabel *and* George and Susan
Healy 1887:9-10; *Belvedere* (MNBedfHi), August 12, 1885; *Lucretia* (MNBedfHi), August 11, 1885; *Young Phoenix* (MNBedfHi), August 11, 1885.
J. B. Vincent
USRC *Bear* (DNA), July 17, 1887; Healy to Secretary of the Treasury, November 26, 1887, Healy Papers, Sheldon Jackson College, Sitka, Alaska; Aldrich 1889:42-44, 77, 188-95.
examples of revenue cutter's support services
Aldrich 1889:78-79; USRC *Bear* (DNA), August 14, 1886, August 4, 1888, June 2, 1889; August 6, 1894; August 15, 1895.
1880 petition for a refuge station
The Chronicle (San Francisco), November 25, 1880.
wreck of the Daniel Webster
Hooper 1884:71-72; Muir 1917:204-8; E. W. Nelson 1899:300; unidentified newspaper clipping (MNBedfHi), scrapbook 17, p. 25.
"ground as fine as matches"
The Daily Post (San Francisco), October 7, 1882.
lease meterological station
The Chronicle (San Francisco), November 25, 1888.
overland surveys for whalers' escape route
Murphy 1968:19-37.
Healy's proposal for a refuge station
Healy to Sumner Kimball, July 24, 1884, and petition to the Secretary of the Treasury, November 6, 1884, Refuge Station correspondence, Records of U.S. Coast Guard, Record Group 26 (DNA) (hereafter referred to as Refuge Station correspondence); Healy 1889:10-11.
Knowles's proposal
Knowles to General Superintendent of Lifesaving Service, November 10, 1884, Refuge Station correspondence.
Eliza, Jane Gray, *and* Bounding Billow
The Weekly Bulletin (San Francisco), September 12, 1888; *The Chronicle* (San Francisco), September 7 and 15, 1888; Wreck Reports, Atlantic Companies, New York, New York.
Loss of Jane Gray, Young Phoenix, *and* Fleetwing
The Chronicle (San Francisco), September 7, 1888; Wreck Reports, San Francisco District, Records of Bureau of Customs, Record Group 36, (DNA); Brower, ms., 333; *Triton* (MNBedf), August 3, 1888; *Fleetwing* (MNBedf), August 4, 1888.
The *Jane Gray* was found capsized by USS *Thetis*. She was righted and towed to Port Clarence for jury repairs. In an earlier article (Bockstoce 1977b) I stated incorrectly that the *Young Phoenix* had collided with the *Mary and Susan*.
"regular picnic"
Triton (SCS), August 3, 1888.
loss of Ino
Brower, ms., 333 and 336-40.
damage to Reindeer, John Howland *and other vessels*
Whalemen's Shipping List (New Bedford), September 11, 1888; USRC *Bear* (DNA), August 4, 1888.
Bear *returns with shipwrecked sailors*
Whalemen's Shipping List (New Bedford), September 11, 1888; *The Chronicle* (San Francisco), September 7, 1888; *The Weekly Bulletin* (San Francisco), September 12, 1888.
vessels trapped in autumn of 1888
The Examiner (San Francisco), October 20 and 28, 1888; *William Baylies* (RP), October 1, 1888; *Rosario* (MShaK), September 26-28, 1888; unidentified newspaper clipping, scrapbook, p. 59; Healy papers

(CsmH); *The Sun* (Washington, D. C.), October 31, 1897.
The trapped vessels were the *Hunter, Sea Breeze, Ocean, Andrew Hicks, Lancer, William Baylies, Hidalgo, Rosario, Eliza, John P. West,* and "big" *Ohio.*

"like a rat in a hole"
Unidentified newspaper clipping, scrapbook, p. 59, Healy Papers (CsmH).

1888 clamor for refuge station and appropriation
For instance: *The Examiner* (San Francisco), October 28, 1888; *The Chronicle* (San Francisco), October 28, 1888; *The Sun* (Baltimore), June 4, 1889; Isaac B. Thompkins to R. T. Davis, December 12, 1888 and passim, Refuge Station correspondence; H. R. 12215, 50th Congress, 2d session.

Borden, Bayne and the refuge station
For a more comprehensive review of the station's history, see Bockstoce 1979.

candidates for superintendent
Refuge Station correspondence, passim.

build refuge station
USRC *Bear* (DNA), July 27-August 16, 1889.

Brower on Borden
Brower, ms., 383.

Borden complains about Stevenson's appointment
Borden to Secretary of the Treasury, September 1, 1890, Refuge Station correspondence.

Borden and Bayne in conspiracy
Borden, Refuge Station journal (CtY), November 21, 1890; Leander Stevenson to Healy, February 14, 1891, Refuge Station correspondence.

Borden threatens Stevenson
Stevenson to Sheldon Jackson, June 30, 1891, Sheldon Jackson Papers (PPPrHi); Stevenson to Healy, July 19, 1892, Refuge Station correspondence.

Borden in regard to Stevenson's "insubordination"
Borden to Secretary of the Treasury, February 17, 1891, Refuge Station correspondence.

Borden attempts to explain sale of Silver Wave
Mary Healy diary (CsmH), 1891 passim; proclamation to Secretary of the Treasury from whaleship owners (CsmH), December 15, 1892; Borden to Secretary of the Treasury, August 23, 1891, April 14, 1892, Refuge Station correspondence.

sale of Silver Wave *and events of winter of 1891-92*
Statements to Healy from Peter Lieb, Conrad Siem, and George F. Smith, 1892. Refuge Station correspondence; Borden to Stevenson, October 1, 1891, Refuge Station correspondence; Borden to Secretary of the Treasury, September 16, 1891, Refuge Station correspondence; Station Log, Point Barrow Refuge Station (DNA), Record Group 26; Langford file, papers incoming, December 5, 1891, Sheldon Jackson Collection (TxAuCH).

call for Borden's ouster
petition to Secretary of the Treasury, February 29, 1892, Refuge Station correspondence.

events of summer of 1892
Healy to Secretary of the Treasury, March 10 and September 22, 1892, Refuge Station correspondence; Sheldon Jackson to Martin Gantz, January 26, 1892, Refuge Station correspondence; Borden, Refuge Station journal (CtY), July 29, 1892; Healy to L. G. Sheppard, September 29, 1892, Alaska file, Revenue Cutter Service, Record Group 26 (DNA).

support for refuge station wanes
Knowles to Perkins, February 4, 1896, Healy to Secretary of the Treasury, September 25, 1892, Refuge Station correspondence.

Government sells station
C. S. Hamlin to Hooper, March 27 and December 23, 1896 passim, Refuge Station correspondence.

Pacific Steam Whaling Company abandons Point Barrow operation
Brower, ms., 390.

Brower and McIlhenny organize relief
McIlhenny, journal (PC), 48-59; Brower, ms., 525.

send Tilton and Walker south
McIlhenny, journal (PC), 60-68; *The Chronicle* (San Francisco), April 18, 1898.

gather food at Point Barrow
McIlhenny, journal (PC), 42; McIlhenny, diary (PC), May 4, 1888; Brower, ms., 517.

work on Navarch *at Point Barrow*
McIlhenny, Journal (PC), 89-95.

expedition north to Point Barrow
U.S. Treasury Department 1899.

conditions at Point Barrow
Brower, ms., 517 and 526.

Driggs's opinion of relief expedition
Driggs to Joshua Kimber, November 20, 1899, Alaska Missionary Papers (TxAuCh).

ships escape from Point Barrow, 1898
U.S. Treasury Department 1899:131-34.

Chapter 14

search for wintering sites eash of Herschel Island
I. O. Stringer diary (CaOTCHAr), May 5, 1893; *Narwhal* (PC), July 21, 1894; *The Chronicle* (San Francisco), October 3, 1893; *The Weekly Bulletin* (San Francisco), October 4, 1893, June 21, 1894; *The Morning Call* (San Francisco), November 12, 1893.

Balaena *and* Grampus *at Balaena Bay, 1895-96*
A. J. Allen 1978:20; author's field notes; I. O. Stringer diary (CaOTCHAr), September 1, 1896.

merchant skipper scheme
Bodfish 1936:87-88; I. O. Stringer diary (CaOTCHAr), May 5, 1893; A. J. Allen 1978:46-47.

establish shore whaling Eskimos at Cape Bathurst
Russell 1898:151; Frank Russell diary (DSI), August 14, 1894.

1895 season
Henry Koenig diary (AkU), August 24, 1895; *Mary D. Hume* (MH-BA), July-August 1895; West et al. 1965:55-56: USS *Bear* (DNA), 1895.

1896 season
Unidentified newspaper clipping, scrapbook T-4 (MNBedfHi), p. 23; Seale 1946:110; *Mermaid* (MShaK), 1896; Report of U.S. Commissioner for Eduation for 1896, p. 1459; J. A. Cook 1926:109.

1897 season
Chapter 13, this volume.
wintering at Langton Bay, 1897-98
Bodfish 1936:151-58; *Narwhal* (MedDHi), 1897-98; *Beluga (MNBedfHi), 1897-98; author's field notes.*
deterioration in crews
The Evanglist, April 30, 1891; *Balaena* (MH-BA), 1901; *Narwhal* (MH-BA), May 20, 1904; *Grampus* (MH-BA), March 25, 1899.
Baillie Island's drawbacks
J. A. Cook 1926:215; A. J. Allen 1978:103; *Grampus* (MH-BA), September 17 and 22, October 1, 1899.
winter of 1898-99 at Baillie Island
Bodfish 1936:164-74; *Beluga* (MNBedfHi), 1898-99; unidentified newspaper clipping, scrapbook 2 (MNBedfHi), November 4, 1905.
Sophia Sutherland
Bodfish, 1936:164-174; *Beluga* (MNBedfHi), 1898-1899; unidentified newspaper clipping, scrapbook 2 (MNBedfHi), November 4, 1905.
Penelope
Unidentified newspaper clipping, scrapbook T-4 (MNBedfHi), p. 170; John Bertonccini, sketchbook (MShaK); *Narwhal* (MH-BA), August 17, 1902; Stefansson 1922:58-59; Nuligak 1971:31.
winter of 1900-1901 at Baillie Island
Bodfish 1936:178-90; *Beluga* (MNBedfHi), 1900-1901; I. O. Stringer diary (CaOTCHAr), August 4-9, 1900; Anderson 1924;506-7; Stone 1901:86; Anderson 1913:186; Cook and Pederson 1937:183-84; Innis 1977:357; Anonymous 1902.
1901 season
Bodfish 1936:191; J. A. Cook 1926:234.
"greatest of marine lotteries"
Unidentified newspaper clipping, scrapbook 16 (MNBedfHi), p. 218.
gasoline auxiliaries
Unidentified newspaper clipping, scrapbook 2 (MNBedfHi), p. 112.
winter of 1905-6
Bodfish 1936:226; *William Baylies* (MNBedfHi), August, 1905; J. A. Cook 1926:285-337; *The Examiner* (San Francisco), November 4, 1905; Faber 1916.
Klengenberg
Stefansson 1922:48-56; Stefansson 1912:306; MacInnes 1932:61, 239; Brower, ms., 554, 638; *Bowhead* (CaACG), August 10, 1906; Joseph Bernard, ms. (AkU), II:303.
McKenna
A. J. Allen 1978:140; J. W. Kelly to M. A. Healy, August 9, 1892, Alaska File, RG 26 (DNA); Sheldon Jackson 1902:122; *Polar Bear* (PC), June 29, 1912.
1906 season
Stefansson 1922:57; *Alexander* (MedDHi), 1906; Stefansson 1909a:26-27.
"whaling grounds . . . need a rest"
The Morning Mercury (New Bedford), March 3, 1906.
farthest cruises
Buxton, ms. (PC), October 27 and November 14, 1897; A. J. Stone diary (CU-BANC); Faber 1916;139-42; *The Chronicle* (San Francisco), October 28-29, 1899, November 3, 1902, October 31, 1904; *Beluga* (MNBedfHi), October 7, 1902; Vilhjalmur Stefansson, diary (NhD), August 10, 1906; Ross and MacIver 1982:map 10.
depletion of bowhead stock
The Herald (New York), February 2, 1908; Anonymous [1906]:163.
change in fashion
Ewing, 1981:113.
1908 season
Tilton 1969:261; J. A. Cook 1926:337; unidentified newspaper clipping, scrapbook 2 (MNBedfHi), p. 246.
1909 season
Stefansson 1912:106; *New Westminster Columbian* (Victoria, B.C.), March 10, 1909.
whalebone trust
Unidentified newspaper clippings (MNBedfHi), scrapbook T-4, pp. 2, 78, and 380.

Bibliography

Aldrich, Herbert L.
1889 *Arctic Alaska and Siberia, or Eight Months with the Arctic Whalemen.* Chicago and New York: Rand McNally and Company.

Alexander, Vera
1974 "Primary Productivity Regions of the Nearshore Beaufort Sea, with Reference to Potential Holes of Ice Biota," in *The Coast and Shelf of the Beaufort Sea.* Washington, D.C.: Arctic Institute of North America.

Allen, Arthur James
1978 *A Whaler and Trader in the Arctic.* Anchorage: Alaska Northwest Publishing Company.

Allen, Everett S.
1973 *Children of the Light: The Rise and Fall of New Bedford Whaling and The Death of the Arctic Fleet.* Boston: Little, Brown and Company.

Allen, Joel Asaph
1880 *History of North American Pinnipeds.* U.S. Geological and Geographical Survey of the Territories, Misc. Publication 12, Washington, D.C.: Government Printing Office.

Allen, William H.
n.d. *Excerpts from Journal of a Voyage in the Ship* Bengal, *of New London, Connecticut, by Wm. Allen, 3rd Mate.* Typescript mimeograph. New Bedford, Mass.: Old Dartmouth Historical Society.

Amsden, Charles W.
1979 "Hard Times: A Case Study from Northern Alaska and Implications for Arctic Prehistory," in *Thule Eskimo Culture: An Anthropological Retrospective,* ed. Allen P. McCartney. Archaeological Survey of Canada, no. 88, National Museum of Man, Mercury Series, Ottawa.

Anderson, Rudolph Martin
1913 "Notes on Muskoxen," in J. A. Allen, "Ontogenic and Other Variations in Muskoxen, with a Systematic Review of the Muskox Group, Recent and Extinct," *Memoirs of the American Museum of Natural History,* n.s., vol. 1, part 4.
1924 "Report on the Natural History Collections of the Expedition," in Vilhjalmur Stefansson, *My Life with the Eskimo.* New York: Macmillan Company.

Anonymous
1858 "Whaling and Whaling Grounds," *The Nautical Magazine and Naval Chronicle,* vol. 27.

Anonymous
1868 Letter to George R. Philips, Esq., New Bedford, 12 mo. 1868 [prob. by Humphrey Seabury].

Anonymous
1902 "The Musk-Ox," *Zoological Society Bulletin,* New York Zoological Society.

Anonymous
1906 "Whaling in the North Pacific," in *Fisheries: A World Industry.* Unidentified broadside, Old Dartmouth Historical Society Collection.

Ansel, Willits D.
1978 *The Whaleboat: A Study of Design, Construction and Use from 1850 to 1970.* Mystic, Conn.: Mystic Seaport, Inc.

Apollonio, Spencer
1971 "The Arctic: Deep Freeze Fountain of Life," *Oceans Magazine,* vol. 4, no. 1

Ashley, Clifford W.
1926 *The Yankee Whaler.* Boston and New York: Houghton Mifflin Company.

Bailey, Alfred M., and Russell W. Hendee
1926 "Notes on the Mammals of Northwestern Alaska," *Journal of Mammology,* vol. 7, no. 1

Bailey, George W.
1880 Letter to Hon. John Sherman, Secretary of the Treasury, in letter from the Secretary of the Treasury, U.S. Senate, 46th Congress, 2d Session, Ex. Doc. no. 132.
1880a *Report Upon Alaska and Its People.* U.S. Revenue Marine Service. Washington, D.C.: Government Printing Office.

Baker, Marcus, annd others
1893 "An Undiscovered Island Off the Northern Coast of Alaska," *National Geographic Magazine,* vol. 4.
1906 *Geographic Dictionary of Alaska,* 2d edition, Department of the Interior, U.S. Geological Survey. Washington, D.C.: Government Printing Office.

Baker, William A.
1977 "The Design and Construction of Steam Whalers," in John Bockstoce, *Steam Whaling in the Western Arctic.* New Bedford, Mass.: Old Dartmouth Historical Society.

Baragwanaih, J.
1908 "Wintering on a Whaler," *Outdoor Life* (December).

Barker, F. A.
1870–71 Journal. Typescript, Turnbull Library, Auckland, New Zealand.

Barnett, Don G.
1976 "Long Range Ice Forecasting for Alaska's North Coast," *Sea Technology,* vol. 17, no. 7.

Bean, Tarleton H.
1902 "A Naturalist's Adventures," in *The White World,* ed. Rudolf

Kersting. New York: Lewis, Scribner and Company.

Beane, J. F.
1905 *From Forecastle to Cabin.* New York: Editor Publishing Company.

Bee, James W., and E. Raymond Hall
1956 *Mammals of Northern Alaska on the Arctic Slope.* Lawrence, Kan.: University of Kansas Museum of Natural History.

Beechey, Frederick William
1831 *Narrative of a Voyage to the Pacific and Beering's Strait.* London: Henry Colburn and Richard Bentley.

Bernard, Joseph F.
n.d. *Arctic Voyages of the Schooner* Teddy Bear. Manuscript, Archives of the University of Alaska, Fairbanks.

Bertholf, Ellsworth P.
1899 "The Rescue of the Whalers," *Harpers New Monthly Magazine,* vol. 99, no. 589.

Bertonccini, John
n.d. Diary, Liebes Collection, California Academy of Sciences, San Francisco.

Bertrand, Kenneth J.
1971 *Americans in Antarctica, 1775–1948.* New York: American Geographical Society.

Berzin, A. A., and A. A. Rovnin
n.d. "The Distribution and Migrations of Whales in the Northeastern Part of the Pacific, Chukchi and Bering Seas." Translation, Language Service Division, Office of International Fisheries, Marine Fisheries Service, National Oceanic and Atmospheric Administration, U.S. Department of Commerce, Washington, D.C.

Bezrukov, P. L., ed.
1964 *Geographical Description of the Bering Sea.* Jerusalem: Israel Program for Scientific Translations.

Blomkvist, E. E.
1972 "A Russian Scientific Expedition to California and Alaska, 1839–1849," trans. Basil Dmytryshyn and E. A. P. Crownhart-Vaughan, *Oregon Historical Quarterly,* vol. 73, no. 2.

Bockstoce, John R.
1975 "Contracts Between American Whalemen and the Copper Eskimos," *Arctic,* vol. 28, no. 4.
1977a *Steam Whaling in the Western Arctic.* New Bedford, Mass.: Old Dartmouth Historical Society.
1977b "The Arctic Whaling Disaster of 1897," *Prologue: The Journal of the National Archives,* Spring.
1977c "Eskimo Whaling in Alaska," *Alaska Magazine,* vol. 43, no. 9.
1977d *Eskimos of Northwest Alaska in the Early Nineteenth Century.* Oxford: Pitt Rivers Museum, Monograph Series, no. 1.
1979 "Arctic Castaway: The Stormy History of the Point Barrow Refuge Station," *Prologue: The Journal of the National Archives,* vol. 11, no. 3.
1980 "A Preliminary Estimate of the Reduction of the Western Arctic Bowhead Whale Population by the Pelagic Whaling Industry: 1848–1915," *Marine Fisheries Review,* vol. 42, no. 9-10.
1980b "The Consumption of Caribou by Whalemen at Herschel Island, Yukon Territory, 1890—1908," *Arctic and Alpine Research,* vol. 12, no. 3.
1980c "Battle of the Bowheads," *Natural History,* vol. 89, no. 5.

Bockstoce, John R., and Charles F. Batchelder
1978a "A Gazetteer of Whaler's Place Names for the Bering Strait Region and the Western Arctic," *Names,* vol. 26, no. 3, pp. 258-70.
1978b "A Chronological List of Commercial Wintering Voyages to the Bering Strait Region and Western Arctic of North America, 1850–1910," *American Neptune,* vol. 38, no. 2, pp. 81-91.

Bockstoce, John R., and Daniel B. Botkin
1982 "The Harvest of Pacific Walruses by the Pelagic Whaling Industry, 1848–1914," *Arctic and Alpine Research,* vol. 14, no. 3, pp. 183-88.
1983 "The Historical Status and Reduction of the Western Arctic Bowhead Whale *(Balaena mysticetus)* Population by the Pelagic Whaling Industry, 1848–1914," *Scientific Reports of the International Whaling Commission,* Special Issue, no. 5, pp. 107-41.

Bodfish, Hartson H.
1936 *Chasing the Bowhead.* Cambridge, Mass.: Harvard University Press.
1967 "A Letter from Herschel Island—1891," *The Dukes County [Mass.] Intelligencer,* vol. 8, no. 3.

Bogoras, Waldemar
1909 "The Chukchee," *Memoirs of the American Museum of Natural History,* vol. 11.

Bogoslovskaya, L. S., et al.
1982 *The Bowhead Whale Off Chukotka: Migrations and Aboriginal Whaling.* Report of the International Whaling Commission, vol. 32.

Bompas, W.C.
"The Esquimaux of the Mackenzie River." Manuscript, CMS, C.Ç. 1./0., 11 and 12. Church Missionary Society, London.

Borden, Gilbert B.
Journal, Aug. 15, 1889–Aug. 23, 1891, Ms S-265, Yale University.

Bourke, Robert H.
1983 "Currents, Fronts and Fine Structure in the Marginal Ice Zone of the Chukchi Sea," *Polar Record,* vol. 21, no. 135, pp. 569-75.

Braham, Howard W.; Floyd E. Durham; Gordon H. Jarrell; and Stephen Leatherwood
1980 "Ingutuk: Preliminary Evaluation of a Morphological Variation of the Bowhead Whale, *Balaena mysticetus,*" *Marine Fisheries Review,* vol. 42, no. 9-10.

Braham, Howard W.; Mark A. Fraker; and Bruce D. Krogman
1980 "Spring Migration of the Western Arctic Population of Bowhead Whales," *Marine Fisheries Review,* vol. 42, no. 9-10.

Brandt, George E.
1929 "Frozen Toes But Not Cold Feet," *Proceedings,* U.S. Naval Institute, vol. 55.

Brower, Charles D.
1942 *Fifty Years Below Zero, A Lifetime of Adventure In the Far North.* New York: Dodd, Mead and Company.
Brower, William A., Jr., et al.
1977 *Climatic Atlas of the Outer Continental Shelf Waters and Coastal Regions of Alaska.* U.S. Department of the Interior, Washington, D.C.
Brown, James Templeman
1883 *The Whale Fishery and Its Appliances.* Great International Fisheries Exhibition, London. Washington, D.C.: Government Printing Office.
1887 "The Whalemen, Vessels, and Boats, Apparatus and Methods of the Whale Fishery," in *The Fisheries and Fishery Industries of the United States.* ed. George Brown Goode, vol. 2, section 5. Washington, D.C.: Government Printing Office.
Brueggeman, John J.
1982 "Early Spring Distribution of Bowhead Whales in the Bering Sea," *Journal of Wildlife Management,* vol. 46, no. 4.
Bulloch, James D.
1959 *The Secret Service of the Confederate States in Europe.* New York: Thomas Yoseloff.
Burch, Ernest S., Jr.
1976 "Overland Travel Routes in Northwest Alaska," *Anthropological Papers of the University of Alaska,* vol. 18, no. 1.
1980 "Traditional Eskimo Societies in Northwest Alaska," in *Alaska Native Culture and History,* ed. Yoshinobu Kotani and William B. Workman. Senri Ethnological Studies 4, National Museum of Ethnology, Senri, Osaka, Japan.
1981 *The Traditional Eskimo Hunters of Point Hope, Alaska: 1800–1875.* Barrow, Alaska: North Slope Borough.
Burns, John J.
1965 *The Walrus in Alaska: Its Ecology and Management.* Juneau: Alaska Department of Fish and Game.
1970 "Remarks on the Distribution and Natural History of Pagophilic Pinnipeds in the Bering and Chukchi Seas," *Journal of Mammology,* vol. 51, no. 3.
Burns, John J.; Francis H. Fay; and Lewis H. Shapiro
1980 *The Relationship of Marine Mammal Distributions, Densities and Activities to Sea Ice Conditions.* Fairbanks: BLM/NOAA Outer Continental Shelf Environmental Assessment Project.
Burns, Walter Noble
1913 *A Year With a Whaler.* New York: Outing Publishing Company.
Burstyn, Harold L.
1957 *At the Sign of the Quadrant.* Marine Historical Association, no. 32, Mystic, Connecticut.
Burwash, L. T.
1931 *Canada's Western Arctic.* Ottawa: Department of the Interior.
Butler, H. G.
1919 "Terrible Experiences in the Arctic," in *The Wide World Magazine,* vol. 43, no. 255, pp. 163-68.
Butler, R. G.
1891 "Where the Ice Never Melts," *Scribner's Magazine,* vol. 9.
Buxton, N. G.
Manuscript, E. A. McIlhenny Collection, Private Collection.
Cameron, Agnes Deans
1910 *The New North.* New York and London: D. Appleton and Company.
Carroll, Geoffrey M., and John R. Smithhisler
1980 "Observations of Bowhead Whales During Spring Migration," *Marine Fisheries Review,* vol. 42, no. 9-10.
Chatfield, Thomas
n.d. *Reminiscences of Captain Thomas Chatfield. Cotuit, Massachusetts.* Typescript, Old Dartmouth Historical Society, New Bedford.
Church, Albert Cook
1938 *Whaleships and Whaling.* New York: W. W. Norton Company.
Clark, A. Howard
1887a "History and Present Condition of the Fishery in the Whale Fishery," in *The Fisheries and Fishery Industries of the United States,* ed. George Brown Goode, vol. 2, section 5, part 15. Washington, D.C.: Government Printing Office.
1887b "The Pacific Walrus Fishery," in *The Fisheries and Fishery Industries of the United States,* ed. George Brown Goode, vol. 2, section 5, part 17. Washington, D.C.: Government Printing Office.
1887c "The Antarctic Fur-Seal and Sea-Elephant Industries," in *The Fisheries and Fishery Industries of the United States,* ed. George Brown Goode, vol. 2, section 3, part 18. Washington, D.C.: Government Printing Office.
1887d "The American Whale Fishery, 1877–1886," *Science,* vol. 9, no. 217.
Clark, Henry W.
1972 *History of Alaska.* Freeport, N.Y.: Books for Libraries Press.
Coachman, L. K.; K. Aagaard; and R. B. Tripp
1975 *Bering Strait: The Regional Physical Oceanography.* Seattle and London: University of Washington Press.
Coachman, L. K., and K. Aagaard
1981 "Reevaluation of Water Transports in the Vicinity of Bering Strait," in *Eastern Bering Sea Shelf,* vol 1. Oceanography and Resources, Office of Marine Pollution Assessment, National Oceanic and Atmospheric Administration, Department of Commerce, Washington, D.C.
Cole, Allan B.
1947 "The Ringgold-Rodgers-Brooke Expedition to Japan and the North Pacific, 1853–1859," *Pacific Historical Review,* vol. 16, pp. 152-62.
Cole, Allan B., ed.
1968 *Yankee Surveyors in the Shogun's Seas.* New York: Greenwood Press.
Collins, J. W.
1891 "Report on the Fisheries of the Pacific Coast," from *Report of Commissioner of Fish and Fisheries for 1888.* Washington, D.C.: Government Printing Office.
Collinson, Richard
1889 *Journal of H.M.S. Enterprise.* London: Sampson Low, Marston, Searle, and Rivington, Ltd.
Cook, Adrian

1975 *The Alabama Claims: American Politics and Anglo-American Relations, 1865–1872.* Ithaca and London: Cornell University Press.

Cook, John A.

1926 *Pursuing the Whale.* Boston: Houghton Mifflin Company.

Cook, John A., and Samson S. Pederson

1937 *Thar She Blows.* Boston: Chapman and Grimes.

Crawford, Richard William

1981 "Whalers from the Golden Gate: A History of the San Francisco Whaling Industry, 1822–1908," M.A. thesis, San Diego State University.

Dall, William H.

1870 *Alaska and Its Resources.* Boston: Lee and Shepard Company.

1881a "Notes on Alaska and the Vicinity of Bering Strait," *The American Journal of Science,* no. 122, vol. 21, 3d series.

1881b "United States Survey Operations in Neighborhood of Bering Strait," *Proceedings of the Royal Geographical Society,* n.s., vol. 3, no. 1.

1881c "On the So-called Chukchi and Namallo People of Eastern Siberia," *American Naturalist,* vol. 15.

1882 "Report on the Currents and Temperatures in the Bering Sea and Adjacent Waters," Appendix 16 in *Report of the Superintendent of the U.S. Coast and Geodetic Survey, showing the progress of the work during the fiscal year ending with June 1880.* Washington, D.C.: Government Printing Office.

1899 "How Long a Whale May Carry a Harpoon," *National Geographic Magazine,* vol. 10, no. 4.

Davidson, George

1868 "Scientific Expedition to Alaska," *Lippincott's Magazine,* vol. 2.

Davis, William M.

1926 *Nimrod of the Sea, or the American Whaleman.* Boston: Charles E. Lauriat Company.

Davoll, Edward W.

1981 *The Captain's Specific Orders on the Commencement of a Whale Voyage to His Officers and Crew.* Historical Sketch, no. 81. New Bedford: Old Dartmouth Historical Society.

Daws, Gavan

1968 *Shoal of Time: A History of the Hawaiian Islands.* Honolulu: University Press of Hawaii.

DeWindt, Harry

1899 *Through the Gold Fields of Alaska to Bering Strait.* London: Chatto and Windus.

1904 *From Paris to New York by Land.* London: George Newnes, Ltd.

Dmytryshyn, Basil, and E. A. P. Crownhart-Vaughan

1979 *The End of Russian America: Captain P. N. Golovin's Last Report, 1862.* Portland: Oregon Historical Society; distributed by the University of Washington Press.

Doty, William Furman

1900a "Log Book, St. Lawrence Island," in Sheldon Jackson, *Ninth Annual Report on Introduction of Domestic Reindeer into Alaska . . . 1899.* Washington, D.C.: Government Printing Office.

1900b "The Eskimo on St. Lawrence Island, Alaska," in Sheldon Jackson, *Ninth Annual Report on Introduction of Domestic Reindeer into Alaska . . . 1899.* Washington, D.C.: Government Printing Office.

Driggs, John B.

1905 *Short Sketches from Oldest America.* Philadelphia: George W. Jacobs and Company.

Du Pasquier, Thierry

1982 *Les Baleiniers Français au XIX ième Siècle (1814–1868).* Grenoble: Terre et Mer.

Dulles, Foster Rhea

1934 *Lowered Boats.* London: George G. Harrap and Company, Ltd.

Dunbar, Moira

1967 "The Monthly and Extreme Limits of Ice in the Bering Sea," in *Physics of Snow and Ice: Proceedings of the International Conference on Low Temperature Science, 1966,* vol. 1, part 1, ed. Hirobumi Oura. Institute of Low Temperature Science, Hokkaido University, Sapporo, Japan.

Elliott, Henry W.

1886 *An Arctic Province: Alaska and the Seal Islands.* London: Sampson, Low, Marston, Searle, and Rivington.

Ellis, Leonard Bolles

1892 *History of New Bedford and Its Vicinity, 1602–1892.* Syracuse, N.Y.: D. Mason and Company.

Ewing, Elizabeth

1981 *Dress and Undress: A History of Women's Underwear.* London: Bibliophile.

Faber, Kurt

1916 *Unter Eskimos und Walfischfangern.* Stuttgart: Robert Lutz.

Fay, Francis H.

1957 "History and Present Status of the Pacific Walrus Population," in *Transactions of the Twenty-second North American Wildlife Conference,* Wildlife Management Institute, Washington D.C.

1982 *Ecology and Biology of the Pacific Walrus,* North American Fauna, no. 74, U.S. Department of the Interior, Washington, D.C.

Fay, Francis H.; Howard M. Feder; and Samuel W. Stoker

1977 "An Estimation of the Impact of the Pacific Walrus Population on Its Food Resources in the Bering Sea," National Technical Information Service, no. PB 273 505, Springfield, Virginia.

Foote, Don Charles

1965 *Exploration and Resource Utilization in Northwestern Arctic Alaska before 1855,* Ph.D. dissertation, Department of Geography, McGill University, Montreal.

Forbes, R. J.

1958 "Petroleum," in *A History of Technology,* vol. 5, ed. Charles Singer. New York: Oxford University Press.

Fraker, Mark A., and John R. Bockstoce

1980 "Summer Distribution of Bowhead Whales in the Eastern Beaufort Sea," *Marine Fisheries Review,* vol. 42, no. 9-10.

Friis, Herman R.

1971 "A Brief Review of Official United States Geographical Exploration of the Arctic Realm . . . of the North Pacific . . . prior to 1914." Typescript. Paper delivered at California

Historical Society, San Francisco.
Frouin, Charles
1978 *Journal de Bord 1852–1856, Charles Frouin, Chirurgien du Baleinier* L'Espadon. Paris: Editions France-Empire.
Gelett, Charles Wetherby
1917 *A Life on the Ocean.* The Advertiser Historical Series, no. 3. Honolulu: Hawaiian Gazette Company, Ltd.
Gilbert, Benjamin F.
1965 "The Confederate Raider *Shenandoah,*" *Journal of the West,* vol. 4, no. 2.
Godsell, Philip H.
1941 "Pirate Days in Arctic Waters," *Forest and Outdoors,* vol. 37.
1951 *Arctic Trader.* London: Robert Hale Ltd.
Graham, Angus
1935 *The Golden Grindstone.* London: Chatto and Windus.
Grainger, E. H.
1974 "Nutrients in the Southern Beaufort Sea," in *The Coast and Shelf of the Beaufort Sea,* Arctic Institute of North America, Washington, D.C.
Great Britain Naval Intelligence Division
1943 *Pacific Islands.* Vol. 2 (Eastern Pacific), B. R. 519B (restricted), Geographical Handbook Series.
Green, Paul, aided by Abbe Abbott
1959 *I Am Eskimo, Aknik My Name.* Juneau, Alaska: Northwest Publishing Company.
Habershaw, Frederick
1883 "Bringing Whale Oil from the Pacific to New York," *Bulletin of the United States Fish Commission.* Vol. 2 for 1882. Washington, D.C.: Government Printing Office.
Hackett, Frank W.
1882 *The Geneva Award Acts . . . .* Boston: Little, Brown Company.
Hadley, Jack
1915 "Whaling Off the Alaskan Coast," *Bulletin of the American Geographical Society,* vol. 47, no. 12.
Hall, Elton W.
1981 "Panoramic Views of Whaling by Benjamin Russell," Old Dartmouth Historical Society Sketch, no. 80, New Bedford, Mass.
1982 *Sperm Whaling from New Bedford: Clifford W. Ashley's Photographs of the Bark Sunbeam in 1904.* New Bedford, Mass.: Old Dartmouth Historical Society.
Hall, Henry
1884 *Report on the Shipbuilding Industry of the United States.* Washington, D.C.: Government Printing Office.
Hammerich, L. L.
1958 "The Western Eskimo Dialects," *Proceedings of the Thirty-second International Congress of Americanists, Copenhagen, 1956.* Copenhagen: Munksgaard.
Hanna, G. Dallas
1920 "Mammals of the St. Matthew Islands, Bering Sea," *Journal of Mammalogy,* vol. 1, pp. 118-22.
Hare, Lloyd C. M.
1960 *Salted Tories, the Story of the Whaling Fleets of San Francisco.* Mystic, Conn.: Marine Historical Association, Inc.
Harlow, Vincent T.
1964 *The Founding of the Second British Empire, 1763–1793.* London: Longmans.
Harrington, John J.
1866–67 *The Esquimaux.* Vol. 1: Port Clarence, Russian America and Plover Bay, Eastern Siberia.
Harrison, Alfred H.
1908 *In Search of a Polar Continent.* London: Edward Arnold.
Healy, M. A.
1889 *Report of the Cruise of the Revenue Marine Steamer* Corwin *in the Arctic Ocean in the Year 1884.* Washington, D.C.: Government Printing Office.
1887 *Report of the Cruise of the Revenue Marine Steamer* Corwin *in the Arctic Ocean in the Year 1885.* Washington, D.C.: Government Printing Office.
Hegarty, Reginald B.
1959 *Returns of Whaling Vessels Sailing from American Ports.* New Bedford, Mass.: Old Dartmouth Historical Society.
Hegemann, P. F. A.
1880 "Bemerkungen uber die Windverhaltnisse in der Umgebung der Bering-Strasse," *Annalen der Hydrographie und Maritimen Meteorologie,* vol. 8.
1890 "Das Eis und die Stromungsverhaltnisse des Beringmeers der Beringstrasse und des Nordlich Davon Belegenen Eismeeres," *Annalen der Hydrographie und Maritimen Meteorologie,* vol. 18, no. 11.
Hendee, Russell W.
1930 "The Hunt on the Flaw," *Atlantic Monthly,* vol. 145, no. 3.
Henderson, David A.
1972 *Men and Whales at Scammon's Lagoon.* Los Angeles: Dawson's Book Shop.
1984 "Nineteenth Century Gray Whaling: Grounds, Catches and Kills, Practices and Depletion of the Whale Population," in *The Gray Whale,* ed. Mary Lou Jones, Steven L. Swartz, and Steven Leatherwood. Orlando, Fla: Academic Press, Inc.
Herendeen, Edward Perry, with Marcus Baker and A. W. Greely
1893 "An Undiscovered Island off the Northern Coast of Alaska," *National Geographic Magazine,* vol. 5.
Hinckley, Ted C.
1972 *The Americanization of Alaska, 1867–1897.* Palo Alto, Calif.: Pacific Books.
Hohman, Elmo Paul
1928 *The American Whaleman: A Study of Life and Labor in the Whaling Industry.* New York: Longmans, Green and Company.
Holmes, Rev. Lewis
1857 *The Arctic Whaleman, or Winter in the Arctic Ocean.* Boston: Wentworth and Company.
Hooper, Calvin Leighton
1881 *Report of the Cruise of the U.S. Revenue Steamer* Corwin *in the Arctic Ocean, 1880.* Washington, D.C.: Government Printing Office.
1884 *Report of the Cruise of the U.S. Revenue Steamer* Thomas Corwin *in the Arctic Ocean, 1881.* Washington, D.C.:

Government Printing Office.
1885 *Report of the Cruise of the U.S. Revenue Steamer* Thomas Corwin *in the Arctic Ocean in 1881.* Washington, D.C.: James Anglim and Company.

Hopkins, David M., ed.
1967 *The Bering Land Bridge.* Stanford, Calif.: Stanford University Press.

Horan, James D., ed.
1960 *C.S.S.* Shenandoah: *The Memoirs of Lieutenant Commanding James I. Waddell.* New York: Crown Publishers, Inc.

Howay, F. W.
1973 "A List of Trading Vessels in the Maritime Fur Trade." Kingston, Ontario: Limestone Press.

Howerton, Joseph B.
1971 *Whaling Logs in the Maury Log Collection: A Resource for Geographical Research.* Draft paper prepared for the Conference on the National Archives and Research in Historical Geography, Washington, D.C.

Hughes, Charles Campbell
1960 *An Eskimo Village in the Modern World.* Ithaca: Cornell University Press.

Hunt, Cornelius E.
1867 *The* Shenandoah, *or the Last Confederate Cruiser.* New York: G. W. Carleton and Company.

Innis, Harold A.
1977 *Fur Trade in Canada.* Rev. ed. Toronto and Buffalo: University of Toronto Press.

Jackson, Gordon
1978 *The British Whaling Trade.* London: Adam and Charles Black.

Jackson, Sheldon
1896 *Report on Introduction of Domestic Reindeer into Alaska, 1896.* Washington, D.C.: Government Printing Office.
1896a "The Arctic Cruise of the U.S. Revenue Cutter 'Bear,'" *National Geographic Magazine,* vol. 7, no. 1.
1900 *Ninth Annual Report on Introduction of Domestic Reindeer into Alaska . . . 1899.* Washington, D.C.: Government Printing Office.
1902 *Eleventh Annual Report on Introductionn of Domestic Reindeer into Alaska . . . 1901.* 57th Congress, 1st Session, Senate Document no. 98, Washington, D.C.: Government Printing Office.

Jarvis, David H.
1899 *Alaska Coast Pilot Notes . . . as Far as Point Barrow.* U.S. Coast and Geodetic Survey, Bulletin 40. Washington, D.C.: Government Printing Office.

Jenkins, J. T.
1921 *A History of the Whale Fisheries.* London: H. F. annd G. Witherby.

Jenness, Diamond
1964 *Eskimo Administration. II: Canada.* Technical Paper, no. 14, Arctic Institute of North America, Montreal.

Jernegan, Jared
n.d. *Account of the Sea Life of Captain Jared Jernegan.* Manuscript, Dukes County Historical Society, Edgartown, Mass.

Johnson, Martin W.
1963 "Arctic Ocean Plankton," in *Proceedings of the Arctic Basin Symposium, October 1962.* Washington, D.C.: Arctic Institute of North America.

Jones, A. G. E.
1981 "The British Southern Whale and Seal Fisheries," *The Great Circle. Journal of the Australian Association for Maritime History,* vol. 3, no. 1, Nedlands, Western Australia.

Jones, Tim
1979 "Return of the Pacific Walrus: Too Much of a Good Thing?" *Alaska Magazine* (September).

Jones, [William Benjamin] Bill
1927 *The Argonauts of Siberia.* Philadelphia: Dorrance and Company.

Kellett, Henry
1850 "Narratives of the Proceedings of Captain Kellett," Great Britain Parliament, House of Commons, Sessional Papers, Accounts and Papers, Arctic Expeditions, vol. 35, no. 107.

Kelly, John W.
1890 "Ethnographical Memoranda Concerning the Eskimos of Arctic Alaska and Siberia," *Society of Alaskan Natural History and Ethnology,* Bulletin no. 3, Sitka.
1897 "Report of John W. Kelly, Purchasing Agent," in Sheldon Jackson, *Report on Introduction of Domestic Reindeer into Alaska 1897,* Senate Document 30, 55th Congress, 2d Session. Washington, D.C.: Government Printing Office.

Kilian, Bernhard
1983 *The Voyage of the Polar Bear: Whaling and Trading in the North Pacific and Western Arctic, 1913–1914,* ed. John R. Bockstoce, Old Dartmouth Historical Society and Alaska Historical Commission.

King, Judith E.
1974 *Seals of the World.* London: British Museum (Natural History).

Kovacs, Austin, and Malcolm Mellor
1974 "Sea Ice Morphology and Ice as a Geologic Agent in the Southern Beaufort Sea," in *The Coast and Shelf of the Beaufort Sea.* Washington, D.C.: Arctic Institute of North America.

Krause, Arthur, and Aurel Krause
1881 "Die wissenschaftliche Expedition der Bremer geographischen Gesellschaft nach den Kustengebeiten an der Beringsstrasse," *Deutsche Geographische Blatter,* vol. 4, Bremen.
1882 "Die Expedition der Bremer geographischen Gesellschaft nach der Tschuktschen-Halbinsel," *Deutsche Geographische Blatter,* vol. 5, Bremen.

Krech, Shepard, III
1979a "Interethnic Relations in the Lower Mackenzie River Region," *Arctic Anthropology,* vol. 16, p. 2
1979b "The Nakotcho Kutchin: A Tenth Aboriginal Kutchin Band?" *Journal of Anthropological Research* (University of New Mexico), vol. 35, no. 1.

Kugler, Richard C.
1971 "The Penetration of the Pacific by American Whalemen in the Nineteenth Century," in *The Opening of the Pacific—Image and Reality,* Maritime Monographs and Reports, no.

2, National Maritime Museum, London.
1980 "The Whale Oil Trade, 1750–1775," in *Seafaring in Colonial Massachusetts,* Publications of the Colonial Society of Massachusetts, vol. 52, Boston.

Kushner, Howard I.
1972 "Hell-Ships: Yankee Whaling Along the Coasts of Russian-America," *New England Quarterly* (March), pp. 81-95.

Larsen, Helge, and Froelich Rainey
1948 *Ipiutak and the Arctic Whale Hunting Culture.* Anthropological Papers of the American Museum of Natural History, vol. 42, part 1, New York.

Lawrence, Mary Chipman
1966 *The Captain's Best Mate: The Journal of Mary Chipman Lawrence on the Whaler Addison, 1856–1860,* ed. Stanton Garner. Providence, R.I.: Brown University Press.

Laws, R. M.
1979 "Southern Elephant Seal," in *Mammals in the Seas,* vol. 2. Rome: Food and Agriculture Organization of the United Nations.

Leavitt, George B.
1899 Letter quoted in U.S. Treasury Department, Division of Revenue Cutter Service, Doc. 2101, *The Cruise of the U.S. Revenue Cutter* Bear *and the Overland Expedition for the Relief of the Whalers in the Arctic Ocean from Nov. 27, 1897 to Sept. 13, 1898,* pp. 122-23. Washington, D.C.: Government Printing Office.

Lebedev, Dimitri, and Vadim Grekov
1967 "Geographical Exploration by the Russians," in *The Pacific Basin: A History of Its Geographical Exploration,* ed. Herman Friis. New York: American Geographical Society.

Le Boeuf, B. J.
1979 "Northern Elephant Seal," in *Mammals in the Seas,* vol. 2, Rome: Food and Agriculure Organization of the United Nations.

Leet, Robert Edward
1974 *American Whalers in the Western Arctic, 1879–1914.* M.A. thesis, Department of History, University of San Francisco.

Leffingwell, Ernest de K.
1919 *The Canning River Region, Northern Alaska.* U.S. Geological Survey Professional Paper, no. 109. Washington, D.C.: Government Printing Office.

Lerrigo, P. H. J.
1902 "Abstract of Daily Journal on St. Lawrence Island Kept by P. H. J. Lerrigo, M. D.," in Sheldon Jackson, *Eleventh Annual Report on Introduction of Domestic Reindeer into Alaska . . . 1901,* Senate, 57th Congress, 1st Session, Doc. 98. Washington, D.C.: Government Printing Office.

Lindeman, Moritz
1899 *Die Gegenwartige Eismeer-Fischerei und der Walfang.* Berlin: Otto Salle.

Lytle, Thomas G.
1984 *Harpoons and Other Whalecraft.* New Bedford, Mass.: Old Dartmouth Historical Society.

MacInnes, Tom
1932 *Klengenberg of the Arctic.* London: Jonathan Cape.

McIlhenny, Edward Avery
n.d. Personal papers, Manuscript, Private Collection.

McRoy, C. Peter, and John J. Goering
1974 "The Influence of Ice on the Primary Productivity of the Bering Sea," in *Oceanography of the Bering Sea,* ed. D. W. Hood and E. J. Kelley, Institute of Marine Science, University of Alaska, Fairbanks.
1976 "Annual Budget of Primary Production in the Bering Sea," *Marine Science Communications,* vol. 2, no. 5, pp. 255-67.

Madsen, Charles, and John Scott Douglas
1957 *Arctic Trader.* New York: Dodd, Mead and Company.

Maguire, Rochfort
1854–55 "Proceedings of Commander Maguire . . . ," Great Britain Parliament, House of Commons, Sessional Papers, Accounts and Papers. *Papers Relative to the Recent Arctic Expeditions in Search of Sir John Franklin,* vol. 35, no. 1898.

Marine Insurance Company
n.d. Records of Ships' Repairs. Manuscript, Old Dartmouth Historical Society, New Bedford, Mass.

Marquette, Willman M., and John R. Bockstoce
1980 "Historical Shore-based Catch of Bowhead Whales in the Bering, Chukchi, and Beaufort Seas," *Marine Fisheries Review,* vol. 42, no. 9-10.

Masik, August, and Isobel Hutchinson
1935 *Arctic Nights' Entertainments.* London: Blackie and Son, Ltd.

Mason, John Thompson
1898 "The Last of the Confederate Cruisers," *Century Magazine* (August).

Mason, Otis Tufton
1902 "Aboriginal American Harpoons," *Report of U.S. National Museum, 1900.* Washington, D.C.: Government Printing Office.

Mason, W. S.
ca. 1900 *The Frozen Northland: Life with the Eskimo in His Own Country.* Cincinnati: Jennings and Graham; New York: Eaton and Manis.

Mathews, L. Harrison
1978 *The Natural History of the Whale.* New York: Columbia University Press.

Matthews, John
n.d. Journal 1850–1855, aboard HMS *Plover.* Typescript, Royal Geographical Society, London.

Maury, Matthew Fontaine
1851 *Explanations and Sailing Directions to Accompany the Wind and Current Charts . . . ,* 3d ed. Washington, D.C.

Mikkelsen, Ejnar
1909 *Conquering the Arctic Ice.* London: William Heineman.

Mjelde, Michael Jay
1970 *Glory of the Seas.* Middletown, Connecticut.

Moment, David
1957 *The Business of Whaling in America in the 1850s.* Manuscript, Harvard Business School, Cambridge, Mass.

Moore, Golda Pauline
1934 "Hawaii During the Whaling Era, 1820–1880," M.A. thesis, University of Hawaii.

Morgan, Murray
1948 *Dixie Raider: The Saga of the C.S.S. Shenandoah*, New York: E. P. Dutton.
Mountain, D. G.
1974 "Preliminary Analysis of Beaufort Shelf Circulation in Summer," in *The Coast and Shelf of the Beaufort Sea*, Arctic Institute of North America, Arlington, Virginia.
Muir, John
1917 *The Cruise of the Corwin*. Boston: Houghton Mifflin and Company.
Munger, James F.
1967 *Two Years in the Pacific and Arctic Oceans and China*. Fairfield, Wash.: Ye Galleon Press (reprint of 1852 journal of the same title; Vernon, N.Y.: J. R. Howlett, Printer).
Murdoch, John
1885 "Natural History," in Patrick Henry Ray, *Report of the International Polar Expedition to Point Barrow, Alaska*. Washington, D.C.: Government Printing Office.
1891 "Whale Catching at Point Barrow," *Popular Science Monthly*, vol. 38, no. 4.
1892 *Ethnological Results of the Point Barrow Expedition*. Ninth Annual Report of the Bureau of Ethnology. Washington, D.C.: Government Printing Office.
Murphy, John Francis
1968 "Cutter Captain: The Life and Times of John C. Cantwell," Ph.D. dissertation, University of Connecticut at Storrs.
Nelson, Edward William
1887 *Report upon Natural History Collections Made in Alaska between the Years 1877 and 1881*. U.S. Army Signal Service. Washington, D.C.: Government Printing Office.
1899 *The Eskimo About Bering Strait*. Eighteenth Annual Report of the Bureau of American Ethnology. Washington, D.C.: Government Printing Office.
Nelson, E. W., and F. W. True.
1887 "Mammals of Northern Alaska," in *Report Upon Natural History Collections Made in Alaska between the Years 1877 and 1881*. U.S. Army Signal Service. Washington, D.C.: Government Printing Office.
Newhall, Charles L.
1981 *The Adventures of Jack, or A Life on the Wave*. Fairfield, Wash.: Ye Galleon Press.
Nordenskiold, A. E.
1882 *The Voyage of the Vega around Asia and Europe*. New York: Macmillan Company.
n.d. A. E. Nordenskiold Papers, Stockholm University Library, Library of the Royal Academy of Sciences.
Nourse, J. E.
1884 *American Explorations in the Ice Zones*. 3d edition. Boston: D. Lothrop and Company.
Nuligak
1971 *I, Nuligak*, ed. M. Metayer. New York: Pocket Books.
Official Records of the Union and Confederate Navies
var. Washington, D.C.: Government Printing Office.
Old Dartmouth Historical Society
var. *Old Dartmouth Historical Society Sketches*. New Bedford, Mass.
Orth, Donald J.
1971 *Dictionary of Alaska Place Names*. Geological Survey, Professional Paper 567. Washington, D.C.: Government Printing Office.
Osbon, Bradley S.
1902 "The Perils of Polar Whaling," in *The White World*, ed. Rudolf Kersting. New York: Lewis, Scribner, and Company.
Otis, Harrison G.
1880 "Report of Special Agent H. G. Otis upon the Illicit Traffic in Rum and Fire-Arms in Alaska," in *Letter from the Secretary of the Treasury, March 30, 1880*, 46th Congress, 2d Session, U.S. Senate, Ex. Doc. no. 132.

Page, James
1900 *Ice and Ice Movements in Bering Sea During the Spring Months*. H.O. Publication 116, Hydrographic Office. Washington, D.C.: Government Printing Office.
Peake, Frank A.
1966 *The Bishop Who Ate His Boots: A Biography of Isaac O. Stringer*. Ottawa: The Anglican Church of Canada.
Pease, Z. W., and George A. Hough
1889 *New Bedford Massachusetts: Its History, Industries, Institutions, and Attractions*, ed. William L. Sayer. New Bedford, Mass.: Board of Trade.
Pedersen, Theodore
1944 "Call All Hands," *Alaska Sportsman*, vol. 10, no. 4.
Pedersen, C. T.
1952 "My First Whale," *The Beaver* (September).
Penniman, T. K.
n.d. *Pictures of Ivory and Other Animal Teeth, Bone and Antler*. Occasional Papers on Technology, no. 5, Pitt Rivers Museum, University of Oxford, Oxford.
Petitot, R. P. E.
1876 *Monographie des Esquimaux Tchiglit du Mackenzie et de Anderson*. Paris: Ernest Leroux.
Pivorunas, August
1979 "The Feeding Mechanisms of Baleen Whales," *American Scientist*, vol. 67, no. 4.
Poole, Dorothy Cottle
1971 "Vineyard Whalemen in the Arctic," *The Dukes County* [Mass.] *Intelligencer*, vol. 13, no. 1.
Purrington, Philip F.
1961 "Mr. Bulloch's War with New Bedford," *The Bulletin* (Spring), Old Dartmouth Historical Society, New Bedford, Mass.
Rainey, Froelich G.
1940 "Eskimo Method of Capturing Bowhead Whales," *Journal of Mammology*, vol. 21, no. 3.
1947 *The Whale Hunters of Tigara*. Anthropological Papers of the American Museum of Natural History, vol. 41, part 2, New York.
Ray, Dorothy Jean
1975a *The Eskimos of Bering Strait, 1650–1898*. Seattle: University of Washington Press.
1975b "Early Maritime Trade with the Eskimo of Bering Strait and

the Introduction of Firearms," *Arctic Anthropology,* vol. 12, no. 1.

Ray, Patrick Henry

1885 *Report of the International Polar Expedition to Point Barrow, Alaska.* Washington, D.C.: Government Printing Office.

Ronnberg, Erik A. R., Jr.

1974 "Copper Sheathing of Whaleships," *Nautical Research Journal* (Washington, D.C.), vol. 20, no. 4.

Ross, W. Gillies, and Anne MacIver

1982 *Distribution of the Kills of Bowhead Whales and Other Sea Mammals by Davis Strait Whalers, 1829–1910.* Arctic Pilot Project (January), Lennoxville, Quebec.

Rosse, Irving C.

1884 "Medical and Anthropological Notes on Alaska," in *Cruise of the Revenue Steamer* Corwin *in Alaska and the N.W. Arctic Ocean in 1881: Notes and Memoranda . . . .* Washington, D.C.: Government Printing Office.

Royal Geographical Society

1881 "Proceedings of Foreign Societies," *Proceedings of the Royal Geographical Society,* n.s., vol. 3, no. 12.

Roys, Thomas W.

1851 Letter to M. F. Maury, January 15, in Matthew Fontaine Maury, *Explanations and Sailing Directions to Accompany the Wind and Current Charts . . . ,* 6th edition. Philadelphia: E. C. and J. Biddle.

n.d. *The Voyages of Thomas Welcome Roys.* Manuscript, Suffolk County Whaling Museum, Sag Harbor, New York.

1854 "Description of Whales," Manuscript holograph, Mariners Museum, Newport News, Virginia.

Russell, Frank

1898 *Explorations in the Far North.* Iowa City: University of Iowa Press.

n.d. Journal of Frank Russell, April 26, 1893—August 19, 1894, Anthropological Archives, Smithsonian Institution.

Sambrotto, R. N.; J. J. Goering; and C. P. McRoy

1984 "Large Yearly Production of Phytoplankton in the Western Bering Strait," *Science,* vol. 225 (14 September), pp. 1147-49.

Scammon, Charles M.

1874 *The Marine Mammals of the North-Western Coast of North America.* San Francisco: John H. Carmany and Company.

Scharf, J. Thomas

1977 *History of the Confederate States' Navy.* New York: Fairfax-Crown.

Schmitt, Frederick P.; Cornelis de Jong; and Frank H. Winter

1980 *Thomas Welcome Roys: America's Pioneer of Modern Whaling.* Published for the Mariners Museum by the University Press of Virginia.

Seale, Alvin

1946 *The Quest for the Golden Cloak and Other Experiences of a Field Naturalist.* San Francisco: Stanford University Press.

Seemann, Berthold

1853 *Narrative of the Voyage of H.M.S. Herald During the Years 1845–1851. . . .* London: Reeve and Company.

Seward, William H.

1852 "Commerce in the Pacific Ocean: Speech of William H. Seward in the Senate of the United States, July 29, 1852." Washington: Buell and Blanchard.

Shklovsky, I. W.

1916 *In Far North-East Siberia.* London: Macmillan Company.

Simpson, Edward

1890 *Report of Ice and Ice Movements in Bering Sea and the Arctic Basin.* U.S. Hydrographic Office Report, No. 92, Washington, D.C.

Simpson, John

1875 "Observations on the Western Eskimo . . . ," in *Arctic Geography and Ethnology.* London: Royal Geographical Society.

Smith, Gaddis

1978 "Whaling History and the Courts," *The Log of Mystic Seaport,* vol. 30, no. 3, pp. 67-80.

Smith, George W.

n.d. *The Autobiography of an Old Whaleman, Captain George W. Smith.* Manuscript, Old Dartmouth Historical Society, New Bedford, Mass.

Smith, N. Byron

1851-53 "Account of a Whaling Voyage Aboard the *Nile* of Greenport, 1851–1853," Manuscript, Library of Congress, Washington, D.C.

Snow, H. J.

1910 *In Forbidden Seas.* London: Edward Arnold.

Sonnenfeld, Joseph

1957 "Changes in Subsistence among the Barrow Eskimo," Ph.D. dissertation, Johns Hopkins University, Baltimore, Maryland.

Spengemann, Friedrich

1952 *Sudseefahrer.* Bremen.

Stackpole, Edouard A.

1953 *The Sea-Hunters: The New England Whalemen During Two Centuries, 1635–1835.* Philadelphia: J. B. Lippincott Company.

Starbuck, Alexander

1964 *History of the American Whale Fishery.* Originally part 4 of the Report of the U.S. Commission on Fish and Fisheries, Washington, 1878. New York: Argosy-Antiquarian Ltd.

Starokadomskiy, L. M.

1976 *Charting the Russian Northern Sea Route.* Montreal: Arctic Institute of North America, McGill-Queen's University Press.

Stefansson, Vilhjalmur

1909 "The Eskimo Trade Jargon of Herschel Island," *American Anthropologist,* n.s., vol. 2.

1909a "The North Coast of Alaska Badly Charted," *Bulletin of the American Geographical Society,* vol. 41, no. 5.

1912 *My Life with the Eskimo.* New York: Harper and Brothers.

1914 *Prehistoric and Present Commerce among the Arctic Coast Eskimo.* Museum Bulletin 6, Anthropological Series no. 3, Geological Survey, Department of Mines, Ottawa.

1922 *Hunters of the Great North.* New York: Harcourt, Brace and Company.

1934 "An Eskimo Discovery of an Island North of Alaska,"

Geographical Review, vol. 24, no. 1.
Stern, Philip Van Doren
1962 *The Confederate Navy: A Pictorial History.* Garden City, N.Y.: Doubleday.
Stevenson, Charles H.
1903 "Fish Oils, Fats and Waxes," U.S. Fish Commission Report of 1902, U.S. Commission of Fish and Fisheries. Washington, D.C.: Government Printing Office.
1904 "Aquatic Products in Arts and Industries," U.S. Commission of Fish and Fisheries, part 28, *Report of the Commissioner for the Year Ending June 30, 1902.* Washington, D.C.: Government Printing Office.
1904a "Utilization of the Skins of Aquatic Mammals," U.S. Commission of Fish and Fisheries, *Report of the Commissioner for the Year Ending June 30, 1902.* Washington, D.C.: Government Printing Office.
Stewart, Ethel G.
1955 "Fort McPherson and the Peel River Area," M.A. thesis, Department of History, Queen's University, Kingston, Ontario.
Stockton, Charles H.
1890 "The Arctic Cruise of the U.S.S. *Thetis* in the Summer and Autumn of 1889," *National Geographic Magazine,* vol. 2.
Stone, A. J.
1900 "Some Results of a Natural History Journey to Northern British Columbia, Alaska, and the Northwest Territory, in the Interest of the American Museum of Natural History," *Bulletin of the American Museum of Natural History,* vol. 13, no. 5.
1901 Letter of Feb. 28, quoted in J. A. Allen, "The Musk-Oxen of Arctic America and Greenland," *Bulletin of the American Museum of Natural History,* vol. 14.
Swenson, Olaf
1944 *Northwest of the World.* New York: Dodd, Mead and Company.
Swithinbank, Charles
1960 *Ice Atlas of Arctic Canada.* Ottawa: Defence Research Board.
Taylor, Nathaniel W.
1929 *Life on a Whaler, or Antarctic Adventures in the Isle of Desolation,* ed. Howard Palmer. New London: New London Historical Society.
Thiercelin, Dr.
1866 *Journal d'un Baleinier: Voyages en Oceanie.* Paris: L. Hachette.
Thornton, Harrison R.
1931 *Among the Eskimos of Wales, Alaska.* Baltimore and London: John Hopkins University Press.
Thrum, Thomas G.
1913 "Honolulu's Share in the Pacific Whaling Industry of By-Gone Days," *The Hawaiian Almanac and Annual for 1913* (Honolulu), vol. 39.
Tikhmenev, P. A.
1978 *A History of the Russian-American Company,* trans. and ed. Richard A. Pierce and Alton S. Donnelly. Seattle and London: University of Washington Press.
Tilton, George Fred
1969 *"Cap'n George Fred" Himself.* Edgartown, Mass.: Dukes County Historical Society.
Tolmachoff, Innokenty P.
1949 *Siberian Passage.* New Brunswick, Rutgers University Press.
Tower, Walter S.
1907 *A History of the American Whale Fishery.* Publications of the University of Pennsylvania, Series in Political Economy and Public Law, no. 20, Philadelphia.
Tressler, D. K.
1923 *Marine Products of Commerce: The Acquisition, Handling, Biological Aspects and the Science and Technology of Their Preparations and Preservation.* New York: Chemical Catalog Company.
Twain, Mark
1966 *Mark Twain's Letters from Hawaii,* ed. A. Grove Day. Honolulu: University Press of Hawaii.
United States Coast and Geodetic Survey
1926 *United States Coast Pilot, Alaska Part II, Yakutat Bay to Arctic Ocean,* 2d edition. Washington, D.C.: Government Printing Office.
United States Commission of Education
1894 *Report of the Commissioner of Education for the Year 1890–91.* Washington, D.C.: Government Printing Office.
United States Commission of Fish and Fisheries
1892 *Report of the Commissioner for 1888.* Washington, D.C.: Government Printing Office.
1904 *Report of the Commissioner for the Year Ending June 30, 1902.* Washington, D.C.: Government Printing Office.
United States Treasury Department, Division of Revenue Cutter Service
1899 *Report of the Cruise of the U.S. Revenue Cutter* Bear *and the Overland Expedition for the Relief of the Whalers in the Arctic Ocean from November 27, 1897 to September 13, 1898.* Washington, D.C.: Government Printing Office.
United States
1870 *Correspondence Concerning Claims Against Great Britain Transmitted to the Senate of the United States.* Washington, D.C.: Government Printing Office.
1872 *The Case of the United States, to be Laid before the Tribunal of Arbitration, to be Convened at Geneva . . . ,* 42d Congress, 2d Session, Ex. Doc. no. 31, Washington, D.C.: Government Printing Office.
Usher, Peter
1971 "The Canadian Western Arctic: A Century of Change," *Anthropologica,* n.s., vol. 13, nos. 1 and 2.
Uzanne, Octave
1898 *Fashions in Paris, 1797–1897.* London: William Heinemann.
Vermilyea, Lucius H.
1958 "A Whaler's Letter," *American Neptune,* vol. 18, no. 4.
Waddell, James I.
n.d. Confederate Document 51, Waddell Papers, no. 80. Copy [typescript] of Waddell's manuscript, U.S. Naval Museum, Washington, D.C.

Wadhams, Peter
1976 "Oil and Ice in the Beaufort Sea," *Polar Record*, vol. 18, no. 114.
Wardman, George
1884 *A Trip to Alaska. . . .* San Francisco: Samuel Carson and Company.
Waugh, Norah
1968 *The Cut of Women's Clothes, 1600–1930.* New York: Theatre Arts Books.
Webster, Bayard
1979 "Mystery of Bering Sea Solved," *New York Times*, June 26.
Wells, Roger and John W. Kelly
1890 *English-Eskimo and Eskimo-English Vocabularies. . . .* Bureau of Education, Circular of Information, no. 2. Washington, D.C.: Government Printing Office.
West, Captain Ellsworth Luce, and Eleanor Ransom Mayhew
1965 *Captain's Papers, a Log of Whaling and Other Sea Experiences.* Barre, Mass.: Barre Publishers.
Whittaker, C. E.
n.d. *Arctic Eskimo.* London: Seeley, Service and Company, Ltd.
Whittle, William C.
1910 *Cruises of the Confederate States Steamers "Shenandoah" and "Nashville."* n.p.
Whymper, Frederick
1869 "Russian America, or 'Alaska': The Natives of the Youkon River and Adjacent Country," *Transactions of the Ethnological Society of London*, n.s., vol. 7.
Wilkinson, David
n.d. *Whaling in Many Seas and Cast Adrift in Siberia.* London: Henry J. Drane.
Williams, Harold
1964 *One Whaling Family.* Boston: Houghton Mifflin Company.
Williamson, Harold F., and Arnold R. Daum
1959 *The American Petroleum Industry: The Age of Illumination, 1859–1899.* Evanston, Ill.: Northwestern University Press.
Winchester, James W., and Charles C. Bates
1957 "Meteorological Conditions and the Associated Sea Ice Distribution in the Chukchi Sea During the Summer of 1955," *Journal of Atmospheric and Terrestrial Physics*, Special Supplement.
Woldt, A.
1884 *Capitain Jacobsen's Reise an der Nordwestkuste Amerikas 1881–1883. . . .* Leipzig: Max Spohr.
Wood, Dennis
n.d. *Abstracts of Whaling Voyages.* Manuscript, New Bedford Free Public Library, New Bedford, Mass.
Woolfe, Henry D.
1893 "The Seventh District," in *Report on Population and Resources of Alaska at the Eleventh Census: 1890*, House of Representatives, 52d Congress, 1st Session, Resources Doc. no. 340, part 7. Washington, D.C.: Government Printing Office.
Works Projects Administration
1940 *Ship Registers of New Bedford, Massachusetts.* National Archives Project, Washington, D.C.
Wrangell, Ferdinand Petrovich, Baron
1840 *Narrative of an Expedition to the Polar Sea in the Years 1820, 1821, 1822, and 1823*, ed. [Sir] Edward Sabine. London: James Madden and Company.
Wright, Allen A.
1976 *Prelude to Bonanza: The Discovery and Exploration of the Yukon.* Sydney, B.C.: Gray's Publishing Ltd.
Wursig, B.; C. W. Clark; and others
1982 *Behavior, Disturbance Responses and Feeding of Bowhead Whales (Balaena Mysticetus) in the Beaufort Sea, 1980–1981.* Bryan, Texas: LGL Ecological Research Associates, Inc.
Zagoskin, L.A.
1967 *Lieutenant Zagoskin's Travels in Russian America, 1842–1844. . . .* Toronto: Arctic Institute of North America and University of Toronto Press.
Zaslow, Morris
1971 *The Opening of the Canadian North, 1870–1914.* Toronto: McClelland and Stewart.
Zubov, N. N.
1947 *Dinamicheskaya Okeanologiya.* Moscow: Gidrometeoizdat.

Index